AF522296

MILESTINES IN PETROLOGY AND FUTURE PERSPECTIVES

MILESTINES IN PETROLOGY AND FUTURE PERSPECTIVES

Ganesh Prasad

RANDOM PUBLICATIONS
NEW DELHI - 110 002 (INDIA)

Milestines in Petrology and Future Perspectives

ISBN 978-93-51117-08-7

Published in 2015 in India by

RANDOM PUBLICATIONS

4376-A/4B, Gali Murari Lal, Ansari Road

New Delhi-110 002

Phone: +9111-43580356, 23289044

E-mail: randomexports@gmail.com; sales@randompublications.com; info@randompublications.com

Reprinted 2019

Type Setting by: Friends Media, Delhi-110089

Digitally Printed at: Replika Press Pvt. Ltd.

Preface

Petrology is the branch of geology that studies the origin, composition, distribution and structure of rocks. In the oil industry, lithology, or more specifically mud logging, is the graphic representation of geological formations being drilled through, and drawn on a log called a mud log. As the cuttings are circulated out of the borehole they are sampled, examined (typically under a 10x microscope) and tested chemically when needed. Petrology utilizes the classical fields of mineralogy, petrography, optical mineralogy, and chemical analyses to describe the composition and texture of rocks. Modern petrologists also include the principles of geochemistry and geophysics through the studies of geochemical trends and cycles and the use of thermodynamic data and experiments to better understand the origins of rocks. There are three branches of petrology, corresponding to the three types of rocks: igneous, metamorphic, and sedimentary, and another dealing with experimental techniques.

Petrology has evolved dramatically over the past century from a science primarily concerned with the diversity of rocks to a science investigating the origin of rocks. Fundamental to all petrologic studies are robust field observations, careful sample collection and detailed petrographic study. Once the foundations of a solid petrologic study have been built, we can turn to an impressive arsenal of new and innovative approaches to coax a mountain of information from the most minuscule of crystals. One of the fastest growing disciplines used extensively by petrologists is isotope geology. Radiogenic isotope geology has its origins rooted in the discovery of radioactivity at the turn of the century and has blossomed into a field that provides a common denominator for all earth science studies; Time. The impact of isotope geology in the field of petrology over the past few decades has been phenomenal and answers to frequently asked questions, such as when did the magma form and what is its origin, are routinely addressed with an increasing array of geochronometers and isotope tracing techniques. Many significant advances have been made in the field of isotope geology in recent years that have virtually revolutionized the way we approach isotopic studies of igneous, metamorphic and sedimentary rocks. Overall, many research directions have evolved towards the analysis of increasingly smaller portions of rocks and mineral grains with the expectation of at least comparable if not greatly improved precision and accuracy. The impact of these recent advances has been most profound in the field of geochronology. We have come to expect age determinations for minerals and rocks with uncertainties of less the 1-2 m.y. regardless of their antiquity. With regard to U-Pb geochronology of minerals, in particular zircon, there are a bewildering number of approaches now in use including Isotope Dilution Thermal Ionization Mass Spectrometry [IDTIMS], Sensitive High Resolution Ion MicroProbe [SHRIMP], Inductively Coupled

Plasma Mass Spectrometry [ICP-MS], Vapour Digestion Technique [VDT], and Thermal Evaporation [TE]. In addition, there are a variety of chemical (not isotopic) U-Pb dating techniques (Th-U-total Pb isochron method commonly utilizing an electron microprobe [CHIME], Energy-dispersive Miniprobe Multi-element Analyser method [EMMA]) that can be successfully applied to obtain a date for minerals with high U and/or Th contents but devoid of common lead. Some of these techniques, such as CHIME, EMMA, ICP-MS, and TE, provide rapid and economical approaches for age determination at the expense of precision and occasionally accuracy. On the other hand, some techniques are much better suited where spatial resolution is essential [CHIME, EMMA, ICP-MS]. The ideal combination of micron-scale spatial resolution and high-precision [e.g. SHRIMP] can be achieved but generally at much greater cost. Therefore, it is important to not only understand the principles, methodology and attendant error evaluation procedures used for each technique, but it is incumbent on the researcher to select the most appropriate technique that suits the geological question at hand. Another exciting recent development in the field of isotope geology that will impact the future of petrology is the enormous potential of the new generation multi-collector inductively coupled plasma mass spectrometers [MC-ICP-MS]. The future looks very promising for profiling studies where both U-Pb age and Hf isotopic compositions are determined across individual growth bands of zoned garnet, limonite and zircon crystals.

Some observers believe that oil shales will be mined and utilized as a major source of fuel in the future as petroleum and natural gas supplies inevitably dwindle. This valuable book is designed to give students a balanced and comprehensive coverage of these new advances, as well as a firm grounding in the classical aspects of igneous and metamorphic petrology.

I thank all members of my team who have helped in the preparation of the book. My special thanks go to "Random Publications" who have published the book.

— Ganesh Prasad

Contents

Chapter 1

Chemical Compound of Petroleum Products

A chemical compound is a pure chemical substance consisting of two or more different chemical elements that can be separated into simpler substances by chemical reactions. Chemical compounds have a unique and defined chemical structure; they consist of a fixed ratio of atoms that are held together in a defined spatial arrangement by chemical bonds.

Chemical compounds can be molecular compounds held together by covalent bonds, salts held together by ionic bonds, intermetallic compounds held together by metallic bonds, or complexes held together by coordinate covalent bonds. Pure chemical elements are not considered chemical compounds, even if they consist of molecules which contain only multiple atoms of a single element (such as H_2, S_8, etc.), which are called diatomic molecules or polyatomic molecules.

Definitions

There are exceptions to the definition above, and large amounts of the solid chemical matter familiar on Earth do not have simple formulas. Certain crystalline compounds are called "non-stoichiometric" because they vary in composition due to either the presence of foreign elements trapped within the crystal structure or a deficit or excess of the constituent elements. Such non-stoichiometric compounds form most of the crust and mantle of the Earth.

Other compounds regarded as chemically identical may have varying amounts of heavy or light isotopes of the constituent elements, which will make the ratio of elements by mass vary slightly.

Elementary Concepts

Characteristic properties of compounds:

1. Elements in a compound are present in a definite proportion.

Example: 2 atoms of hydrogen + 1 atom of oxygen becomes 1 molecule of compound-water.

2. *Compounds have a definite set of properties*

Elements comprising a compound do not retain their original properties.

Example: hydrogen (element, which is combustible and non-supporter of combustion) + oxygen (element, which is non-combustible and supporter of combustion) becomes water (compound, which is non-combustible and non-supporter of combustion)

Valency is the number of hydrogen atoms which can combine with one atom of the element forming a compound.

Compounds Compared to Mixtures

The physical and chemical properties of compounds are different from those of their constituent elements. This is one of the main criteria for distinguishing a compound from a mixture of elements or other substances because a mixture's properties are generally closely related to and dependent on the properties of its constituents. Another criterion for distinguishing a compound from a mixture is that the constituents of a mixture can usually be separated by simple, mechanical means such as filtering, evaporation, or use of a magnetic force, but the components of a compound can only be separated by a chemical reaction. Conversely, mixtures can be created by mechanical means alone, but a compound can only be created (either from elements or from other compounds, or a combination of the two) by a chemical reaction.

Some mixtures are so intimately combined that they have some properties similar to compounds and may easily be mistaken for compounds. One example is alloys. Alloys are made mechanically, most commonly by heating the constituent metals to a liquid state, mixing them thoroughly, and then cooling the mixture quickly so that the constituents are trapped in the base metal. Other examples of compound-like mixtures include intermetallic compounds and solutions of alkali metals in a liquid form of ammonia.

Formula

Chemists describe compounds using formulas in various formats. For compounds that exist as molecules, the formula for the molecular unit is shown. For polymeric materials, such as minerals and many metal oxides, the empirical formula is normally given, e.g. NaCl for table salt. The elements in a chemical formula are normally listed in a specific order, called the Hill system. In this system, the carbon atoms (if there are any) are usually listed first, any hydrogen atoms are listed next, and all other elements follow in alphabetical order. If the formula contains no carbon, then all of the elements, including hydrogen, are listed alphabetically. There are, however, several important exceptions to the normal rules. For ionic compounds, the positive ion is almost always listed first and the negative ion is listed second. For oxides, oxygen is usually listed last.

Organic acids generally follow the normal rules with C and H coming first in the formula. For example, the formula for trifluoroacetic acid is usually written as $C_2HF_3O_2$. More descriptive formulas can convey structural information, such as writing the formula for trifluoroacetic acid as CF_3CO_2H.

On the other hand, the chemical formulas for most inorganic acids and bases are exceptions to the normal rules. They are written according to the rules for ionic compounds (positive first, negative second), but they also follow rules that emphasize their Arrhenius definitions. Specifically, the formula for most inorganic acids begins with hydrogen and the formula for most bases ends with the hydroxide ion (OH^-). Formulas for inorganic compounds do not often convey structural information, as illustrated by the common use of the formula H_2SO_4 for a molecule (sulfuric acid) that contains no H-S bonds. A more descriptive presentation would be $O_2S(OH)_2$, but it is almost never written this way.

Phases and Thermal Properties

Compounds may have several possible phases. All compounds can exist as solids, at least at low enough temperatures. Molecular compounds may also exist as liquids, gases, and, in some cases, even plasmas. All compounds decompose upon applying heat. The temperature at which such fragmentation occurs is often called the decomposition temperature. Decomposition temperatures are not sharp and depend on the rate of heating.

CAS Number

Every chemical substance, including chemical compounds, that has been described in the literature carries a unique numerical identifier, its CAS number.

Chemical Substance

In chemistry, a chemical substance is a form of matter that has constant chemical composition and characteristic properties. It cannot be separated into components by physical separation methods, i.e. without breaking chemical bonds. They can be solids, liquids or gases.

Chemical substances are often called 'pure' to set them apart from mixtures. A common example of a chemical substance is pure water; it has the same properties and the same ratio of hydrogen to oxygen whether it is isolated from a river or made in a laboratory. Other chemical substances commonly encountered in pure form are diamond (carbon), gold, table salt (sodium chloride) and refined sugar (sucrose). However, simple or seemingly pure substances found in nature can in fact be mixtures of chemical substances. For example, tap water may contain small amounts of dissolved sodium chloride and compounds containing iron, calcium and many other chemical substances.

Chemical substances exist as solids, liquids, gasses, or plasma and may change between these phases of matter with changes in temperature or pressure. Chemical reactions convert one chemical substance into another. Forms of energy, such as light and heat, are not considered to be matter, and thus they are not "substances" in this regard.

Definition

Chemical substances (also called pure substances) are often defined as "any material with a definite chemical composition" in most introductory general chemistry textbooks. According to this definition a chemical substance can either be a pure chemical element or a pure chemical compound. But, there are exceptions to this definition; a pure substance can also be defined as a form of matter that has both definite composition and distinct properties.

The chemical substance index published by CAS also includes several alloys of uncertain composition. Non-stoichiometric compounds are a special case (in inorganic chemistry) that violates the law of constant composition, and for them, it is sometimes difficult to draw the line between a mixture and a compound, as in the case of palladium hydride. Broader definitions of chemicals or chemical substances can be found, for example: "the term 'chemical substance' means any organic or inorganic substance of a particular molecular identity, including – (i) any combination of such substances occurring in whole or in part as a result of a chemical reaction or occurring in nature"

History

The concept of a "chemical substance" became firmly established in the late eighteenth century after work by the chemist Joseph Proust on the composition of some pure chemical compounds such as basic copper carbonate. He deduced that, "All samples of a compound have the same composition; that is, all samples have the same proportions, by mass, of the elements present in the compound." This is now known as the law of constant composition.

Later with the advancement of methods for chemical synthesis particularly in the realm of organic chemistry; the discovery of many more chemical elements and new techniques in the realm of analytical chemistry used for isolation and purification of elements and compounds from chemicals that led to the establishment of modern chemistry, the concept was defined as is found in most chemistry textbooks. However, there are some controversies regarding this definition mainly because the large number of chemical substances reported in chemistry literature need to be indexed.

Chemical Elements

An element is a chemical substance that is made up of a particular kind of atoms and hence cannot be broken down or transformed by a chemical reaction into a different element, though it can be transmutated into another element through a nuclear reaction. This is so, because all of the atoms in a sample of an element have the same number of protons, though they may be different isotopes, with differing numbers of neutrons.

There are about 120 known elements, about 80 of which are stable – that is, they do not change by radioactive decay into other elements. However, the number of chemical substances that are elements can be more than 120, because some elements can occur as more than a single chemical substance (allotropes).

For instance, oxygen exists as both diatomic oxygen (O_2) and ozone (O_3). The majority of elements are classified as metals. These are elements with a characteristic lustre such as iron, copper, and gold. Metals typically conduct electricity and heat well, and they are malleable and ductile. Around a dozen elements, such as carbon, nitrogen, and oxygen, are classified as non-metals. Non-metals lack the metallic properties described above, they also have a high electronegativity and a tendency to form negative ions. Certain elements such as silicon sometimes resemble metals and sometimes resemble non-metals, and are known as metalloids.

Chemical Compounds

A pure chemical compound is a chemical substance that is composed of a particular set of molecules or ions. Two or more elements combined into one substance through a chemical reaction form a chemical compound. All compounds are substances, but not all substances are compounds.

A chemical compound can be either atoms bonded together in molecules or crystals in which atoms, molecules or ions form a crystalline lattice. Compounds based primarily on carbon and hydrogen atoms are called organic compounds, and all others are called inorganic compounds. Compounds containing bonds between carbon and a metal are called organometallic compounds.

Compounds in which components share electrons are known as covalent compounds. Compounds consisting of oppositely charged ions are known as ionic compounds, or salts.

In organic chemistry, there can be more than one chemical compound with the same composition and molecular weight. Generally, these are called isomers. Isomers usually have substantially different chemical properties, may be isolated and do not spontaneously convert to each other. A common example is glucose vs. fructose. The former is an aldehyde, the latter is a ketone. Their interconversion requires either enzymatic or acid-base catalysis. However, there are also tautomers, where isomerization occurs spontaneously, such that a pure substance cannot be isolated into its tautomers. A common example is glucose, which has open-chain and ring forms. One cannot manufacture pure open-chain glucose because glucose spontaneously cyclizes to the hemiacetal form.

Substances versus Mixtures

All matter consists of various elements and chemical compounds, but these are often intimately mixed together. Mixtures contain more than one chemical substance, and they do not have a fixed composition. In principle, they can be separated into the component substances by purely mechanical processes. Butter, soil and wood are common examples of mixtures.

Grey iron metal and yellow sulfur are both chemical elements, and they can be mixed together in any ratio to form a yellow-grey mixture. No chemical process occurs, and the material can be identified as a mixture by the fact that the sulfur and the iron can be separated by a mechanical process, such as using a magnet to attract the iron away from the sulfur.

In contrast, if iron and sulfur are heated together in a certain ratio (1 atom of iron for each atom of sulfur, or by weight, 56 grams (1 mol) of iron to 32 grams (1 mol) of sulfur), a chemical reaction takes place and a new substance is formed, the compound iron(II) sulfide, with chemical formula FeS. The resulting compound has all the properties of a chemical substance and is not a mixture. Iron(II) sulfide has its own distinct properties such as melting point and solubility, and the two elements cannot be separated using normal mechanical processes; a magnet will be unable to recover the iron, since there is no metallic iron present in the compound.

Chemicals versus Chemical Substances

While the term *chemical substance* is a precise technical term that is synonymous with "chemical" for professional chemists, the meaning of the word *chemical* varies for non-chemists within the English speaking world or those using English. For industries, government and society in general in some countries, the word *chemical* includes a wider class of substances that contain many mixtures of such chemical substances, often finding application in many vocations. In countries that require a list of ingredients in products, the "chemicals" listed would be equated with "chemical substances".

Within the chemical industry, manufactured "chemicals" are chemical substances, which can be classified by production volume into bulk chemicals, fine chemicals and chemicals found in research only. Bulk chemicals are produced in very large quantities, usually with highly optimized continuous processes and to a relatively low price. Fine chemicals are produced at a high cost in small quantities for special low-volume applications such as biocides, pharmaceuticals and speciality chemicals for technical applications.

Research chemicals are produced individually for research, such as when searching for synthetic routes or screening substances for pharmaceutical activity. In effect, their price per gram is very high, although they are not sold. The cause of the difference in production volume is the complexity of the molecular structure of the chemical. Bulk chemicals are usually much less complex. While fine chemicals may be more complex, many of them are simple enough to be sold as "building blocks" in the synthesis of more complex molecules targeted for single use, as named above.

The *production* of a chemical includes not only its synthesis but also its purification to eliminate by-products and impurities involved in the synthesis. The last step in production should be the analysis of batch lots of chemicals in order to identify and quantify the percentages of impurities for the buyer of the chemicals. The required purity and analysis depends on the application, but higher tolerance of impurities is usually expected in the production of bulk chemicals.

Thus, the user of the chemical in the US might chose between the bulk or "technical grade" with higher amounts of impurities or a much purer "pharmaceutical grade" (labeled "USP", United States Pharmacopeia).

Naming and Indexing

Every chemical substance has one or more systematic names, usually named according to the IUPAC rules for naming. An alternative system is used by the Chemical Abstracts Service (CAS).

Many compounds are also known by their more common, simpler names, many of which predate the systematic name. For example, the long-known sugar glucose is now systematically named 6-(hydroxymethyl)oxane-2,3,4,5-tetrol. Natural products and pharmaceuticals are also given simpler names, for example the mild pain-killer Naproxen is the more common name for the chemical compound (S)-6-methoxy-α-methyl-2-naphthaleneacetic acid.

Chemists frequently refer to chemical compounds using chemical formulae or molecular structure of the compound. There has been a phenomenal growth in the number of chemical compounds being synthesized (or isolated), and then reported in the scientific literature by professional chemists around the world. An enormous number of chemical compounds are possible through the chemical combination of the known chemical elements.

As of May 2011, about sixty million chemical compounds are known. The names of many of these compounds are often nontrivial and hence not very easy to remember or cite accurately. Also it is difficult to keep the track of them in the literature. Several international organizations like IUPAC and CAS have initiated steps to make such tasks easier. CAS provides the abstracting services of the chemical literature, and provides a numerical identifier, known as CAS registry number to each chemical substance that has been reported in the chemical literature (such as chemistry journals and patents).

This information is compiled as a database and is popularly known as the Chemical substances index. Other computer-friendly systems that have been developed for substance information, are: SMILES and the International Chemical Identifier or InChI.

Table: *Identification of a typical chemical substance*

Common name	*Systematic name*	*Chemical formula*	*Chemical structure*	*CAS registry number*	*InChI*
alcohol, or ethyl alcohol	ethanol	C_2H_5OH	CH_3CH_2OH (skeletal: ⁄⁀OH)	[64-17-5]	1/C2H6O/c1-2-3/ h3H,2H2,1H3

Isolation, purification, characterization, and identification

Often a pure substance needs to be isolated from a mixture, for example from a natural source (where a sample often contains numerous chemical substances) or after a chemical reaction (which often give mixtures of chemical substances).

Chemical Nomenclature

A chemical nomenclature is a set of rules to generate systematic names for chemical compounds. The nomenclature used most frequently worldwide is the one created and developed by the International Union of Pure and Applied Chemistry (IUPAC).

The IUPAC's rules for naming organic and inorganic compounds are contained in two publications, known as the *Blue Book* and the *Red Book* respectively. A third publication, known as the *Green Book*, describes the recommendations for the use of symbols for physical quantities (in association with the IUPAP), while a fourth, the *Gold Book*, contains the definitions of a large number of technical terms used in chemistry.

Similar compendia exist for biochemistry (the *White Book*, in association with the IUBMB), analytical chemistry (the *Orange Book*), macromolecular chemistry (the *Purple Book*) and clinical chemistry (the *Silver Book*). These "color books" are supplemented by shorter recommendations for specific circumstances which are published from time to time in the journal *Pure and Applied Chemistry*.

Aims of Chemical Nomenclature

The primary function of chemical nomenclature is to ensure that a spoken or written chemical name leaves no ambiguity as to what chemical compound the name refers: each chemical name should refer to a single substance. A less important aim is to ensure that each substance has a single name, although the number of acceptable names is limited.

Preferably, the name also conveys some information about the structure or chemistry of a compound. CAS numbers form an extreme example of names that do not perform this function: each CAS number refers to a single compound but none contain information about the structure. The form of nomenclature used depends on the audience to which it is addressed. As such, no single *correct* form exists, but rather there are different forms that are more or less appropriate in different circumstances.

A common name will often suffice to identify a chemical compound in a particular set of circumstances. To be more generally applicable, the name should indicate at least the chemical formula. To be more specific still, the three-dimensional arrangement of the atoms may need to be specified. In a few specific circumstances (such as the construction of large indices), it becomes necessary to ensure that each compound has a unique name: this requires the addition of extra rules to the standard IUPAC system (the CAS system is the most commonly used in this context), at the expense of having names which are longer and less familiar to most readers. Another system gaining popularity is the International Chemical Identifier (InChI)—while InChI symbols are not human-readable, they contain complete information about substance structure. That makes them more general than CAS numbers.

The IUPAC system is often criticized for the above failures when they become relevant (for example in differing reactivity of sulfur allotropes which IUPAC doesn't distinguish). While IUPAC has a human-readable advantage over CAS numbering, it would be difficult to claim that the IUPAC names for some larger, relevant molecules (such as rapamycin) are human-readable, and so most researchers simply use the informal names.

Differing Aims of Lexicography and Chemical Nomenclature

It is generally understood that the aims of normal English lexicography versus chemical nomenclature vary and are to an extent at odds. Dictionaries of English words on the web or otherwise collect and report the meanings of words as their uses appear and change over time. Chemical nomenclature on the other hand (with IUPAC nomenclature as the best example) is necessarily more restrictive: it aims to standardize communication and practice so that when a chemical term is used, it has a fixed meaning relating to chemical structure, and thereby, can give insights into chemical properties and derived molecular functions.

These differing aims can have profound effects on valid understanding in chemistry, especially with regard to chemical classes made popular for perceived health benefits; this is particularly true in the new media age of the web, where word meanings can rapidly change. A discussion related to the term polyphenol is a clear example (where various web definitions and common uses of the word are at odds with its chemical nomenclature, which connects polyphenol structure and bioactivity).

History

The nomenclature of alchemy is rich in description, but does not effectively meet the aims outlined above. Opinions differ whether this was deliberate on the part of the early practitioners of alchemy or whether it was a consequence of the particular (and often esoteric) theoretical framework in which they worked. While both explanations are probably valid to some extent, it is remarkable that the first "modern" system of chemical nomenclature appeared at the same time as the distinction (by Lavoisier) between elements and compounds, in the late eighteenth century. The French chemist Louis-Bernard Guyton de Morveau published his recommendations in 1782, hoping that his "constant method of denomination" would "help the intelligence and relieve the memory". The system was refined in collaboration with Berthollet, de Fourcroy and Lavoisier, and promoted by the latter in a textbook which would survive long after his death at the guillotine in 1794. The project was also espoused by Jons Jakob Berzelius, who adapted the ideas for the German-speaking world.

The recommendations of Guyton covered only what would be today known as inorganic compounds. With the massive expansion of organic chemistry in the mid-nineteenth century and the greater understanding of the structure of organic compounds, the need for a less *ad hoc* system of nomenclature was felt just as the theoretical tools became available to make this possible. An international conference was convened in Geneva in 1892 by the national chemical societies, from which the first widely accepted proposals for standardization arose. A commission was set up in 1913 by the Council of the International Association of Chemical Societies, but its work was interrupted by World War I. After the war, the task passed to the newly formed International Union of Pure and Applied Chemistry, which first appointed commissions for organic, inorganic and biochemical nomenclature in 1921 and continues to do so to this day.

Types of Nomenclature

Organic Chemistry

- Substitutive name
- Functional class name, also known as a radicofunctional name

- Conjunctive name
- Additive name
- Subtractive name
- Multiplicative name
- Fusion name
- Hantzsch–Widman name
- Replacement name

Inorganic Chemistry

Compositional Nomenclature:

Examples of compositional names are:

- PCl_5 phosphorus pentachloride
- N_2O_4 dinitrogen tetraoxide.

An alternative method uses the oxidation state on the metal in place of suffices, e.g.:

- $SnCl_2$, tin(II) chloride as an alternative to tin dichloride.

Generally this system, known as Stock nomenclature or international nomenclature, is preferred over the prefix system for ionic compounds.

Substitutive Nomenclature: This naming method generally follows established IUPAC organic nomenclature. Hydrides of the main group elements (groups 13–17) are given -ane base names, e.g. borane (BH_3), oxidane (H_2O), phosphane (PH_3) (the name phosphine is also in common use, but is not recommended by IUPAC). The compound PCl_3 would be named substitutively as trichlorophosphane.

Additive Nomenclature: This naming method has been developed principally for coordination compounds although it can be more widely applied. An example of its application is:

- $[CoCl(NH_3)_5]Cl_2$ pentaamminechloridocobalt(III) chloride.

Note that ligands such as chloride become chlorido- rather than chloro- as in substitutive naming.

Prices of Elements and their Compounds

This table lists the elements by their name and gives some recent historical prices for them and their commonly traded compounds. The first of the two price columns shows the price in US dollars per kg of the compound specified. The second shows the USD price per kg of element in the compound. Dates on prices facilitate forex conversion.

Elements of only short half-lives are unlikely to be traded on a secondary market outside of nuclear lab production, and were therefore excluded from the main table. The current cutoff excludes elements with atomic number greater than 105. They have no isotope with a half-life greater than 1.5 hours and are listed separately later.

These elements above 105 were excluded from the above table:

Table: *Highly unstable elements*

Name	*Symbol*	*No.*
Bohrium	Bh	107
Copernicium	Cn	112
Darmstadtium	Ds	110
Hassium	Hs	108
Meitnerium	Mt	109
Roentgenium	Rg	111
Seaborgium	Sg	106
Ununhexium	Uuh	116
Ununoctium	Uuo	118
Ununpentium	Uup	115
Ununquadium	Uuq	114
Ununseptium	Uus	117
Ununtrium	Uut	113

Chemical Safety Signs

In our everyday, we use chemicals for a variety of our jobs. This includes work at office work and also household work as well. Even when it comes to bathroom cleaning or polishing our kitchen tiles we are using chemicals. This is the reason that one realizes that as we come in contact with chemicals so often, we need to safely use chemicals. Also one set of chemicals need to be put to use for a particular task only and we should never take the liberty of using it for other purposes.

Need for Chemical Safety Signs

When using a particular chemical there are a particular set of instructions that we need to follow. These may be in terms of handling, storing or using the chemicals, care has to be taken to protect oneself. Many times some side effects may happen due to some reaction with a chemical, thus it becomes essential that we tale precaution all the while. Chemicals can be very reactive at times and need to be stored in particular types of containers. Proper storage is one of the key elements towards safe handling of chemicals. This can lead to an unwanted reaction at times, thus there is always a particular reason for a chemical to be in a particular material of a container. On the container a label must be pasted that this particular chemical has corrosive properties; this is very essential when the chemical is being taken out to be put in another container. This is very critical towards the safety of the people dealing with the chemical.

Improper labeling can also lead to life threatening accidents. The label should very clearly mention the contents of the container and the concentration levels for the same, and it should talk about the first aid steps in case an accident has occurred.

Also clear instructions need to be given as to the storage steps to be taken for the chemical keeping in mind the climatic conditions of a place, as there could be reactions in case of temperature changes. The entire exercise is solely based in providing safety to the lives of the people working in the area.

Chemical Identification Signs

Make sure that your hazardous chemical storage areas are marked with a chemical identification sign.

Signs are laminated for durability and superior chemical resistance.

Chemical Element

A chemical element is a pure chemical substance consisting of one type of atom distinguished by its atomic number, which is the number of protons in its nucleus. Familiar examples of elements include carbon, oxygen, aluminum, iron, copper, gold, mercury, and lead.

As of May 2011, 118 elements have been identified, the latest being ununoctium in 2002. Of the 118 known elements, only the first 92 are believed to occur naturally on Earth. Of these, 80 are stable or essentially so, while the others are radioactive, decaying into lighter elements over various timescales from fractions of a second to billions of years. Additional elements, of higher atomic numbers than those naturally occurring, have been produced artificially as the products of nuclear reactions.

Hydrogen and helium are by far the most abundant elements in the universe. However, oxygen is the most abundant element in the Earth's crust, making up almost half of its mass. Although all known chemical matter is composed of these various elements, chemical matter itself constitutes only about 15% of the total matter in the universe.

The remainder is dark matter, which is not composed of chemical elements as we know them, since it does not contain protons, neutrons or electrons. The chemical elements are thought to have been produced by various cosmic processes, including hydrogen, helium (and smaller amounts of lithium, beryllium and boron) created during the Big Bang and cosmic-ray spallation. Production of heavier elements from carbon to the very heaviest elements proceeds by stellar nucleosynthesis, and these elements are made available for nebular solar system and planetary formation by supernovae and other cataclysmic cosmic events, which blast these newly created elements from their stars into space.

While most elements are generally viewed as stable, natural transformation of one element to another also occurs in the present time, through decay of radioactive elements as well as other nuclear processes such as cosmic ray bombardment and natural nuclear fission of the nuclei of various heavy elements.

When two distinct elements are chemically combined, with the atoms held together by chemical bonds, the result is termed a chemical compound. Two thirds of the chemical elements occur on Earth only as compounds, and in many cases in the remaining third, often the compound forms are the most common.

Chemical compounds may be composed of elements combined in exact whole-number ratios of atoms, as in water, table salt, and such minerals as quartz, calcite, and some ores. However, chemical bonding of many types of elements results in crystalline solids and metallic alloys for which exact chemical formulas do not exist. Most of the solid substance of the Earth is of this latter type: the atoms that are form the substance of the Earth's crust, mantle, and inner core form chemical compounds of many compositions, but that do not have precise empirical formulas.

In all of these presentations, the physical and chemical properties of the pure individual elements are not apparent. This is true even for elements that occur in uncombined form, if they occur in mixtures, which most do. While all of the 94 naturally occurring elements have been identified in mineral samples from the Earth's crust, only a small minority of elements are found as recognizable, relative pure minerals. Among the more common of such "native elements" are copper, silver, gold, carbon (as coal, graphite, or diamonds), sulfur, and mercury. All but a few of the most inert elements, such as noble gases and noble metals, are usually found on Earth in chemically combined form, as chemical compounds.

While about 32 of the chemical elements occur on Earth in native uncombined form, many of these occur as mixtures. For example, atmospheric air is primarily a mixture of nitrogen, oxygen, and argon. Native solid elements also usually occur in various mixtures, such alloys of iron and nickel.

The history of discovery and use of the chemical elements began with the numerous primitive human societies that found native elements like copper and gold, and extracted (smelted) iron and a few other metals from their ores. Alchemists and chemists subsequently identified many more, with nearly all of the naturally-occurring elements known by 1900. The properties of the chemical elements are often summarized using the periodic table that organizes the elements by increasing atomic number into rows ("periods") in which the columns ("groups") share recurring ("periodic") physical and chemical properties. Either in its pure forms or in various chemical compounds or mixtures, almost every element has at least one important human use. Save for short half-lived radioactive elements, all of the elements through uranium, and also americium, are now available industrially, most to high degrees of purity.

Around two dozen of the elements are essential to various kinds of biological life. Most rare elements on Earth are not needed by life (exceptions being selenium and iodine), while a few quite common ones (aluminum and titanium) are not used. Most organisms share element needs, but there are a few differences between plants and animals. For example, ocean algae use bromine but land plants and animals seem to need none.

All animals require sodium, but some plants do not. Plants need boron and silicon, but animals may not (or may need ultra-small amounts). Just six elements—carbon, hydrogen, nitrogen, oxygen, calcium, and phosphorus—make up almost 99% of the mass of a human body. In addition to the six major elements that compose most of the human body, humans require consumption of at least a dozen more elements.

Description

The lightest of the chemical elements are hydrogen and helium, both created by Big Bang nucleosynthesis during the first 20 minutes of the universe in a ratio of around 3:1 by mass (approximately 12:1 by number of atoms). Almost all other elements found in nature, including some further hydrogen and helium created since then, were made by various natural or (at times) artificial methods of nucleosynthesis. On

Earth, small amounts of new atoms are naturally produced in nucleogenic reactions, or in cosmogenic processes, such as cosmic ray spallation. New atoms are also naturally produced on Earth as radiogenic daughter isotopes of ongoing radioactive decay processes such as alpha decay, beta decay, spontaneous fission, cluster decay, and other rarer modes of decay.

Of the 94 naturally occurring elements, those with atomic numbers 1 through 40 are all considered to be stable isotopes. Elements with atomic numbers 41 through 82 are apparently stable (except technetium, element 43 and promethium, element 61) but theoretically unstable, and thus possibly mildly radioactive. The half-lives of elements 41 through 82 are so long however that their radioactive decay has yet to be detected by experiment. These "theoretical radionuclides" have half-lives at least 100 million times longer than the estimated age of the universe. Elements with atomic numbers 83 through 94 are unstable to the point that their radioactive decay can be detected. Some of these elements, notably thorium (atomic number 90) and uranium (atomic number 92), have one or more isotopes with half-lives long enough to survive as remnants of the explosive stellar nucleosynthesis that produced the heavy elements before the formation of our solar system. For example, at over 1.9×10 years, over a billion times longer than the current estimated age of the universe, bismuth-209 (atomic number 83) has the longest known alpha decay half-life of any naturally occurring element. The very heaviest elements (those beyond plutonium, atomic number 94) undergo radioactive decay with half-lives so short that they have only been observed as the result of experimental observation.

As of 2010, there are 118 known elements (in this context, "known" means observed well-enough, even from just a few decay products, to have been differentiated from any other element). Of these 118 elements, 94 occur naturally on Earth. Six of these occur in extreme trace quantities: technetium, atomic number 43; promethium, number 61; astatine, number 85; francium, number 87; neptunium, number 93; and plutonium, number 94. These 94 elements have been detected in the universe at large, in the spectra of stars and also supernovae, where short-lived radioactive elements are newly being made. The first 94 elements have been detected directly on Earth as primordial nuclides present from the formation of the solar system, or as naturally-occurring fission or transmutation products of uranium and thorium.

The remaining 24 heavier elements, not found today either on Earth or in astronomical spectra, have been derived artificially. All of the heavy elements that are derived solely through artificial means are radioactive, with very short half-lives; if any atoms of these elements were present at the formation of Earth, they are extremely likely to have already decayed, and if present in novae, have been in quantities too small to have been noted.

Technetium was the first purportedly non-naturally occurring element to be synthesized, in 1937, although trace amounts of technetium have since been found in nature (and also the element may have been discovered naturally in 1925). This pattern of artificial production and later natural discovery has been repeated with several other radioactive naturally-occurring rare elements.

Lists of the elements are available by name, by symbol, by atomic number, by density, by melting point, and by boiling point as well as Ionization energies of the elements. The nuclides of stable and radioactive elements are also available as a list of nuclides, sorted by length of half-life for those that are unstable. One of the most convenient, and certainly the most traditional presentation of the elements, is in form of periodic table, which groups elements with similar chemical properties (and usually also similar electronic structures) together.

Atomic Number

The atomic number of an element is equal to the number of protons that defines the element. For example, all carbon atoms contain 6 protons in their nucleus; so the atomic number of carbon is 6. Carbon atoms may have different numbers of neutrons; atoms of the same element having different numbers of neutrons are known as isotopes of the element.

The number of protons in the atomic nucleus also determines its electric charge, which in turn determines the number of electrons of the atom in its non-ionized state. The electrons are placed into atomic orbitals which determine the atom's various chemical properties. The number of neutrons in a nucleus usually has very little effect on an elements' chemical properties (except in the case of hydrogen and deuterium).

Thus, all carbon isotopes have nearly identical chemical properties because they all have six protons and six electrons, even though carbon atoms may differ in number of neutrons. It is for this reason that atomic number rather than mass number or atomic weight is considered the identifying characteristic of a chemical element. The symbol for atomic number is Z.

Atomic Mass and Atomic Weight

The mass number of an element, *A*, is the number of nucleons (protons and neutrons) in the atomic nucleus. Different isotopes of a given element are distinguished by their mass numbers, which are conventionally written as a super-index on the left hand side of the atomic symbol (e.g., U). The mass number is always a simple whole number and has units of "nucleons." An example of use of a mass number is "magnesium-24," which has 24 nucleons (12 protons and 12 neutrons).

Whereas the mass number simply counts the total number of neutrons and protons and is thus a natural (or whole) number, the atomic mass of a single isotope is a real number. In general, it differs in value when expressed in u for a given nuclide (or isotope) slightly from the mass number, since the mass of the protons and neutrons is not exactly 1 u, the electrons contribute a lesser share to the atomic mass as neutron number exceeds proton number, and (finally) because of the nuclear binding energy.

For example, the atomic weight of chlorine-35 to five significant digits is 34.969 u and that of chlorine-37 is 36.966 u. However, the atomic mass in u of pure isotope atoms is quite close (always within 1%) to its simple mass number. The only exception to the atomic mass of an isotope atom not being a natural number is C, which has a mass of exactly 12 by definition, because u is *defined* as 1/12 of the mass of a free neutral carbon-12 atom in the ground state.

The relative atomic mass (historically and commonly also called "atomic weight") of an element is the *average* of the atomic masses of all the chemical element's isotopes as found in a particular environment, weighted by isotopic abundance, relative to the atomic mass unit (u). This number may be a fraction that is *not* close to a whole number, due to the averaging process. For example, the relative atomic mass of chlorine is 35.453 u, which differs greatly from a whole number due to being made of an average of 76% chlorine-35 and 24% chlorine-37. Whenever a relative atomic mass value differs by more than 1% from a whole number, it is due to this averaging effect resulting from significant amounts of more than one isotope being naturally present in the sample of the element in question.

Isotopes

Isotopes are atoms of the same element (that is, with the same number of protons in their atomic nucleus), but having *different* numbers of neutrons. Most (66 of 94) naturally occurring elements have more than one stable isotope. Thus, for example, there are three main isotopes of carbon. All carbon atoms have 6 protons in the nucleus, but they can have either 6, 7, or 8 neutrons. Since the mass numbers of these are 12, 13 and 14 respectively, the three isotopes of carbon are known as carbon-12, carbon-13, and carbon-14, often abbreviated to ^{12}C, ^{13}C, and ^{14}C. Carbon in everyday life and in chemistry is a mixture of ^{12}C, ^{13}C, and (a very small fraction of) ^{14}C atoms.

Except in the case of the isotopes of hydrogen (which differ greatly from each other in relative mass—enough to cause chemical effects), the isotopes of the various elements are typically chemically nearly indistinguishable from each other. For example, the three naturally-occurring isotopes of carbon have essentially the same chemical properties, but different nuclear properties. In this example, carbon-12 and carbon-13 are stable atoms, but carbon-14 is unstable; it is radioactive, undergoing beta decay into nitrogen-14.

All of the elements have some isotopes that are radioactive (radioisotopes), although not all of these radioisotopes occur naturally. The radioisotopes typically decay into other elements upon radiating an alpha or beta particle. If an element has isotopes that are not radioactive, they are termed "stable." All of the known stable isotopes occur naturally. The many radioisotopes that are not found in nature have been characterized from being artificially made. Certain elements have no stable isotopes and are composed *only* of radioactive isotopes: specifically the elements without any stable isotopes are technetium (atomic number 43), promethium (atomic number 61), and all observed elements with atomic numbers greater than 82.

Of the 80 elements with at least one stable isotope, 26 have only one stable isotope, and the mean number of stable isotopes for the 80 stable elements is 3.1 stable isotopes per element. The largest number of stable isotopes that occur for an element is 10 (for tin, element 50).

Allotropes

Atoms of pure elements may bond to each other chemically in more than one way, allowing the pure element to exist in multiple structures (spacial arrangements of atoms), known as allotropes, which differ in their properties.

For example, carbon can be found as diamond, which has a tetrahedral structure around each carbon atom; graphite, which has layers of carbon atoms with a hexagonal structure stacked on top of each other; graphene, which is a single layer of graphite that is incredibly strong; fullerenes, which have nearly spherical shapes; and carbon nanotubes, which are tubes with a hexagonal structure (even these may differ from each other in electrical properties). The ability for an element to exist in one of many structural forms is known as 'allotropy'.

The standard state, or reference state, of an element is defined as its thermodynamically most stable state at 1 bar at a given temperature (typically at 298.15 K). In thermochemistry, an element is defined to have an enthalpy of formation of zero in its standard state. For example, the reference state for carbon is graphite, because it is more stable than the other allotropes.

Properties

Several kinds of descriptive categorizations can be applied broadly to the elements, including consideration of their general physical and chemical properties, their states of matter under familiar conditions, their melting and boiling points, their densities, their crystal structures as solids, and their origins.

General Properties

Several terms are commonly used to characterize the general physical and chemical properties of the chemical elements. A first distinction is between the metals, which readily conduct electricity, and the nonmetals, which do not, with a small group (the *metalloids*) having intermediate properties, often behaving as semiconductors.

A more refined classification is often shown in colored presentations of the periodic table; this system restricts the terms "metal" and "nonmetal" to only certain of the more broadly defined metals and nonmetals, adding additional terms for certain sets of the more broadly viewed metals and nonmetals.

The version of this classification used in the periodic tables presented here includes: actinides, alkali metals, alkaline earth metals, halogens, lanthanides, metals (or "other metals"), metalloids, noble gases, nonmetals (or "other nonmetals"), and transition metals. In this system, the alkali metals, alkaline earth metals, and transition metals, as well as the lanthanides and the actinides, are special groups of the metals viewed in a broader sense. Similarly, the halogens and the noble gases are nonmetals, viewed in the broader sense. In some presentations, the halogens are not distinguished, with astatine identified as a metalloid and the others identified as nonmetals.

States of Matter

Another commonly used basic distinction among the elements is their state of matter (phase), solid, liquid, or gas, at a selected standard temperature and pressure (STP). Most of the elements are solids at conventional temperatures and atmospheric pressure, while several are gases. Only bromine and mercury are liquids at 0 degrees Celsius (32 degrees Fahrenheit) and normal atmospheric pressure; caesium and gallium are solids at that temperature, but melt at 28.4 °C (83.2 °F) and 29.8 °C (85.6 °F), respectively.

Melting and Boiling Points

Melting and boiling points, typically expressed in degrees Celsius at a pressure of one atmosphere, are commonly used in characterizing the various elements. While known for most elements, either or both of these measurements is still undetermined for some of the radioactive elements available in only tiny quantities. Since helium remains a liquid even at absolute zero at atmospheric pressure, it has only a boiling point, and not a melting point, in conventional presentations.

Densities

The density at a selected standard temperature and pressure (STP) is frequently used in characterizing the elements. Density is often expressed in grams per cubic centimeter (g/cm^3). Since several elements are gases at commonly encountered temperatures, their densities are usually stated for their gaseous forms; when liquefied or solidified, the gaseous elements have densities similar to those of the other elements.

When an element has allotropes with different densities, one representative allotrope is typically selected in summary presentations, while densities for each allotrope can be stated where more detail is provided. For example, the three familiar allotropes of carbon (amorphous carbon, graphite, and diamond) have densities of 1.8–2.1, 2.267, and 3.515 g/cm^3 respectively.

Crystal Structures

The elements studied to date as solid samples have eight kinds of crystal structures: cubic, body-centered cubic, face-centered cubic, hexagonal, monoclinic, orthorhombic, rhombohedral, and tetragonal. For some of the synthetically produced transuranic elements, available samples have been too small to determine crystal structures.

Occurrence and Origin on Earth

The elements may also be categorized by their origin on Earth, with the first 94 considered naturally occurring, and those with atomic numbers beyond 94 being synthetic (produced technologically, but not known to occur naturally).

Of the naturally occurring elements, 84 are considered primordial, either stable or long-persisting, and the remaining 10 are transient elements with half lives too short to have been present at the beginning of the solar system. These are produced either "recurrently" (often found) or "incidentally" (very rare) as decay products or through other natural nuclear reactionprocesses. The 91 regularly occurring natural elements include the 80 stable or essentially stable, primordial elements (from hydrogen through lead, omitting technetium and promethium); bismuth, thorium, uranium, and plutonium (radioactive but still remaining from primordial times); and the 7 transiently existing but recurrently produced decay products of thorium, uranium, and plutonium (polonium, astatine, radon, francium, radium, actinium, and protactinium).

Three additional naturally occurring elements, technetium, promethium, and neptunium, are only incidentally occurring, present in natural materials only as transiently existing atoms produced from uranium or other heavy elements by rare nuclear processes. Note that helium is recurrently produced naturally from radioactive decay, but little if any primordial helium still exists at the Earth's surface, since this light gas readily escapes from the atmosphere into outer space.

The Periodic Table

The properties of the chemical elements are often summarized using the periodic table, which powerfully and elegantly organizes the elements by increasing atomic number into rows ("periods") in which the columns ("groups") share recurring ("periodic") physical and chemical properties. The current standard table contains 118 confirmed elements as of April 10, 2010.

Although earlier precursors to this presentation exist, its invention is generally credited to Russian chemist Dmitri Mendeleev in 1869, who intended the table to illustrate recurring trends in the properties of the elements. The layout of the table has been refined and extended over time, as new elements have been discovered, and new theoretical models have been developed to explain chemical behavior.

Use of the periodic table is now ubiquitous within the academic discipline of chemistry, providing an extremely useful framework to classify, systematize and compare all the many different forms of chemical behavior. The table has also found wide application in physics, geology, biology, materials science, engineering, agriculture, medicine, nutrition, environmental health, and astronomy. Its principles are especially important in chemical engineering.

Nomenclature and Symbols

The various chemical elements are formally identified by their unique atomic numbers, by their accepted names, and by their symbols.

Atomic Numbers

The known elements have atomic numbers from 1 through 118, conventionally presented as Arabic numerals. Since the elements can be uniquely sequenced by atomic number, conventionally from lowest to hightest (as in a periodic table), sets of elements are sometimes specified by such notation as "through", "beyond", or "from ... through", as in "through iron", "beyond uranium", or "from lanthanum through lutetium". The terms "light" and "heavy" are sometimes also used informally to indicate relative atomic numbers (not densities!), as in "lighter than carbon" or "heavier than lead", although technically the weight or mass of atoms of an element (their atomic weights or atomic masses) do not always increase monotonically with their atomic numbers.

Element Names

The naming of various substances now known as elements precedes the atomic theory of matter, as names were given locally by various cultures to various minerals, metals, compounds, alloys, mixtures, and other materials, although at the time it was not known which chemicals were elements and which compounds. As they were identified as elements, the existing names for anciently-known elements (*e.g.*, gold, mercury, iron) were kept in most countries. National differences emerged over the names of elements either for convenience, linguistic niceties, or nationalism. For a few illustrative examples: German speakers use "Wasserstoff" (water substance) for "hydrogen", "Sauerstoff" (acid substance) for "oxygen" and "Stickstoff" (smothering substance) for "nitrogen", while English and some romance languages use "sodium" for "natrium" and "potassium" for "kalium", and the French, Italians, Greeks, Portuguese and Poles prefer "azote/azot/azoto" (from roots meaning "no life") for "nitrogen".

For purposes of international communication and trade, the official names of the chemical elements both ancient and more recently recognized are decided by the International Union of Pure and Applied Chemistry (IUPAC), which has decided on a sort of international English language, drawing on traditional English names even when an element's chemical symbol is based on a Latin or other traditional word, for example adopting "gold" rather than "aurum" as the name for the 79th element (Au). IUPAC prefers the British spellings "aluminium" and "caesium" over the U.S. spellings "aluminum" and "cesium", and the U.S. "sulfur" over the British "sulphur". However, elements that are practical to sell in bulk in many countries often still have locally used national names, and countries whose national language does not use the Latin alphabet are likely to use the IUPAC element names.

According to IUPAC, chemical elements are not proper nouns in English; consequently, the full name of an element is not routinely capitalized in English, even if derived from a proper noun, as in californium and einsteinium. Isotope names of chemical elements are also uncapitalized if written out, *e.g.*, carbon-12 or uranium-235. Chemical element *symbols* are always capitalized.

In the second half of the twentieth century, physics laboratories became able to produce nuclei of chemical elements with half-lives too short for an appreciable amount of them to exist at any time. These are also named by IUPAC, which generally adopts the name chosen by the discoverer. This practice can lead to the controversial question of which research group actually discovered an element, a question that has delayed naming of elements with atomic number of 104 and higher for a considerable time.

Precursors of such controversies involved the nationalistic namings of elements in the late 19th century. For example, *lutetium* was named in reference to Paris, France. The Germans were reluctant to relinquish naming rights to the French, often calling it *cassiopeium*. Similarly, the British discoverer of *niobium* originally named it *columbium,* in reference to the New World. It was used extensively as such by American publications prior to international standardization.

Chemical Symbols

For the listing of current and not used Chemical symbols, and other symbols that look like chemical symbols, see List of elements by symbol.

Specific Chemical Elements

Before chemistry became a science, alchemists had designed arcane symbols for both metals and common compounds. These were however used as abbreviations in diagrams or procedures; there was no concept of atoms combining to form molecules. With his advances in the atomic theory of matter, John Dalton devised his own simpler symbols, based on circles, which were to be used to depict molecules.

The current system of chemical notation was invented by Berzelius. In this typographical system chemical symbols are not used as mere abbreviations – though each consists of letters of the Latin alphabet – they are symbols intended to be used by peoples of all languages and alphabets. The first of these symbols were intended to be fully universal; since Latin was the common language of science at that time, they were abbreviations based on the Latin names of metals – Cu comes from Cuprum, Fe comes from Ferrum, Ag from Argentum. The symbols were not followed by a period (full stop) as abbreviations were. Later chemical elements were also assigned unique chemical symbols, based on the name of the element, but not necessarily in English. For example, sodium has the chemical symbol 'Na' after the Latin *natrium*. The same applies to "W" (wolfram) for tungsten, "Fe" (ferrum) for iron, "Hg" (hydrargyrum) for mercury, "Sn" (stannum) for tin, "K" (kalium) for potassium, "Au" (aurum) for gold, "Ag" (argentum) for silver, "Pb" (plumbum) for lead, "Cu" (Cuprum) for copper, and "Sb" (stibium) for antimony.

Chemical symbols are understood internationally when element names might need to be translated. There are sometimes differences; for example, the Germans have used "J" instead of "I" for iodine, so the character would not be confused with a Roman numeral.

The first letter of a chemical symbol is always capitalized, as in the preceding examples, and the subsequent letters, if any, are always lower case (small letters). Thus, the symbols for californium or einsteinium are Cf and Es.

General Chemical Symbols

There are also symbols for series of chemical elements, for comparative formulas. These are one capital letter in length, and the letters are reserved so they are not permitted to be given for the names of specific elements. For example, an "X" is used to indicate a variable group amongst a class of compounds (though usually a halogen), while "R" is used for a radical, meaning a compound structure such as a hydrocarbon chain. The letter "Q" is reserved for "heat" in a chemical reaction. "Y" is also often used as a general chemical symbol, although it is also the symbol of yttrium. "Z" is also frequently used as a general variable group. "L" is used to represent a general ligand in inorganic and organometallic chemistry. "M" is also often used in place of a general metal. At least one additional, two-letter generic chemical symbol is also in informal usage, "Ln" for any lanthanide element.

Isotope Symbols

Isotopes are distinguished by the atomic mass number (total protons and neutrons) for a particular isotope of an element, with this number combined with the pertinent element's symbol. IUPAC prefers that isotope symbols be written in superscript notation when practical, for example ^{12}C and ^{235}U. However, other notations, such as carbon-12 and uranium-235, or C-12 and U-235, are also used.

As a special case, the three naturally occurring isotopes of the element hydrogen are often specified as H for ^{1}H (protium), D for ^{2}H (deuterium), and T for ^{3}H (tritium). This convention is easier to use in chemical equations, replacing the need to write out the mass number for each atom. For example, the formula for heavy water may be written D_2O instead of $^{2}H_2O$.

Origin of the Elements

Only about 4% of the total mass of the universe is made of atoms or ions, and thus represented by chemical elements. This fraction is about 15% of the total matter, with the remainder of the matter (85%) being dark matter. The nature of dark matter is unknown, but it is not composed of atoms of chemical elements because it contains no protons, neutrons, or electrons. (The remaining non-matter part of the mass of the universe is composed of the even more mysterious dark energy).

The universe's 94 naturally occurring chemical elements are thought to have been produced by at least four cosmic processes. Most of the hydrogen and helium in the universe was produced primordially in the first few minutes of the Big Bang. Three recurrently occurring later processes are thought to have produced the remaining elements. Stellar nucleosynthesis, an ongoing process, produces all elements from carbon through iron in atomic number, but little lithium, beryllium, or boron. Elements heavier in atomic number than iron, as heavy as uranium and plutonium, are produced by explosive nucleosynthesis in supernovas and other cataclysmic cosmic events. Cosmic ray spallation (fragmentation) of carbon, nitrogen, and oxygen is important to the production of lithium, beryllium and boron.

During the early phases of the Big Bang, nucleosynthesis of hydrogen nuclei resulted in the production of hydrogen-1 (protonium, ^{1}H) and helium-4 (^{4}He), as well as a smaller amount of deuterium (^{2}H) and very minuscule amounts (on the order of 10^{-10}) of lithium and beryllium. Even smaller amounts of boron may have been produced in the Big Bang, since it has been observed in some very old stars, while carbon has not.

It is generally agreed that no heavier elements than boron were produced in the Big Bang. As a result, the primordial abundance of atoms (or ions) consisted of roughly 75% ^{1}H, 25% ^{4}He, and 0.01%

deuterium, with only tiny traces of lithium, beryllium, and perhaps boron. Subsequent enrichment of galactic halos occurred due to stellar nucleosynthesis and supernova nucleosynthesis. However, the element abundance in intergalactic space can still closely resemble primordial conditions, unless it has been enriched by some means.

On Earth (and elsewhere), trace amounts of various elements continue to be produced from other elements as products of natural transmutation processes. These include some produced by cosmic rays or other nuclear reactions, and others produced as decay products of long-lived primordial nuclides.

For example, trace (but detectable) amounts of carbon-14 (^{14}C) are continually produced in the atmosphere by cosmic rays impacting nitrogen atoms, and argon-40 (^{40}Ar) is continually produced by the decay of primordially occurring but unstable potassium-40 (^{40}K). Also, three primordially occurring but radioactive actinides, thorium, uranium, and plutonium, decay through a series of recurrently produced but unstable radioactive elements such as radium and radon, which are transiently present in any sample of these metals or their ores or compounds. Three other radioactive elements, technetium, promethium, and neptunium, occur only incidentally in natural materials, produced as individual atoms by natural fission of the nuclei of various heavy elements or in other rare nuclear processses.

Human technology has produced various additional elements beyond these first 94, with those through atomic number 118 now known.

Abundance

The following graph (note log scale) shows abundance of elements in our solar system. The table shows the twelve most common elements in our galaxy (estimated spectroscopically), as measured in parts per million, by mass. Nearby galaxies that have evolved along similar lines have a corresponding enrichment of elements heavier than hydrogen and helium. The more distant galaxies are being viewed as they appeared in the past, so their abundances of elements appear closer to the primordial mixture. As physical laws and processes appear common throughout the visible universe, however, it is expected that these galaxies will likewise have evolved similar abundances of elements.

The abundance of the chemical elements on Earth varies from air to crust to ocean, and in various types of life. The abundance of elements in Earth's crust differs from those in the universe mainly in selective loss of the very lightest elements (hydrogen and helium) and also volatile neon, carbon, nitrogen and sulfur, as a result of solar heating in the early formation of the solar system. Aluminum is also far more common in the Earth and Earth's crust than the universe and solar system.

The composition of the human body, by contrast, more closely follows the composition of seawater, save that the human body has additional stores of carbon and nitrogen which are necessary to form the proteins and nucleic acids that are characteristic of living organisms.

Certain kinds of organisms require particular additional elements, for example the magnesium in chlorophyll in green plants, the calcium in mollusc shells, or the iron in the hemoglobin in vertebrate animals' red blood cells.

Element	***Parts per millionby mass***
Hydrogen	739,000
Helium	240,000
Oxygen	10,400
Carbon	4,600
Neon	1,340
Iron	1,090
Nitrogen	960
Silicon	650
Magnesium	580
Sulfur	440
Potassium	210
Nickel	100

History

Evolving Definitions: The concept of an "element" as an undivisible substance has developed through three major historical phases: Classical definitions (such as those of the ancient Greeks), chemical definitions, and atomic definitions.

Classical Definitions

Ancient philosophy posited a set of classical elements to explain observed patterns in nature. These *elements* originally referred to *earth*, *water*, *air* and *fire* rather than the chemical elements of modern science.

The term 'elements' (*stoicheia*) was first used by the Greek philosopher Plato in about 360 BCE, in his dialogue Timaeus, which includes a discussion of the composition of inorganic and organic bodies and is a speculative treatise on chemistry. Plato believed the elements introduced a century earlier by Empedocles were composed of small polyhedral *forms*: tetrahedron (fire), octahedron (air), icosahedron (water), and cube (earth).

Aristotle, c. 350 BCE, also used the term *stoicheia* and added a fifth element called *aether,* which formed the heavens. Aristotle defined an element as:

Element – one of those bodies into which other bodies can decompose, and that itself is not capable of being divided into other.

Chemical Definitions

In 1661, chemist Robert Boyle showed that there were more than just the four classical elements that the ancients had assumed. The first modern list of chemical elements was given in Antoine Lavoisier's 1789 *Elements of Chemistry*, which contained thirty-three elements, including light and caloric. By 1818, Jons Jakob Berzelius had determined atomic weights for forty-five of the forty-nine then-accepted accepted elements. Dmitri Mendeleev had sixty-six elements in his periodic table of 1869.

From Boyle until the early 20th century, an element was defined as a pure substance that could not be decomposed into any simpler substance. Put another way, a chemical element cannot be transformed into other chemical elements by chemical processes. Elements during this time were generally distinguished by their atomic weights, a property measurable with fair accuracy by available analytical techniques.

Atomic Definitions

The 1913 discovery by Henry Moseley that the nuclear charge is the physical basis for an atom's atomic number, further refined when the nature of protons and neutrons became appreciated, eventually led to the current definition of an element, based on atomic number (number of protons per atomic nucleus).

The use of atomic numbers, rather than atomic weights, to distinguish elements has greater predictive value (since these numbers are integers), and also resolves some ambiguities in the chemistry-based view due to varying properties of isotopes and allotropes within the same element. Currently IUPAC defines an element to exist if it has isotopes with a lifetime longer than the 10^{-14} seconds which takes the nucleus to form an electronic cloud.

By 1914, seventy-two elements were known, all naturally occurring. The remaining naturally occurring elements were discovered or isolated is subsequent decades, and various additional elements have also been produced synthetically, with much of that work pioneered by Glenn T. Seaborg. In 1955, element 101 was discovered and named mendelevium in honor of D.I. Mendeleev, the first to arrange the elements in a periodic manner. Most recently, the synthesis of element 118 was reported in October 2006, and the synthesis of element 117 was reported in April 2010.

Discovery and Recognition of Various Elements

Ten materials familiar to various prehistoric cultures are now known to be chemical elements: Carbon, copper, gold, iron, lead, mercury, silver, sulfur, tin, and zinc. Three additional materials now accepted as elements, arsenic, antimony, and bismuth, were recognized as distinct substances prior to 1500 AD. Phosphorus, cobalt, and platinum were isolated before 1750.

Most of the remaining naturally occurring chemical elements were identified and characterized by 1900, including:

- Such now-familiar industrial materials as aluminum, silicon, nickel, chromium, magnesium, and tungsten
- Reactive metals such as lithium, sodium, potassium, and calcium
- The halogens fluorine, chlorine, bromine, and iodine
- Gases such as hydrogen, oxygen, nitrogen, helium, argon, and neon
- Most of the rare-earth elements, including cerium, lanthanum, gadolinium, and neodymium, and
- The more common radioactive elements, including uranium, thorium, radium, and radon.

Elements isolated or produced since 1900 include:

- The three remaining undiscovered regularly occurring stable natural elements: hafnium, lutetium, and rhenium
- Plutonium, first produced synthetically but now also known from a few long-persisting natural occurrences
- The three incidentally occurring natural elements (neptunium, promethium, and technetium), all first produced synthetically but later discovered in trace amounts in certain geological samples
- Three scarcer decay products of uranium or thorium (astatine, francium, and protactinium),

- Various synthetic transuranic elements, beginning with americium, curium, berkelium, and californium

Recently Discovered Elements

The first transuranium element (element with atomic number greater than 92) discovered was neptunium in 1940. As of February 2010, only the elements up to 112, copernicium, have been confirmed as discovered by IUPAC, while claims have been made for synthesis of elements 113, 114, 115, 116, 117 and 118.

The discovery of element 112 was acknowledged in 2009, and the name 'copernicium' and the atomic symbol 'Cn' were suggested for it. The name and symbol were officially endorsed by IUPAC on February 19, 2010. The heaviest element that is believed to have been synthesized to date is element 118, ununoctium, on October 9, 2006, by the Flerov Laboratory of Nuclear Reactions in Dubna, Russia.

Element 117 was the latest element claimed to be discovered, in 2009. IUPAC officially recognized ununquadium and ununhexium, elements 114 and 116, in June 2011.

List of the 118 Known Chemical Elements

The following sortable table includes the 118 known chemical elements, with the names linking to the *Wikipedia* articles on each.

- Atomic number, name, and symbol all serve independently as unique identifiers.
- Names are those accepted by IUPAC; provisional names for recently produced elements not yet formally named are in parentheses.
- Group, period, and block refer to an element's position in the periodic table.
- State of matter *(solid, liquid,* or *gas)* applies at standard temperature and pressure conditions (STP).
- Occurrence distinguishes naturally occurring elements, categorized as either *primordial* or *transient* (from decay), and additional *synthetic* elements that have been produced technologically, but are not known to occur naturally.
- Description summarizes an element's properties using the broad categories commonly presented in periodic tables:

  Actinide, alkali metal, alkaline earth metal, halogen, lanthanide, metal, metalloid, noble gas, non-metal, and *transition metal.*

List of Elements

Atomic no.	*Name*	*Symbol*	*Group*	*Period*	*Block*	*State at STP*	*Occurrence*	*Description*
1	Hydrogen	H	1	1	s	Gas	Primordial	Non-metal
2	Helium	He	18	1	s	Gas	Primordial	Noble gas
3	Lithium	Li	1	2	s	Solid	Primordial	Alkali metal
4	Beryllium	Be	2	2	s	Solid	Primordial	Alkaline earth metal
5	Boron	B	13	2	p	Solid	Primordial	Metalloid
6	Carbon	C	14	2	p	Solid	Primordial	Non-metal
7	Nitrogen	N	15	2	p	Gas	Primordial	Non-metal
8	Oxygen	O	16	2	p	Gas	Primordial	Non-metal
9	Fluorine	F	17	2	p	Gas	Primordial	Halogen
10	Neon	Ne	18	2	p	Gas	Primordial	Noble gas
11	Sodium	Na	1	3	s	Solid	Primordial	Alkali metal
12	Magnesium	Mg	2	3	s	Solid	Primordial	Alkaline earth metal
13	Aluminium	Al	13	3	p	Solid	Primordial	Metal
14	Silicon	Si	14	3	p	Solid	Primordial	Metalloid
15	Phosphorus	P	15	3	p	Solid	Primordial	Non-metal
16	Sulfur	S	16	3	p	Solid	Primordial	Non-metal
17	Chlorine	Cl	17	3	p	Gas	Primordial	Halogen
18	Argon	Ar	18	3	p	Gas	Primordial	Noble gas
19	Potassium	K	1	4	s	Solid	Primordial	Alkali metal
20	Calcium	Ca	2	4	s	Solid	Primordial	Alkaline earth metal
21	Scandium	Sc	3	4	d	Solid	Primordial	Transition metal
22	Titanium	Ti	4	4	d	Solid	Primordial	Transition metal
23	Vanadium	V	5	4	d	Solid	Primordial	Transition metal
24	Chromium	Cr	6	4	d	Solid	Primordial	Transition metal
25	Manganese	Mn	7	4	d	Solid	Primordial	Transition metal
26	Iron	Fe	8	4	d	Solid	Primordial	Transition metal
27	Cobalt	Co	9	4	d	Solid	Primordial	Transition metal
28	Nickel	Ni	10	4	d	Solid	Primordial	Transition metal

Contd...

Atomic no.	*Name*	*Symbol*	*Group*	*Period*	*Block*	*State at STP*	*Occurrence*	*Description*
29	Copper	Cu	11	4	d	Solid	Primordial	Transition metal
30	Zinc	Zn	12	4	d	Solid	Primordial	Transition metal
31	Gallium	Ga	13	4	p	Solid	Primordial	Metal
32	Germanium	Ge	14	4	p	Solid	Primordial	Metalloid
33	Arsenic	As	15	4	p	Solid	Primordial	Metalloid
34	Selenium	Se	16	4	p	Solid	Primordial	Non-metal
35	Bromine	Br	17	4	p	Liquid	Primordial	Halogen
36	Krypton	Kr	18	4	p	Gas	Primordial	Noble gas
37	Rubidium	Rb	1	5	s	Solid	Primordial	Alkali metal
38	Strontium	Sr	2	5	s	Solid	Primordial	Alkaline earth metal
39	Yttrium	Y	3	5	d	Solid	Primordial	Transition metal
40	Zirconium	Zr	4	5	d	Solid	Primordial	Transition metal
41	Niobium	Nb	5	5	d	Solid	Primordial	Transition metal
42	Molybdenum	Mo	6	5	d	Solid	Primordial	Transition metal
43	Technetium	Tc	7	5	d	Solid	Transient	Transition metal
44	Ruthenium	Ru	8	5	d	Solid	Primordial	Transition metal
45	Rhodium	Rh	9	5	d	Solid	Primordial	Transition metal
46	Palladium	Pd	10	5	d	Solid	Primordial	Transition metal
47	Silver	Ag	11	5	d	Solid	Primordial	Transition metal
48	Cadmium	Cd	12	5	d	Solid	Primordial	Transition metal
49	Indium	In	13	5	p	Solid	Primordial	Metal
50	Tin	Sn	14	5	p	Solid	Primordial	Metal
51	Antimony	Sb	15	5	p	Solid	Primordial	Metalloid
52	Tellurium	Te	16	5	p	Solid	Primordial	Metalloid
53	Iodine	I	17	5	p	Solid	Primordial	Halogen
54	Xenon	Xe	18	5	p	Gas	Primordial	Noble gas

Contd...

Atomic no.	*Name*	*Symbol*	*Group*	*Period*	*Block*	*State at STP*	*Occurrence*	*Description*
55	Caesium	Cs	1	6	s	Solid	Primordial	Alkali metal
56	Barium	Ba	2	6	s	Solid	Primordial	Alkaline earth metal
57	Lanthanum	La	3	6	f	Solid	Primordial	Lanthanide
58	Cerium	Ce	3	6	f	Solid	Primordial	Lanthanide
59	Praseodymium		Pr	3	6	f	Solid	Primordial Lanthanide
60	Neodymium	Nd	3	6	f	Solid	Primordial	Lanthanide
61	Promethium	Pm	3	6	f	Solid	Transient	Lanthanide
62	Samarium	Sm	3	6	f	Solid	Primordial	Lanthanide
63	Europium	Eu	3	6	f	Solid	Primordial	Lanthanide
64	Gadolinium	Gd	3	6	f	Solid	Primordial	Lanthanide
65	Terbium	Tb	3	6	f	Solid	Primordial	Lanthanide
66	Dysprosium	Dy	3	6	f	Solid	Primordial	Lanthanide
67	Holmium	Ho	3	6	f	Solid	Primordial	Lanthanide
68	Erbium	Er	3	6	f	Solid	Primordial	Lanthanide
69	Thulium	Tm	3	6	f	Solid	Primordial	Lanthanide
70	Ytterbium	Yb	3	6	f	Solid	Primordial	Lanthanide
71	Lutetium	Lu	3	6	d	Solid	Primordial	Lanthanide
72	Hafnium	Hf	4	6	d	Solid	Primordial	Transition metal
73	Tantalum	Ta	5	6	d	Solid	Primordial	Transition metal
74	Tungsten	W	6	6	d	Solid	Primordial	Transition metal
75	Rhenium	Re	7	6	d	Solid	Primordial	Transition metal
76	Osmium	Os	8	6	d	Solid	Primordial	Transition metal
77	Iridium	Ir	9	6	d	Solid	Primordial	Transition metal
78	Platinum	Pt	10	6	d	Solid	Primordial	Transition metal
79	Gold	Au	11	6	d	Solid	Primordial	Transition metal
80	Mercury	Hg	12	6	d	Liquid	Primordial	Transition metal
81	Thallium	Tl	13	6	p	Solid	Primordial	Metal
82	Lead	Pb	14	6	p	Solid	Primordial	Metal

Contd...

Atomic no.	*Name*	*Symbol*	*Group*	*Period*	*Block*	*State at STP*	*Occurrence*	*Description*
83	Bismuth	Bi	15	6	p	Solid	Primordial	Metal
84	Polonium	Po	16	6	p	Solid	Transient	Metalloid
85	Astatine	At	17	6	p	Solid	Transient	Halogen
86	Radon	Rn	18	6	p	Gas	Transient	Noble gas
87	Francium	Fr	1	7	s	Solid	Transient	Alkali metal
88	Radium	Ra	2	7	s	Solid	Transient	Alkaline earth metal
89	Actinium	Ac	3	7	f	Solid	Transient	Actinide
90	Thorium	Th	3	7	f	Solid	Primordial	Actinide
91	Protactinium	Pa	3	7	f	Solid	Transient	Actinide
92	Uranium	U	3	7	f	Solid	Primordial	Actinide
93	Neptunium	Np	3	7	f	Solid	Transient	Actinide
94	Plutonium	Pu	3	7	f	Solid	Primordial	Actinide
95	Americium	Am	3	7	f	Solid	Synthetic	Actinide
96	Curium	Cm	3	7	f	Solid	Synthetic	Actinide
97	Berkelium	Bk	3	7	f	Solid	Synthetic	Actinide
98	Californium	Cf	3	7	f	Solid	Synthetic	Actinide
99	Einsteinium	Es	3	7	f	Solid	Synthetic	Actinide
100	Fermium	Fm	3	7	f	Solid	Synthetic	Actinide
101	Mendel-evium	Md	3	7	f	Solid	Synthetic	Actinide
102	Nobelium	No	3	7	f	Solid	Synthetic	Actinide
103	Lawrencium	Lr	3	7	d	Solid	Synthetic	Actinide
104	Rutherfordium		Rf	4	7	d	Synthetic	Transition metal
105	Dubnium	Db	5	7	d		Synthetic	Transition metal
106	Seaborgium	Sg	6	7	d		Synthetic	Transition metal
107	Bohrium	Bh	7	7	d		Synthetic	Transition metal
108	Hassium	Hs	8	7	d		Synthetic	Transition metal
109	Meitnerium	Mt	9	7	d		Synthetic	
110	Darmstadtium		Ds	10	7	d		Synthetic
111	Roentgenium	Rg	11	7	d		Synthetic	
112	Copernicium	Cn	12	7	d		Synthetic	Transition metal
113	(Ununtrium)	Uut	13	7	p		Synthetic	

Contd...

Atomic no.	*Name*	*Symbol*	*Group*	*Period*	*Block*	*State at STP*	*Occurrence*	*Description*
114	(Ununquadium)	Uuq	14	7	p		Synthetic	
115	(Ununpentium)	Uup	15	7	p		Synthetic	
116	(Ununhexium)	Uuh	16	7	p		Synthetic	
117	(Ununseptium)	Uus	17	7	p		Synthetic	
118	(Ununoctium)	Uuo	18	7	p		Synthetic	

Fossil Fuel

Fossil fuels are fuels formed by natural processes such as anaerobic decomposition of buried dead organisms. The age of the organisms and their resulting fossil fuels is typically millions of years, and sometimes exceeds 650 million years. The fossil fuels, which contain high percentages of carbon, include coal, petroleum, and natural gas. Fossil fuels range from volatile materials with low carbon:hydrogen ratios like methane, to liquid petroleum to nonvolatile materials composed of almost pure carbon, like anthracite coal.

Methane can be found in hydrocarbon fields, alone, associated with oil, or in the form of methane clathrates. It is generally accepted that they formed from the fossilized remains of dead plants by exposure to heat and pressure in the Earth's crust over millions of years. This biogenic theory was first introduced by Georg Agricola in 1556 and later by Mikhail Lomonosov in the 18th century. It was estimated by the Energy Information Administration that in 2007 primary sources of energy consisted of petroleum 36.0%, coal 27.4%, natural gas 23.0%, amounting to an 86.4% share for fossil fuels in primary energy consumption in the world. Non-fossil sources in 2006 included hydroelectric 6.3%, nuclear 8.5%, and others (geothermal, solar, tide, wind, wood, waste) amounting to 0.9 percent. World energy consumption was growing about 2.3% per year.

Fossil fuels are non-renewable resources because they take millions of years to form, and reserves are being depleted much faster than new ones are being made. The production and use of fossil fuels raise environmental concerns. A global movement toward the generation of renewable energy is therefore under way to help meet increased energy needs.

The burning of fossil fuels produces around 21.3 billion tonnes (21.3 gigatonnes) of carbon dioxide (CO_2) per year, but it is estimated that natural processes can only absorb about half of that amount, so there is a net increase of 10.65 billion tonnes of atmospheric carbon dioxide per year (one tonne of atmospheric carbon is equivalent to 44/12 or 3.7 tonnes of carbon dioxide). Carbon dioxide is one of the greenhouse gases that enhances radiative forcing and contributes to global warming, causing the average surface temperature of the Earth to rise in response, which the vast majority of climate scientists agree will cause major adverse effects.

Origin

Petroleum and natural gas are formed by the anaerobic decomposition of remains of organisms including phytoplankton and zooplankton that settled to the sea (or lake) bottom in large quantities under anoxic conditions, millions of years ago. Over geological time, this organic matter, mixed with mud, got buried under heavy layers of sediment.

The resulting high levels of heat and pressure caused the organic matter to chemically alter, first into a waxy material known as kerogen which is found in oil shales, and then with more heat into liquid and gaseous hydrocarbons in a process known as catagenesis. There is a wide range of renewable, or hydrocarbon, compounds in any given fuel mixture.

The specific mixture of hydrocarbons gives a fuel its characteristic properties, such as boiling point, melting point, density, viscosity, etc. Some fuels like natural gas, for instance, contain only very low boiling, gaseous components. Others such as gasoline or diesel contain much higher boiling components. Terrestrial plants, on the other hand, tend to form coal and methane. Many of the coal fields date to the Carboniferous period of Earth's history. Terrestrial plants also form type III kerogen, a source of natural gas.

Importance

Fossil fuels are of great importance because they can be burned (oxidized to carbon dioxide and water), producing significant amounts of energy per unit weight. The use of coal as a fuel predates recorded history. Coal was used to run furnaces for the melting of metal ore. Semi-solid hydrocarbons from seeps were also burned in ancient times, but these materials were mostly used for waterproofing and embalming.

Commercial exploitation of petroleum, largely as a replacement for oils from animal sources (notably whale oil), for use in oil lamps began in the 19th century.

Natural gas, once flared-off as an unneeded byproduct of petroleum production, is now considered a very valuable resource.

Figure: *A petrochemical refinery in Grangemouth, Scotland, UK*

Heavy crude oil, which is much more viscous than conventional crude oil, and tar sands, where bitumen is found mixed with sand and clay, are becoming more important as sources of fossil fuel. Oil shale and similar materials are sedimentary rocks containing kerogen, a complex mixture of high-molecular weight organic compounds, which yield synthetic crude oil when heated (pyrolyzed). These materials have yet to be exploited commercially. These fuels can be employed in internal combustion engines, fossil fuel power stations and other uses.

Prior to the latter half of the 18th century, windmills and watermills provided the energy needed for industry such as milling flour, sawing wood or pumping water, and burning wood or peat provided domestic heat. The widescale use of fossil fuels, coal at first and petroleum later, to fire steam engines enabled the Industrial Revolution. At the same time, gas lights using natural gas or coal gas were coming into wide use. The invention of the internal combustion engine and its use in automobiles and trucks greatly increased the demand for gasoline and diesel oil, both made from fossil fuels. Other forms of transportation, railways and aircraft, also required fossil fuels. The other major use for fossil fuels is in generating electricity and as feedstock for the petrochemical industry. Tar, a leftover of petroleum extraction, is used in construction of roads.

Figure: *An oil well in the Gulf of Mexico*

Levels and Flows

Levels of primary energy sources are the reserves in the ground. Flows are production. The most important part of primary energy sources are the carbon based fossil energy sources. Coal, oil, and natural gas provided 79.6% of primary energy production during 2002 (in million tonnes of oil equivalent (mtoe)) (34.9+23.5+21.2).

Levels (proved reserves) during 2005-2007

- Coal: 997,748 million short tonnes (905 billion metric tonnes), 4,416 billion barrels (702.1 km^3) of oil equivalent
- Oil: 1,119 billion barrels (177.9 km^3) to 1,317 billion barrels (209.4 km^3)
- Natural gas: 6,183-6,381 trillion cubic feet (175-181 trillion cubic metres), 1,161 billion barrels ($184.6 \times 10^{\triangle 9}$ m^3) of oil equivalent

Flows (daily production) during 2006

- Coal: 18,476,127 short tonnes (16,761,260 metric tonnes), 52,000,000 barrels (8,300,000 m^3) of oil equivalent per day
- Oil: 84,000,000 barrels per day (13,400,000 m^3/d)
- Natural gas: 104,435 billion cubic feet (2,960 billion cubic metres), 19,000,000 barrels (3,000,000 m^3) of oil equivalent per day

Years of production left in the ground with the current proved reserves and flows above

- Coal: 148 years
- Oil: 43 years
- Natural gas: 61 years

Years of production left in the ground with the most optimistic proved reserve estimates (Oil & Gas Journal, World Oil)

- Coal: 417 years
- Oil: 43 years
- Natural gas: 167 years.

The calculation above assumes that the product could be produced at a constant level for that number of years and that all of the proved reserves could be recovered. In reality, consumption of all three resources has been increasing. While this suggests that the resource will be used up more quickly, in reality, the production curve is much more akin to a bell curve. At some point in time, the production of each resource within an area, country, or globally will reach a maximum value, after which, the production will decline until it reaches a point where is no longer economically feasible or physically possible to produce. Hubbert peak theory for detail on this decline curve with regard to petroleum. Note also that proved reserve estimates do not include strategic reserves, which (globally) amount to 4.1 billion more barrels.

The above discussion emphasizes worldwide energy balance. It is also valuable to understand the ratio of reserves to annual consumption (R/C) by region or country. For example, energy policy of the United Kingdom recognizes that Europe's R/C value is 3.0, very low by world standards, and exposes that region to energy vulnerability. Alternatives to fossil fuels are a subject of intense debate worldwide.

Limits and Alternatives

The principle of supply and demand suggests that as hydrocarbon supplies diminish, prices will rise. Therefore higher prices will lead to increased alternative, renewable energy supplies as previously uneconomic sources become sufficiently economical to exploit.

Artificial gasolines and other renewable energy sources currently require more expensive production and processing technologies than conventional petroleum reserves, but may become economically viable in the near future. See Energy development. Different alternative sources of energy include nuclear, hydroelectric, solar, wind, and geothermal.

Environmental Effects

In the United States, more than 90% of greenhouse gas emissions come from the combustion of fossil fuels. Combustion of fossil fuels also produces other air pollutants, such as nitrogen oxides, sulfur dioxide, volatile organic compounds and heavy metals.

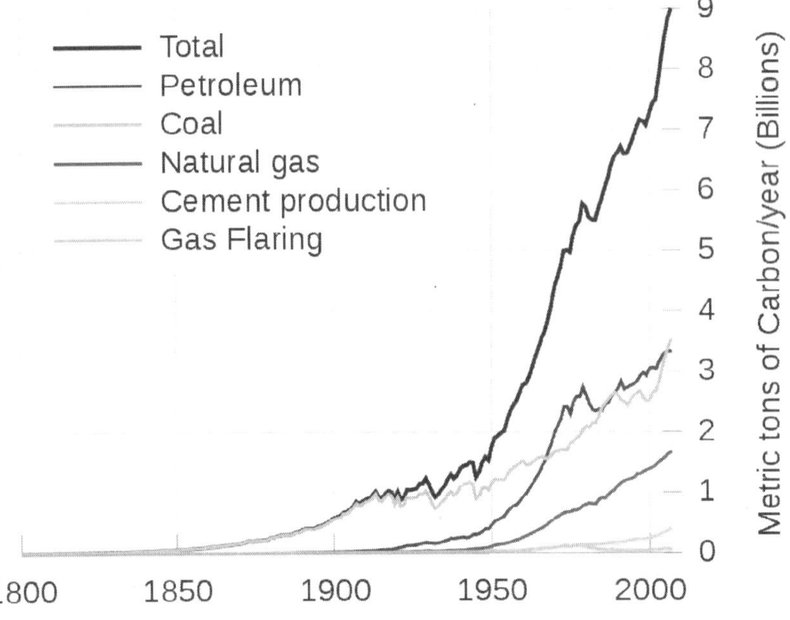

Figure: *Global fossil carbon emission by fuel type, 1800-2007. Note: Carbon only represents 27% of the mass of* CO_2

According to Environment Canada:

> *"The electricity sector is unique among industrial sectors in its very large contribution to emissions associated with nearly all air issues. Electricity generation produces a large share of Canadian nitrogen oxides and sulphur dioxide emissions, which contribute to smog and acid*

rain and the formation of fine particulate matter. It is the largest uncontrolled industrial source of mercury emissions in Canada. Fossil fuel-fired electric power plants also emit carbon dioxide, which may contribute to climate change. In addition, the sector has significant impacts on water and habitat and species. In particular, hydro dams and transmission lines have significant effects on water and biodiversity."

According to U.S. Scientist Jerry Mahlman and USA Today: Mahlman, who crafted the IPCC language used to define levels of scientific certainty, says the new report will lay the blame at the feet of fossil fuels with "virtual certainty," meaning 99% sure. That's a significant jump from "likely," or 66% sure, in the group's last report in 2001, Mahlman says. His role in this year's effort involved spending two months reviewing the more than 1,600 pages of research that went into the new assessment.

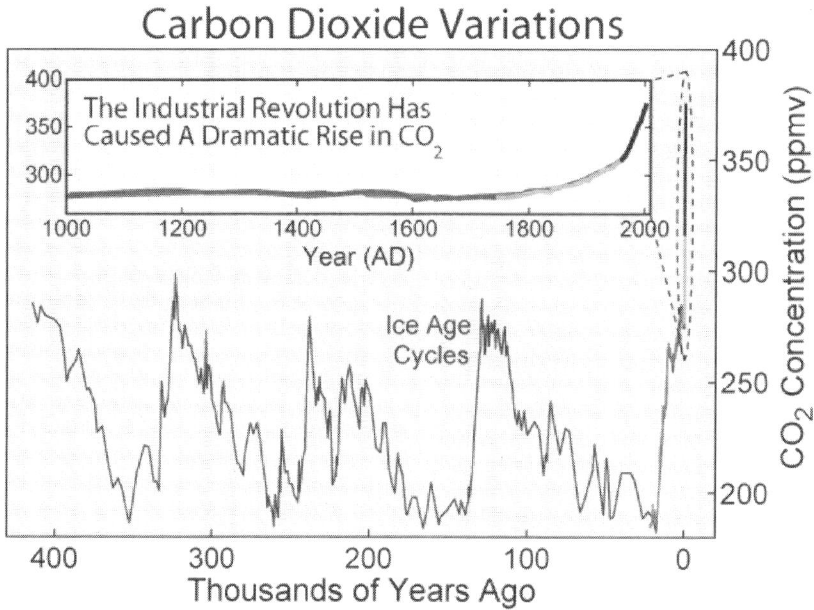

Figure: *Carbon dioxide variations over the last 400,000 years, showing a rise since the industrial revolution.*

Combustion of fossil fuels generates sulfuric, carbonic, and nitric acids, which fall to Earth as acid rain, impacting both natural areas and the built environment. Monuments and sculptures made from marble and limestone are particularly vulnerable, as the acids dissolve calcium carbonate.

Fossil fuels also contain radioactive materials, mainly uranium and thorium, which are released into the atmosphere. In 2000, about 12,000 tonnes of thorium and 5,000 tonnes of uranium were released worldwide from burning coal. It is estimated that during 1982, US coal burning released 155 times as much radioactivity into the atmosphere as the Three Mile Island incident. However, this radioactivity from coal burning is minuscule at each source and has not shown to have any adverse effect on human physiology.

Burning coal also generates large amounts of bottom ash and fly ash. These materials are used in a wide variety of applications, utilizing, for example, about 40% of the US production.

Harvesting, processing, and distributing fossil fuels can also create environmental concerns. Coal mining methods, particularly mountaintop removal and strip mining, have negative environmental impacts, and offshore oil drilling poses a hazard to aquatic organisms. Oil refineries also have negative environmental impacts, including air and water pollution. Transportation of coal requires the use of diesel-powered locomotives, while crude oil is typically transported by tanker ships, each of which requires the combustion of additional fossil fuels.

Environmental regulation uses a variety of approaches to limit these emissions, such as command-and-control (which mandates the amount of pollution or the technology used), economic incentives, or voluntary programs.

An example of such regulation in the USA is the "EPA is implementing policies to reduce airborne mercury emissions. Under regulations issued in 2005, coal-fired power plants will need to reduce their emissions by 70 percent by 2018.".

In economic terms, pollution from fossil fuels is regarded as a negative externality. Taxation is considered one way to make societal costs explicit, in order to 'internalize' the cost of pollution. This aims to make fossil fuels more expensive, thereby reducing their use and the amount of pollution associated with them, along with raising the funds necessary to counteract these factors.

Former CIA Director James Woolsey recently outlined the national security arguments in favor of moving away from fossil fuels.

Chapter 2

Carbon Footprint

A carbon footprint has historically been defined as "the total set of greenhouse gas (GHG) emissions caused by an organization, event, product or person.". However, calculating a carbon footprint which conforms to this definition is often impracticable due to the large amount of data required, which is often costly and time consuming to obtain. A more practicable definition has been suggested, which is gaining acceptance within the field:

> *"A measure of the total amount of carbon dioxide (CO_2) and methane (CH_4) emissions of a defined population, system or activity, considering all relevant sources, sinks and storage within the spatial and temporal boundary of the population, system or activity of interest. Calculated as carbon dioxide equivalent (CO_2e) using the relevant 100-year global warming potential (GWP100)."*

Greenhouse gases can be emitted through transport, land clearance, and the production and consumption of food, fuels, manufactured goods, materials, wood, roads, buildings, and services. For simplicity of reporting, it is often expressed in terms of the amount of carbon dioxide, or its equivalent of other GHGs, emitted.

The concept name of the carbon footprint originates from ecological footprint discussion. The carbon footprint is a subset of the ecological footprint and of the more comprehensive Life Cycle Assessment (LCA).

An individual's, nation's, or organisations carbon footprint can be measured by undertaking a GHG emissions assessment. Once the size of a carbon footprint is known, a strategy can be devised to reduce it, e.g. by technological developments, better process and product management, changed Green Public or Private Procurement (GPP), carbon capture, consumption strategies, and others.

The mitigation of carbon footprints through the development of alternative projects, such as solar or wind energy or reforestation, represents one way of reducing a carbon footprint and is often known as Carbon offsetting.

The main influences on carbon footprints include population, economic output, and energy and carbon intensity of the economy. These factors are the main targets of individuals and businesses in order to decrease carbon footprints. Scholars suggest the most effective way to decrease a carbon footprint is to either decrease the amount of energy needed for production or to decrease the dependence on carbon emitting fuels.

By Asrea

Of Products

Several organizations have calculated carbon footprints of products; The US Environmental Protection Agency has addressed paper, plastic (candy wrappers), glass, cans, computers, carpet and tires. Australia has addressed lumber and other building materials.

Academics in Australia, Korea and the US have addressed paved roads. Companies, nonprofits and academics have addressed manufacture and operation of cars, buses, trains, airplanes, ships and pipelines. The US Postal Service has addressed mailing letters and packages. Carnegie Mellon University has estimated the CO_2 footprints of 46 large sectors of the economy in each of eight countries. Carnegie Mellon, Sweden and the Carbon Trust have addressed foods at home and in restaurants.

The Carbon Trust has worked with UK manufacturers on foods, shirts and detergents, introducing a CO_2 label in March 2007. The label is intended to comply with a new British Publicly Available Specification (i.e. not a standard), PAS 2050, and is being actively piloted by The Carbon Trust and various industrial partners.

Evaluating the package of some products is key to figuring out the carbon footprint. . The key ways to determine the carbon footprint is to look at the materials that were used to make the item. For example, the juice carton is made of Aseptic carton, the beer can is made of aluminum and some water bottles either made of glass or plastic. The larger the size, the larger the footprint will be.

Of Electricity

The following table compares, from peer-reviewed studies of full life cycle emissions and from various other studies, the carbon footprint of various forms of energy generation: Nuclear, Hydro, Coal, Gas, Solar Cell, Peat and Wind generation technology.

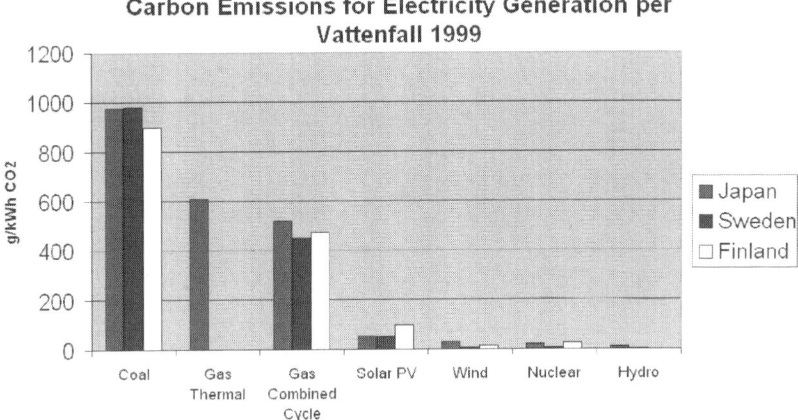

Figure: *The Vattenfall study found renewable and nuclear generation responsible for far less CO2 than fossil fuel generation.*

These studies thus concluded that hydroelectric, wind, and nuclear power always produced the least CO_2 per kilowatt-hour of any other electricity sources. These figures do not allow for emissions due to accidents or terrorism. Renewable electricity generation methods, for example wind power and hydropower, emit no carbon from the operation, but do leave a footprint during construction phase and maintenance during operation.

Of Heat and Various Combined Heat and Power Schemes, Heat Pumps etc ...

The previous table gives the carbon footprint per kilowatt-hour of electricity generated, which is about half the world's man-made CO_2 output. The CO_2 footprint for heat is equally significant and research shows that using waste heat from power generation in combined heat and power district heating, chp/dh has the lowest carbon footprint. much lower than micro-power or heat pumps.

Kyoto Protocol, Carbon Offsetting, and Certificates

Carbon dioxide emissions into the atmosphere, and the emissions of other GHGs, are often associated with the burning of fossil fuels, like natural gas, crude oil and coal.

The Kyoto Protocol defines legally binding targets and timetables for cutting the GHG emissions of industrialized countries that ratified the Kyoto Protocol. Accordingly, from an economic or market perspective, one has to distinguish between a *mandatory market* and a *voluntary market*. Typical for both markets is the trade with emission certificates:

- Certified Emission Reduction (CER)
- Emission Reduction Unit (ERU)
- Verified Emission Reduction (VER).

Mandatory Market Mechanisms

To reach the goals defined in the Kyoto Protocol, with the least economical costs, the following flexible mechanisms were introduced for the mandatory market:

- Clean Development Mechanism (CDM)
- Joint Implementation (JI)
- Emissions trading.

The CDM and JI mechanisms requirements for projects which create a supply of emission reduction instruments, while Emissions Trading allows those instruments to be sold on international markets.

Projects which are compliant with the requirements of the CDM mechanism generate Certified Emissions Reductions (CERs).

Projects which are compliant with the requirements of the JI mechanism generate Emissions Reduction Units (ERUs).

The CERs and ERUs can then be sold through Emissions Trading. The demand for the CERs and ERUs being traded is driven by:

- Shortfalls in national emission reduction obligations under the Kyoto Protocol.
- Shortfalls amongst entities obligated under local emissions reduction schemes.

Nations which have failed to deliver their Kyoto emissions reductions obligations can enter Emissions Trading to purchase CERS and ERUs to cover their treaty shortfalls. Nations and groups of nations can also create local emission reduction schemes which place mandatory carbon dioxide emission targets on entities within their national boundaries.

If the rules of a scheme allow, the obligated entities may be able to cover all or some of any reduction shortfalls by purchasing CERs and ERUs through Emissions Trading. While local emissions reduction schemes have no status under the Kyoto Protocol itself, they play a prominent role in creating the demand for CERs and ERUs, stimulating Emissions Trading and setting a market price for emissions.

A well-known mandatory local emissions trading scheme is the EU Emissions Trading Scheme (EU ETS).

New changes are being made to the trading schemes. The EU Emissions Trading Scheme is set to make some new changes within the next year. The new changes will target the emissions produced by flight travel in and out of the European Union.

Other nations are scheduled to start participating in Emissions Trading Schemes within the next few year. These nations include China, India and the United States.

Voluntary Market Mechanisms

In contrast to the strict rules set out for the mandatory market, the voluntary market provides companies with different options to acquire emissions reductions. A solution, comparable with those developed for the mandatory market, has been developed for the voluntary market, the Verified Emission Reductions (VER). This measure has the great advantage that the projects/activities are managed according to the quality standards set out for CDM/JI projects but the certificates provided are not registered by the governments of the host countries or the Executive Board of the UNO. As such, high quality VERs can be acquired at lower costs for the same project quality. However, at present VERs can not be used in the mandatory market.

The voluntary market in North America is divided between members of the Chicago Climate Exchange and the Over The Counter (OTC) market. The Chicago Climate Exchange is a voluntary yet legally binding cap-and-trade emission scheme whereby members commit to the capped emission reductions and must purchase allowances from other members or offset excess emissions. The OTC market does not involve a legally binding scheme and a wide array of buyers from the public and private spheres, as well as special events that want to go carbon neutral.

There are project developers, wholesalers, brokers, and retailers, as well as carbon funds, in the voluntary market. Some businesses and nonprofits in the voluntary market encompass more than just one of the activities listed above. A report by Ecosystem Marketplace shows that carbon offset prices increase as it moves along the supply chain—from project developer to retailer.

While some mandatory emission reduction schemes exclude forest projects, these projects flourish in the voluntary markets. A major criticism concerns the imprecise nature of GHG sequestration quantification methodologies for forestry projects. However, others note the community co-benefits that forestry projects foster. Project types in the voluntary market range from avoided deforestation, afforestation/ reforestation, industrial gas sequestration, increased energy efficiency, fuel switching, methane capture from coal plants and livestock, and even renewable energy. Renewable Energy Certificates (RECs) sold on the voluntary market are quite controversial due to additionality concerns. Industrial Gas projects receive criticism because such projects only apply to large industrial plants that already have high fixed costs. Siphoning off industrial gas for sequestration is considered picking the low hanging fruit; which is why credits generated from industrial gas projects are the cheapest in the voluntary market.

The size and activity of the voluntary carbon market is difficult to measure. The most comprehensive report on the voluntary carbon markets to date was released by Ecosystem Marketplace and New Carbon Finance in July 2007.

ÆON of Japan is firstly approved by Japanese authority to indicate carbon footprint on three private brand goods in October 2009.

Radiative Forcing

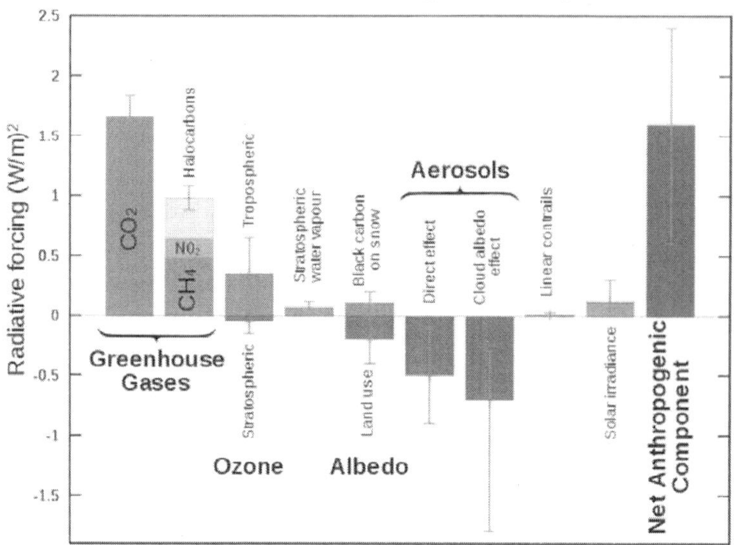

Figure: *2005 radiative forcings as estimated by the IPCC.*

In climate science, radiative forcing is generally defined as the change in net irradiance between different layers of the atmosphere. Typically, radiative forcing is quantified at the tropopause in units of watts per square meter. A positive forcing (more incoming energy) tends to warm the system, while a negative forcing (more outgoing energy) tends to cool it. Sources of radiative forcing include changes in insolation (incident solar radiation) and in concentrations of radiatively active gases and aerosols.

Radiation Balance

The vast majority of the energy which affects Earth's weather comes from the Sun. The planet and its atmosphere absorb and reflect some of the energy, while long-wave energy is radiated back into space. The balance between absorbed and radiated energy determines the average temperature. The planet is warmer than it would be in the absence of the atmosphere: see greenhouse effect. The radiation balance can be altered by factors such as intensity of solar energy, reflection by clouds or gases, absorption by various gases or surfaces, emission of heat by various materials, and other factors related to climate change. Any such alteration is a radiative forcing, and causes a new balance to be reached. In the real world this happens continuously as sunlight hits the surface, clouds and aerosols form, the concentrations of atmospheric gases vary, and seasons alter the ground cover.

IPCC Usage

The term "radiative forcing" has been used in the IPCC Assessments with a specific technical meaning, to denote an externally imposed perturbation in the radiative energy budget of Earth's climate system, which may lead to changes in climate parameters. The exact definition used is:

The radiative forcing of the surface-troposphere system due to the perturbation in or the introduction of an agent (say, a change in greenhouse gas concentrations) is the change in net (down minus up) irradiance (solar plus long-wave; in Wm) at the tropopause AFTER allowing for stratospheric temperatures to readjust to radiative equilibrium, but with surface and tropospheric temperatures and state held fixed at the unperturbed values.

In a subsequent report, the IPCC defines it as:

> *"Radiative forcing is a measure of the influence a factor has in altering the balance of incoming and outgoing energy in the Earth-atmosphere system and is an index of the importance of the factor as a potential climate*

change mechanism. In this report radiative forcing values are for changes relative to preindustrial conditions defined at 1750 and are expressed in watts per square meter (W/m2)."

In simple terms, radiative forcing is "...the rate of energy change per unit area of the globe as measured at the top of the atmosphere." In the context of climate change, the term "forcing" is restricted to changes in the radiation balance of the surface-troposphere system imposed by external factors, with no changes in stratospheric dynamics, no surface and tropospheric feedbacks in operation (i.e., no secondary effects induced because of changes in tropospheric motions or its thermodynamic state), and no dynamically induced changes in the amount and distribution of atmospheric water (vapour, liquid, and solid forms).

Radiative forcing can be used to estimate a subsequent change in equilibrium surface temperature (ΔT_s) arising from that radiative forcing via the equation:

$$\Delta T_s = \lambda \Delta F$$

where λ is the climate sensitivity, usually with units in $K/(W/m^2)$, and ΔF is the radiative forcing. A typical value of λ is 0.8 $K/(W/m^2)$, which gives a warming of 3K for doubling of CO_2.

Example Calculations

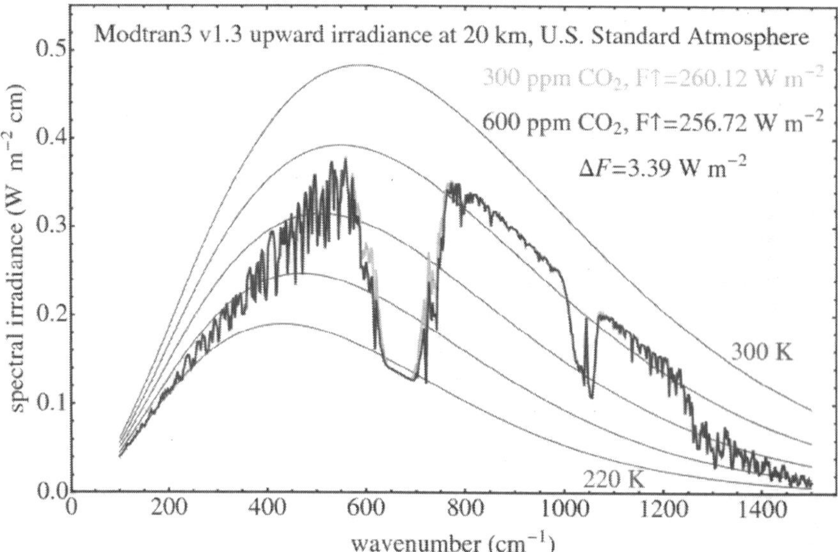

***Figure:** Radiative forcing for doubling CO2, as calculated by radiative transfer code Modtran.*

Radiative forcing (often measured in watts per square meter) can be estimated in different ways for different components. For the case of a change in solar irradiance, the radiative forcing is the change in the solar constant divided by 4 and multiplied by 0.7 to take into account the geometry of the sphere and the amount of reflected sunlight. For a greenhouse gas, such as carbon dioxide, radiative transfer codes that examine each spectral line for atmospheric conditions can be used to calculate the change ΔF as a function of changing concentration. These calculations can often be simplified into an algebraic formulation that is specific to that gas.For instance, the simplified first-order approximation expression for carbon dioxide is:

$$\Delta F = 5.35 \times \ln \frac{C}{C_0}\,\mathrm{W\,m^{-2}}$$

where C is the CO_2 concentration in parts per million by volume and C_0 is the reference concentration. The relationship between carbon dioxide and radiative forcing is logarithmic so that increased concentrations have a progressively smaller warming effect.

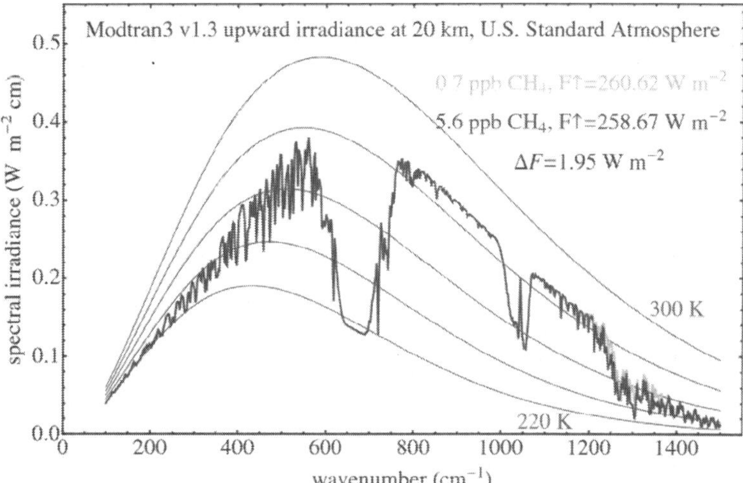

Figure: *Radiative forcing for eight times increase of CH4, as calculated by radiative transfer code Modtran.*

Formulas for other greenhouse gases such as methane, N_2O or CFCs are given in the IPCC reports.

Related Measures

Radiative forcing is intended as a useful way to compare different causes of perturbations in a climate system. Other possible tools can be constructed for the same purpose: for example Shine et al. say "...recent experiments indicate that for changes in absorbing aerosols and ozone, the predictive ability of radiative forcing is much worse... we propose an alternative, the 'adjusted troposphere and stratosphere forcing'. We present GCM calculations showing that it is a significantly more reliable predictor of this GCM's surface temperature change than radiative forcing. It is a candidate to supplement radiative forcing as a metric for comparing different mechanisms...". In this quote, GCM stands for "global circulation model", and the word "predictive" does not refer to the ability of GCMs to forecast climate change. Instead, it refers to the ability of the alternative tool proposed by the authors to help explain the system response.

Global-warming Potential

Global-warming potential (GWP) is a relative measure of how much heat a greenhouse gas traps in the atmosphere. It compares the amount of heat trapped by a certain mass of the gas in question to the amount of heat trapped by a similar mass of carbon dioxide. A GWP is calculated over a specific time interval, commonly 20, 100 or 500 years. GWP is expressed as a factor of carbon dioxide (whose GWP is standardized to 1). For example, the 20 year GWP of methane is 72, which means that if the same mass of methane and carbon dioxide were introduced into the atmosphere, that methane will trap 72 times more heat than the carbon dioxide over the next 20 years.

The substances subject to restrictions under the Kyoto protocol either are rapidly increasing their concentrations in Earth's atmosphere or have a large GWP.

The GWP depends on the following factors:

- the absorption of infrared radiation by a given species
- the spectral location of its absorbing wavelengths
- the atmospheric lifetime of the species.

Thus, a high GWP correlates with a large infrared absorption and a long atmospheric lifetime. The dependence of GWP on the wavelength of absorption is more complicated. Even if a gas absorbs radiation efficiently at a certain wavelength, this may not affect its GWP much if the atmosphere already absorbs most radiation at that wavelength. A gas has the most effect if it absorbs in a "window" of wavelengths where the atmosphere is fairly transparent. The dependence of GWP as a function of wavelength has been found empirically and published as a graph.

Because the GWP of a greenhouse gas depends directly on its infrared spectrum, the use of infrared spectroscopy to study greenhouse gases is centrally important in the effort to understand the impact of human activities on global climate change.

Calculating the Global-warming Potential

Just as radiative forcing provides a simplified means of comparing the various factors that are believed to influence the climate system to one another, global-warming potentials (GWPs) are one type of simplified index based upon radiative properties that can be used to estimate the potential future impacts of emissions of different gases upon the climate system in a relative sense. GWP is based on a number of factors, including the radiative efficiency (infrared-absorbing ability) of each gas relative to that of carbon dioxide, as well as the decay rate of each gas (the amount removed from the atmosphere over a given number of years) relative to that of carbon dioxide.

The radiative forcing capacity (RF) is the amount of energy per unit area, per unit time, absorbed by the greenhouse gas, that would otherwise be lost to space. It can be expressed by the formula:

$$RF = \sum_{n=1}^{100} Abs_i * F_i / (pathlength * density)$$

where the subscript i represents an interval of 10 inverse centimeters. Abs_i represents the integrated infrared absorbance of the sample in that interval, and F_i represents the RF for that interval.

The Intergovernmental Panel on Climate Change (IPCC) provides the generally accepted values for GWP, which changed slightly between 1996 and 2001. An exact definition of how GWP is calculated is to be found in the IPCC's 2001 Third Assessment Report. The GWP is defined as the ratio of the time-integrated radiative forcing from the instantaneous release of 1 kg of a trace substance relative to that of 1 kg of a reference gas:

$$GWP(x) = \frac{\int_0^{TH} a_x \cdot [x(t)] dt}{\int_0^{TH} a_r \cdot [r(t)] dt}$$

where TH is the time horizon over which the calculation is considered; a_x is the radiative efficiency due to a unit increase in atmospheric abundance of the substance (i.e., Wm^{-2} kg^{-1}) and [x(t)] is the time-dependent decay in abundance of the substance following an instantaneous release of it at time t=0.

The denominator contains the corresponding quantities for the reference gas (i.e. CO_2). The radiative efficiencies a_x and a_r are not necessarily constant over time. While the absorption of infrared radiation by many greenhouse gases varies linearly with their abundance, a few important ones display non-linear behaviour for current and likely future abundances (e.g., CO_2, CH_4, and N_2O). For those gases, the relative radiative forcing will depend upon abundance and hence upon the future scenario adopted. Since all GWP calculations are a comparison to CO_2 which is non-linear, all GWP values are affected. Assuming otherwise as is done above will lead to lower GWPs for other gases than a more detailed approach would.

Use in Kyoto Protocol

Under the Kyoto Protocol, the Conference of the Parties decided (decision 2/CP.3) that the values of GWP calculated for the IPCC Second Assessment Report are to be used for converting the various greenhouse gas emissions into comparable CO_2 equivalents when computing overall sources and sinks.

Importance of Time Horizon

Note that a substance's GWP depends on the timespan over which the potential is calculated. A gas which is quickly removed from the atmosphere may initially have a large effect but for longer time periods as it has been removed becomes less important. Thus methane has a potential of 25 over 100 years but 72 over 20 years; conversely sulfur hexafluoride has a GWP of 22,800 over 100 years but 16,300 over 20 years (IPCC TAR). The GWP value depends on how the gas concentration decays over time in the atmosphere. This is often not precisely known and hence the values should not be considered exact. For this reason when quoting a GWP it is important to give a reference to the calculation. The GWP for a mixture of gases can not be determined from the GWP of the constituent gases by any form of simple linear addition. Commonly, a time horizon of 100 years is used by regulators (e.g., the California Air Resources Board).

Values

Carbon dioxide has a GWP of exactly 1 (since it is the baseline unit to which all other greenhouse gases are compared). Although water vapour has a significant influence with regard to absorbing infrared radiation (which is the green house effect; see greenhouse gas), its GWP is not calculated. Its concentration in the atmosphere mainly depends on air temperature. There is no possibility to directly influence atmospheric water vapour concentration.

Carbon Dioxide Equivalent

Carbon dioxide equivalent (CDE) and Equivalent carbon dioxide (or CO_2e) are two related but distinct measures for describing how much global warming a given type and amount of greenhouse gas may cause, using the functionally equivalent amount or concentration of carbon dioxide (CO_2) as the reference.

Carbon dioxide equivalency is a quantity that describes, for a given mixture and amount of greenhouse gas, the amount of CO_2 that would have the same global warming potential (GWP), when measured over a specified timescale (generally, 100 years).

Carbon dioxide equivalency thus reflects the time-integrated radiative forcing of a quantity of *emissions* or rate of greenhouse gas emission - a *flow* into the atmosphere - rather than the instantaneous value of the radiative forcing of the *stock* (concentration) of greenhouse gases *in the atmosphere* described by CO_2e.

The carbon dioxide equivalency for a gas is obtained by multiplying the mass and the GWP of the gas. The following units are commonly used:

- By the UN climate change panel IPCC: billion metric tonnes of CO_2 equivalent (GtCO_2eq).
- In industry: million metric tonnes of carbon dioxide equivalents (MMTCDE).
- For vehicles: g of carbon dioxide equivalents / km (gCDE/km).

For example, the GWP for methane over 100 years is 25 and for nitrous oxide 298. This means that emissions of 1 million metric tonnes of methane and nitrous oxide respectively is equivalent to emissions of 25 and 298 million metric tonnes of carbon dioxide.

Equivalent Carbon Dioxide

Equivalent CO_2 (CO_2e) is the concentration of CO_2 that would cause the same level of radiative forcing as a given type and concentration of greenhouse gas. Examples of such greenhouse gases are methane, perfluorocarbons and nitrous oxide. CO_2e is expressed as parts per million by volume, ppmv.

CO_2e calculation example:

- The radiative forcing for pure CO_2 is approximated by $RF = \alpha ln(C / C_0)$ where C is the present concentration, α is a constant, 5.35 and C_0 the pre-industrial concentration, 278 ppm. Hence the value of CO_2e for an arbitrary gas mixture with a known radiative forcing is given by $C_0 exp(RF / \alpha)$ in ppmv.
- To calculate the radiative forcing for a 1998 gas mixture, IPCC 2001 gives the radiative forcing (relative to 1750) of various gases as: CO_2=1.46 (corresponding to a concentration of 365 ppmv), CH_4=0.48, N_2O=0.15 and other minor gases =0.01 W/

m2. The sum of these is 2.10 W/m2. Inserting this to the above formula, we obtain $CO_2e = 412$ ppmv

Carbon Accounting

Carbon accounting is the accounting process undertaken to measure the amount of carbon dioxide equivalents that will not be released into the atmosphere as a result of Flexible Mechanisms projects under the Kyoto Protocol. These projects thus include (but are not limited to) renewable energy projects and biomass, forage and tree plantations.

Carbon Accounting Software

A number of programs are created in order to assist with carbon accounting:

- Access Dimensions Carbon accounting tool within mid-market financial management software
- BIOMITRE Biomass based climate change, for biomass energy usage may be harmful to the environment, as it will also release additional carbon emissions
- Brighter Planet's Carbon Middleware provides a RESTful carbon emissions calculation API. All calculations include live methodology statements and the calculation models themselves are open source.
- CAMFor Carbon accounting model for forests
- CAMSAT Carbon management self assessment tool, for companies to measure their offset
- CapISA SPM Sustainability Performance Management software.
- Carbon Calculated Carbon Management and Accounting solutions that meet regulatory compliance and industry standards (CRC, EU-ETS, CDP, ISO 14064), along with a platform of emissions data with sources and guidance for consultants.
- Carbon Check is a carbon management software system to gather and report on emissions, the system includes a CO2 emissions calculator and many sustainability reporting functions.
- Carbon Guerrilla SaaS based Carbon Accounting and Business Management tool covering voluntary or compliance based schemes. Over 1,000 sites deployed for CRCEES, EU-ETS, and CDP. Follow us on: Guerrilla Tweet

- CarbonMate On Demand Carbon Accounting software aimed specifically at small and medium sized enterprises.
- Carbon Management Software & Services Directory provides list of Carbon Accounting Software Tools.
- CarbonView Carbon management and accounting platform for enterprises, supply chains and government.
- CBM-CFS3 Operational-Scale Carbon Budget Model of the Canadian Forest Sector
- C-FIX Carbon uptake monitoring through satellite images
- CO2FIX One of the oldest carbon accounting applications available, somewhat outdated
- CarbonLow Emissions Carbon measurement software (iCAT) for businesses to measure their emissions, including scopes 1, 2, 3 and full PAS 2050 product reports from UK Company CarbonLow.
- Carbonetworks Established enterprise level carbon accounting software, run as an internet accessible service
- Clean Air-Cool Planet CA-CP provides a carbon footprint calculator designed specifically for use at colleges and universities under the American College and Universities Presidents' Climate Commitment.
- COMMTracker Environmental Management System Monitors, Tracks and Reports GHG emissions for EPA mandated sites.
- CRedit360 Energy & Carbon Leading companies, including Barclays and Swiss Re use the Carbon Disclosure Project-accredited CRedit360 platform to manage and report their energy and carbon inline with GHG Protocol, CRC, EES, CDP, GRI, ISO14064, NGERS and Climate Registry.
- Enablon Energy and Carbon Management Suite Enablon is one of the world's leading providers of Sustainability Management Software. More than 250 global companies use Enablon to measure and manage their carbon, energy and environmental performance.
- EPS Corp xChange Point: Scopes 1, 2 and 3 carbon accounting and energy management platform for manufacturers.
- FoundationFootprint Enterprise level, web based (SaaS) carbon, energy, water and supply chain emissions management system based on ISO 14064 / GHG Protocol standards.

- GEMIS Global emission model for integrated systems
- GaBi Software by PE International for Product Carbon Footprints
- GORCAM Graz / Oak Ridge carbon accounting model, excel spreadsheet
- Greenhouse Gas Management Institute for greenhouse gas accounting and management training and education
- Forest Vegetation Simulator Developed by the US Forest Service, includes a carbon accounting model
- Real-time web-based carbon accounting software
- HWP Harvested wood products, dead wood carbon assessment
- IPCC Greenhouse Gas Inventory software for the workbook based on IPCC guidelines
- LEAP:Long range Energy Alternatives Planning System: a software tool for energy planning and greenhouse gas mitigation analysis.
- ManageCO2 Carbon Software, Carbon Accounting and Management Reporting Software.
- Nootrol Carbon accounting software for large corporates to manage the emissions within their supply chain.
- Our Impacts SaaS-based, CDP-accredited greenhouse gas accounting software from Ecometrica
- Oracle Environmental Accounting & Reporting greenhouse gas emissions and other environmental reporting directly in Oracle ERP systems.
- SIMAPROLCA, Life cycle assessment based on various accounting systems and that allows different types outputs
- SoFi Software by PE International for Corporate Carbon Footprints / Enterprise Carbon Accounting
- TimberCAM Carbon in wood products assessment
- TEAM Sigma is a global enterprise energy/carbon management package provided by TEAM Energy Auditing Agency Ltd
- TRIRIGA TREES® delivers award-winning carbon accounting software
- Verteego Carbon Scope 3 Carbon Inventory, Accounting, Management and Reporting Enterprise Software.

Chapter 3

Ethylene

Ethylene (IUPAC name: ethene) is a gaseous organic compound with the formula C_2H_4. It is the simplest alkene (older name: *olefin* from its oil-forming property).

Because it contains a carbon-carbon double bond, ethylene is classified as an *unsaturated hydrocarbon.* Ethylene is widely used in industry and is also a plant hormone. Ethylene is the most produced organic compound in the world; global production of ethylene exceeded 107 million tonnes in 2005.

To meet the ever increasing demand for ethylene, sharp increases in production facilities are added globally, particularly in the Persian Gulf countries and in China.

Structure and Properties

This hydrocarbon has four hydrogen atoms bound to a pair of carbon atoms that are connected by a double bond. All six atoms that comprise ethylene are coplanar. The H-C-H angle is 119°, close to the 120° for ideal sp^2 hybridized carbon. The molecule is also relatively rigid: rotation about the C-C bond is a high energy process that requires breaking the π-bond.

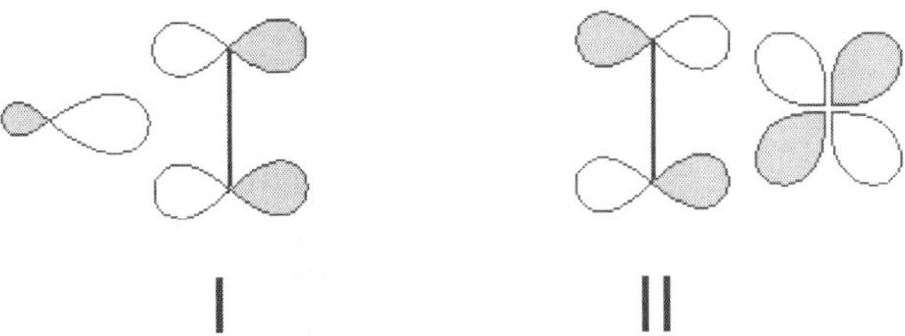

Figure: *Orbital description of bonding between ethylene and a transition metal.*

The π-bond in the ethylene molecule is responsible for its useful reactivity. The double bond is a region of high electron density, thus it is susceptible to attack by electrophiles. Many reactions of ethylene are catalyzed by transition metals, which bind transiently to the ethylene using both the π and π* orbitals.

Being a simple molecule, ethylene is spectroscopically simple. Its UV-vis spectrum is still used as a test of theoretical methods.

Uses

Major industrial reactions of ethylene include in order of scale: 1) polymerization, 2) oxidation, 3) halogenation and hydrohalogenation, 4) alkylation, 5) hydration, 6) oligomerization, and 7) hydroformylation. In the United States and Europe, approximately 90% of ethylene is used to produce three chemical compounds—ethylene oxide, ethylene dichloride, and ethylbenzene—and a variety of kinds of polyethylene.

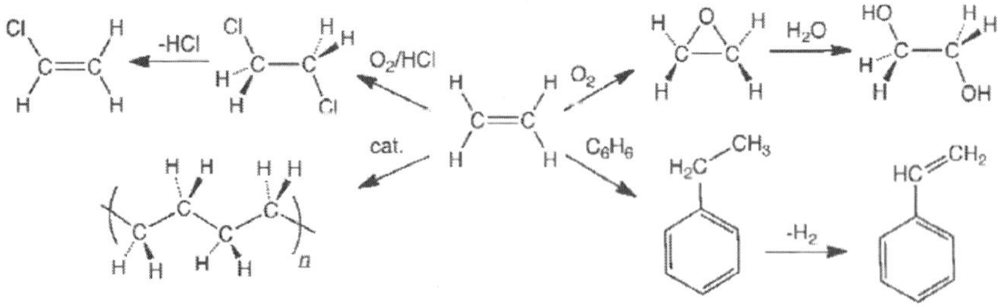

Figure: *Main industrial uses of ethylene. Clockwise from the upper right: its conversions to ethylene oxide, precursor to ethylene glycol, to ethylbenzene, precursor to styrene, to various kinds of polyethylene, to ethylene dichloride, precursor to vinyl chloride.*

Polymerization

Polyethylenes of various types consume more than half of world ethylene supply. Polyethylene, also called *polythene*, is the world's most widely-used plastic, being primarily used to make films used in packaging, carrier bags and trash liners. Linear alpha-olefins, produced by oligomerization (formation of short polymers) are used as precursors, detergents, plasticisers, synthetic lubricants, additives, and also as co-monomers in the production of polyethylenes.

Oxidation

Ethylene is oxidized to produce ethylene oxide, a key raw material in the production of surfactants and detergents by ethoxylation. Ethylene oxide also hydrolyzed to produce ethylene glycol, widely used as an automotive antifreeze as well as higher molecular weight glycols and glycol ethers.

Ethylene undergoes oxidation by palladium to give acetaldehyde. This conversion remains a major industrial process (10M kg/y). The process proceeds via the initial complexation of ethylene to a Pd(II) centre.

Halogenation and hydrohalogenation

Major intermediates from the halogenation and hydrohalogenation of ethylene include ethylene dichloride, ethyl chloride and ethylene dibromide. The addition of chlorine entails "oxychlorination," i.e. chlorine itself is not used. Some products derived from this group are polyvinyl chloride, trichloroethylene, perchloroethylene, methyl chloroform, polyvinylidiene chloride and copolymers, and ethyl bromide.

Alkylation

Major chemical intermediates from the alkylation with ethylene is ethylbenzene, precursor to styrene. Styrene is used principally in polystyrene for packaging and insulation, as well as in styrene-butadiene rubber for tires and footwear. On a smaller scale, ethyltoluene, ethylanilines, 1,4-hexadiene, and aluminium alkyls. Products of these intermediates include polystyrene, unsaturated polyesters and ethylene-propylene terpolymers.

Oxo Reaction

The hydroformylation (oxo reaction) of ethylene results in propionaldehyde, a precursor to propionic acid and n-propyl alcohol.

Hydration

Ethylene can be hydrated to give ethanol, but this method is rarely used industrially.

Niche Uses

An example of a niche use is as an anesthetic agent (in an 85% ethylene/15% oxygen ratio). It can also be used to hasten fruit ripening, as well as a welding gas.

Production

In 2006, global ethylene production was 109 million tonnes. By 2010 ethylene was produced by at least 117 companies in 55 countries.

Ethylene is produced in the petrochemical industry by steam cracking. In this process, gaseous or light liquid hydrocarbons are heated to 750–950 °C, inducing numerous free radical reactions followed by immediate quench to stop these reactions. This process converts large hydrocarbons into smaller ones and introduces unsaturation.

Ethylene is separated from the resulting complex mixture by repeated compression and distillation. In a related process used in oil refineries, high molecular weight hydrocarbons are cracked over zeolite catalysts. Heavier feedstocks, such as naphtha and gas oils require at least two "quench towers" downstream of the cracking furnaces to recirculate pyrolysis-derived gasoline and process water. When cracking a mixture of ethane and propane, only one water quench tower is required.

The areas of an ethylene plant are:

1. steam cracking furnaces:
2. primary and secondary heat recovery with quench;
3. a dilution steam recycle system between the furnaces and the quench system;
4. primary compression of the cracked gas (3 stages of compression);
5. hydrogen sulfide and carbon dioxide removal (acid gas removal);
6. secondary compression (1 or 2 stages);
7. drying of the cracked gas;
8. cryogenic treatment;
9. all of the cold cracked gas stream goes to the demethanizer tower. The overhead stream from the demethanizer tower consists of all the hydrogen and methane that was in the cracked gas stream. Cryogenically (–250 °F (–157 °C)) treating this overhead stream separates hydrogen from methane. Methane recovery is critical to the economical operation of an ethylene plant.
10. the bottom stream from the demethanizer tower goes to the deethanizer tower. The overhead stream from the deethanizer tower consists of all the C_2,'s that were in the cracked gas stream. The C_2 stream contains acetylene, which is explosive above 200 kPa (29 psi). If the partial pressure of acetylene is expected to exceed these values, the C_2 stream is partially hydrogenated. The C_2's then proceed to a C_2 splitter. The product ethylene is taken from the overhead of the tower and the ethane coming from the bottom of the splitter is recycled to the furnaces to be cracked again;
11. the bottom stream from the de-ethanizer tower goes to the depropanizer tower. The overhead stream from the depropanizer tower consists of all the C_3's that were in the cracked gas stream. Before feeding the C_3's to the C_3 splitter, the stream is hydrogenated to convert the methylacetylene and

propadiene (allene) mix. This stream is then sent to the C_3 splitter. The overhead stream from the C_3 splitter is product propylene and the bottom stream is propane which is sent back to the furnaces for cracking or used as fuel.

12. The bottom stream from the depropanizer tower is fed to the debutanizer tower. The overhead stream from the debutanizer is all of the C_4's that were in the cracked gas stream. The bottom stream from the debutanizer (light pyrolysis gasoline) consists of everything in the cracked gas stream that is C_5 or heavier.

Since ethylene production is energy intensive, much effort has been dedicated to recovering heat from the gas leaving the furnaces. Most of the energy recovered from the cracked gas is used to make high pressure (1200 psig) steam. This steam is in turn used to drive the turbines for compressing cracked gas, the propylene refrigeration compressor, and the ethylene refrigeration compressor. An ethylene plant, once running, does not need to import steam to drive its steam turbines. A typical world scale ethylene plant (about 1.5 billion pounds of ethylene per year) uses a 45,000 horsepower (34,000 kW) cracked gas compressor, a 30,000 hp (22,000 kW) propylene compressor, and a 15,000 hp (11,000 kW) ethylene compressor.

Laboratory Synthesis

Ethylene can be produced via ethanol dehydration by heating absolute ethanol with concentrated sulfuric acid or by passing ethanol vapor over hot aluminium oxide catalyst. Interestingly for such a useful compound, ethylene is rarely used in organic synthesis in the laboratory.

Ethylene as a Plant Hormone

Ethylene serves as a hormone in plants. It acts at trace levels throughout the life of the plant by stimulating or regulating the ripening of fruit, the opening of flowers, and the abscission (or shedding) of leaves. Commercial ripening rooms use "catalytic generators" to make ethylene gas from a liquid supply of ethanol. Typically, a gassing level of 500 ppm to 2,000 ppm is used, for 24 to 48 hours. Care must be taken to control carbon dioxide levels in ripening rooms when gassing, as high temperature ripening (68F) has been seen to produce CO2 levels of 10% in 24 hours.

History of Ethylene in Plant Biology

Ethylene has been used in practice since the ancient Egyptians, who would gash figs in order to stimulate ripening (wounding stimulates ethylene production by plant tissues). The ancient Chinese would burn incense in closed rooms to enhance the ripening of pears. In 1864, it was discovered that gas leaks from street lights led to stunting of growth, twisting of plants, and abnormal thickening of stems. In 1901, a Russian scientist named Dimitry Neljubow showed that the active component was ethylene. Doubt discovered that ethylene stimulated abscission in 1917. It wasn't until 1934 that Gane reported that plants synthesize ethylene. In 1935, Crocker proposed that ethylene was the plant hormone responsible for fruit ripening as well as senescence of vegetative tissues.

Ethylene Biosynthesis in Plants

Ethylene is produced from essentially all parts of higher plants, including leaves, stems, roots, flowers, fruits, tubers, and seedlings.

"Ethylene production is regulated by a variety of developmental and environmental factors. During the life of the plant, ethylene production is induced during certain stages of growth such as germination, ripening of fruits, abscission of leaves, and senescence of flowers. Ethylene production can also be induced by a variety of external aspects such as mechanical wounding, environmental stresses, and certain chemicals including auxin and other regulators".

The biosynthesis of the hormone starts with conversion of the amino acid methionine to S-adenosyl-L-methionine (SAM, also called Adomet) by the enzyme Met Adenosyltransferase. SAM is then converted to 1-aminocyclopropane-1-carboxylic-acid (ACC) by the enzyme ACC synthase (ACS); the activity of ACS determines the rate of ethylene production, therefore regulation of this enzyme is key for the ethylene biosynthesis. The final step requires oxygen and involves the action of the enzyme ACC-oxidase (ACO), formerly known as the Ethylene Forming Enzyme (EFE). Ethylene biosynthesis can be induced by endogenous or exogenous ethylene. ACC synthesis increases with high levels of auxins, especially Indole acetic acid (IAA), and cytokinins. ACC synthase is inhibited by abscisic acid.

Ethylene Perception in Plants

Ethylene could be perceived by a transmembrane protein dimer complex. The gene encoding an ethylene receptor has been cloned in *Arabidopsis thaliana* and then in tomato.

Ethylene receptors are encoded by multiple genes in the Arabidopsis and tomato genomes. The gene family comprises five receptors in Arabidopsis and at least six in tomato, most of which have been shown to bind ethylene. DNA sequences for ethylene receptors have also been identified in many other plant species and an ethylene binding protein has even been identified in Cyanobacteria.

Environmental and Biological Triggers of Ethylene

Environmental cues can induce the biosynthesis of the plant hormone. Flooding, drought, chilling, wounding, and pathogen attack can induce ethylene formation in the plant. In flooding, root suffers from lack of oxygen, or anoxia, which leads to the synthesis of 1-aminocyclopropane-1-carboxylic acid (ACC). ACC is transported upwards in the plant and then oxidized in leaves. The product, the ethylene causes epinasty of the leaves.

One speculation recently put forth for epinasty is the downward pointing leaves may act as pump handles in the wind. The ethylene may or may not additionally induce the growth of a valve in the xylem, but the idea would be that the plant would harness the power of the wind to pump out more water from the roots of the plants than would normally happen with transpiration.

Physiological Responses of Plants

Like the other plant hormones, ethylene is considered to have pleiotropic effects. This essentially means that it is thought that at least some of the effects of the hormone are unrelated.

What is actually caused by the gas may depend on the tissue affected as well as environmental conditions. In the evolution of plants, ethylene would simply be a message that was coopted for unrelated uses by plants during different periods of the evolutionary development.

List of Plant Responses to Ethylene

- Seedling triple response, thickening and shortening of hypocotyl with pronounced apical hook.
- In pollination, when the pollen reaches the stigma, the precursor of the ethylene, ACC, is secreted to the petal, the ACC releases ethylene with ACC oxidase.
- Stimulates leaf and flower senescence
- Stimulates senescence of mature xylem cells in preparation for plant use
- Induces leaf abscission
- Induces seed germination
- Induces root hair growth — increasing the efficiency of water and mineral absorption

- Induces the growth of adventitious roots during flooding
- Stimulates epinasty — leaf petiole grows out, leaf hangs down and curls into itself
- Stimulates fruit ripening
- Induces a climacteric rise in respiration in some fruit which causes a release of additional ethylene.
- Affects gravitropism
- Stimulates nutational bending
- Inhibits stem growth and stimulates stem and cell broadening and lateral branch growth outside of seedling stage
- Interference with auxin transport (with high auxin concentrations)
- Inhibits shoot growth and stomatal closing except in some water plants or habitually flooded ones such as some rice varieties, where the opposite occurs (conserving CO_2 and O_2)
- Induces flowering in pineapples
- Inhibits short day induced flower initiation in *Pharbitus nil* and *Chrysanthemum morifolium*

Commercial Issues

Ethylene shortens the shelf life of many fruits by hastening fruit ripening and floral senescence. Ethylene will shorten the shelf life of cut flowers and potted plants by accelerating floral senescence and floral abscission. Flowers and plants which are subjected to stress during shipping, handling, or storage produce ethylene causing a significant reduction in floral display. Flowers affected by ethylene include carnation, geranium, petunia, rose, and many others.

Ethylene can cause significant economic losses for florists, markets, suppliers, and growers. Researchers have developed several ways to inhibit ethylene, including inhibiting ethylene synthesis and inhibiting ethylene perception. Aminoethoxyvinylglycine (AVG), Aminooxyacetic acid (AOA), and silver ions are ethylene inhibitors. Inhibiting ethylene synthesis is less effective for reducing post-harvest losses since ethylene from other sources can still have an effect. By inhibiting ethylene perception, fruits, plants and flowers don't respond to ethylene produced endogenously or from exogenous sources. Inhibitors of ethylene perception include compounds that have a similar shape to ethylene, but do not elicit the ethylene response. One example of an ethylene perception inhibitor is 1-methylcyclopropene (1-MCP).

Commercial growers of bromeliads, including pineapple plants, use ethylene to induce flowering. Plants can be induced to flower either by treatment with the gas in a chamber, or by placing a banana peel next to the plant in an enclosed area.

Chrysanthemum flowering is delayed by ethylene gas and growers have found that carbon dioxide 'burners' and the exhaust fumes from inefficient glasshouse heaters can raise the ethylene concentration to 0.05 vpm causing delay in flowering of commercial crops.

Historical Significance

Many geologists and scholars believe that the famous Greek Oracle at Delphi (the Pythia) went into her trance-like state as an effect of ethylene rising from ground faults.

History

Ethylene appears to have been discovered by Johann Joachim Becher, who obtained it by heating ethanol with sulfuric acid; he mentioned the gas in his *Physica Subterranea* (1669). Joseph Priestley also mentions the gas in his *Experiments and observations relating to the various branches of natural philosophy: with a continuation of the observations on air* (1779), where he reports that Jan Ingenhousz saw ethylene synthesized in the same way by a Mr. Enée in Amsterdam in 1777 and that Ingenhousz subsequently produced the gas himself. The properties of ethylene were studied in 1795 by four Dutch chemists, Johann Rudolph Deimann, Adrien Paets van Troostwyck, Anthoni Lauwerenburgh and Nicolas Bondt, who found that it differed from hydrogen gas and that it contained both carbon and hydrogen. This group also discovered that ethylene could be combined with chlorine to produce the *oil of the Dutch chemists*, 1,2-dichloroethane; this discovery gave ethylene the name used for it at that time, *olefiant gas* (oil-making gas.)

In the mid-19th century, the suffix *-ene* (an Ancient Greek root added to the end of female names meaning "daughter of") was widely used to refer to a molecule or part thereof that contained one fewer hydrogen atoms than the molecule being modified. Thus, *ethylene* (C_2H_4) was the "daughter of ethyl" (C_2H_5). The name ethylene was used in this sense as early as 1852.

In 1866, the German chemist August Wilhelm von Hofmann proposed a system of hydrocarbon nomenclature in which the suffixes -ane, -ene, -ine, -one, and -une were used to denote the hydrocarbons with 0, 2, 4, 6, and 8 fewer hydrogens than their parent alkane. In this system, ethylene became *ethene*. Hofmann's system eventually became the basis for the Geneva nomenclature approved by the International Congress of Chemists in 1892, which remains at the core of the IUPAC nomenclature. However, by that time, the name ethylene was deeply entrenched, and it remains in wide use today, especially in the chemical industry.

Nomenclature

The 1979 IUPAC nomenclature rules made an exception for retaining the non-systematic name ethylene, however, this decision was reversed in the 1993 rules so the IUPAC name is now *ethene*.

Safety

Like all hydrocarbons, ethylene is an asphyxiant and combustible. It has been used as an anesthetic.

Propene

Propene, also known as propylene or methylethylene, is an unsaturated organic compound having the chemical formula C_3H_6. It has one double bond, and is the second simplest member of the alkene class of hydrocarbons, and it is also second in natural abundance.

Properties

At room temperature and atmospheric pressure, propene is a gas, and as with many other alkenes, it is also colourless with a weak but unpleasant smell. Propene has a higher density and boiling point than ethylene due to its greater size. It has a slightly lower boiling point than propane and is thus more volatile. It lacks strongly polar bonds, yet the molecule has a small dipole moment due to its reduced symmetry (its point group is C_s).

Propene has the same empirical formula as cyclopropane but their atoms are connected in different ways, making these molecules structural isomers.

Production

Propene is produced from fossil fuels—petroleum, natural gas, and, to a much lesser extent, coal. Propene is a byproduct of oil refining and natural gas processing. During oil refining, ethylene, propene, and other compounds are produced by as a result of cracking larger hydrocarbon molecules to produce hydrocarbons more in demand. A major source of propene is cracking intended to produce ethylene, but it also results from refinery cracking producing other products. Propene can be separated by fractional distillation from hydrocarbon mixtures obtained from cracking and other refining processes; refinery-grade propene is about 50 to 70%. Another important petrochemical source of propene is propane dehydrogenation. This route is popular in regions, such as the Middle East, where there is an abundance of propane from oil/gas operations. Less common propene sources are the Fischer-Tropsch process, metathesis of ethylene or an ethylene/butene mixture, and catalytic conversion of methanol.

Propene production has remained static at around 35 million tonnes (Europe and North America only) from 2000 to 2008, but it has been increasing in East Asia, most notably Singapore and China. Total world production of propene is currently about half that of ethylene.

Uses

Propene is the second most important starting product in the petrochemical industry after ethylene. It is the raw material for a wide variety of products. Manufacturers of the plastic polypropylene account for nearly two thirds of all demand. Polypropylene is, for example, needed for the production of films, packaging, caps and closures as well as for other applications. In the year 2008 the worldwide sales of propene reached a value of over 90 billion US dollars.

Propene and benzene are converted to acetone and phenol via the cumene process. Propene is also used to produce isopropanol (propan-2-ol), acrylonitrile, propylene oxide (epoxypropane) and epichlorohydrin.

Reactions

Propene resembles other alkenes in that it undergoes addition reactions relatively easily at room temperature. The relative weakness of its double bond (which is less strong than two single bonds) explains its tendency to react with substances that can achieve this transformation. Alkene reactions include: 1) polymerization, 2) oxidation, 3) halogenation and hydrohalogenation, 4) alkylation, 5) hydration, 6) oligomerization, and 7) hydroformylation.

Environmental Safety

Propene is produced naturally by vegetation, particularly certain tree species. It is also a product of combustion, from forest fires and cigarette smoke to motor vehicle and aircraft exhaust. It is an impurity in some heating gases. Observed concentrations have been in the range of 0.1-4.8 parts per billion (ppb) in rural air, 4-10.5 ppb in urban air, and 7-260 ppb in industrial air samples.

In the United States and some European countries a Threshold Limit Value of 500 parts per million (ppm) was established for occupational (8-hour time-weighted average) exposure. It is considered a volatile organic compound (VOC) and emissions are regulated by many governments, but it is not listed by the U.S.

Environmental Protection Agency (EPA) as a hazardous air pollutant under the Clean Air Act. It has a relatively short half-life in the atmosphere, and is not expected to bioaccumulate, based on a calculated bioconcentration factor of 13.18 using a log K_{ow} value of 1.77.

Propene has low acute toxicity from inhalation. Inhalation of the gas can cause anesthetic effects and at very high concentrations, unconsciousness. However, the asphyxiation limit for humans is about 10 times higher (23%) than the lower flammability level.

Aromaticity

In organic chemistry, Aromaticity is a chemical property in which a conjugated ring of unsaturated bonds, lone pairs, or empty orbitals exhibit a stabilization stronger than would be expected by the stabilization of conjugation alone.

The earliest use of the term was in an article by August Wilhelm Hofmann in 1855. There is no general relationship between aromaticity as a chemical property and the olfactory properties of such compounds.

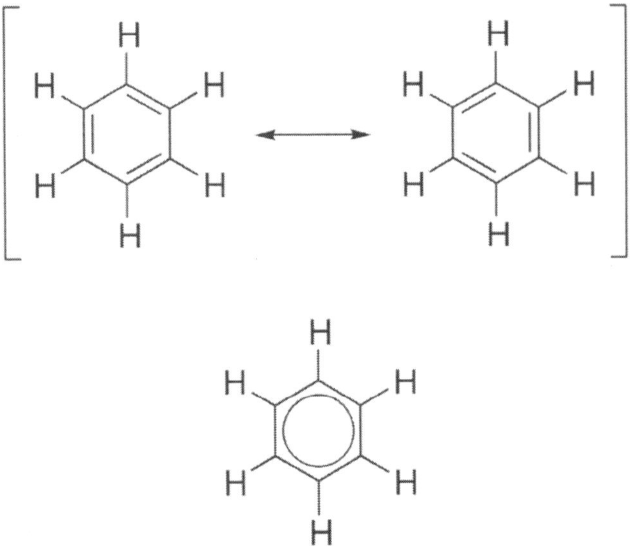

Figure 3: *Two different resonance forms of benzene (top) combine to produce an average structure (bottom)*

Aromaticity can also be considered a manifestation of cyclic delocalization and of resonance. This is usually considered to be because electrons are free to cycle around circular arrangements of atoms which are alternately single- and double-bonded to one another. These bonds may be seen as a hybrid of a single bond and a double bond, each bond in the ring identical to every other.

This commonly seen model of aromatic rings, namely the idea that benzene was formed from a six-membered carbon ring with alternating single and double bonds (cyclohexatriene), was developed by Kekulé.

The model for benzene consists of two resonance forms, which corresponds to the double and single bonds superimposing to give rise to six one-and-a-half bonds.

Benzene is a more stable molecule than would be expected without accounting for charge delocalization.

Theory

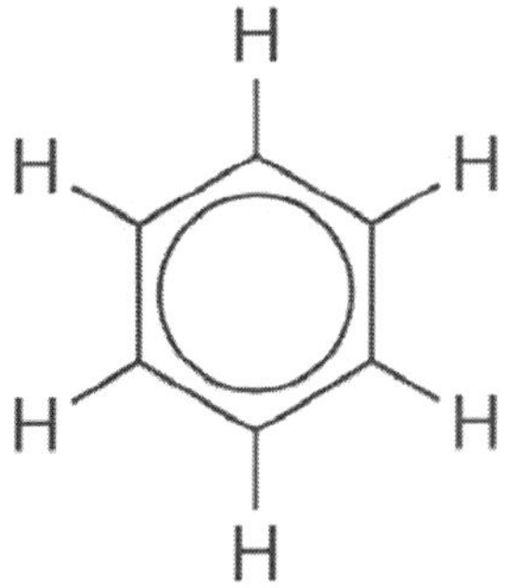

As is standard for resonance diagrams, a double-headed arrow is used to indicate that the two structures are not distinct entities, but merely hypothetical possibilities. Neither is an accurate representation of the *actual* compound, which is best represented by a hybrid (average) of these structures, which can be seen at right. A C=C bond is shorter than a C"C bond, but benzene is perfectly hexagonal—all six carbon-carbon bonds have the same length, intermediate between that of a single and that of a double bond.

A better representation is that of the circular π bond (Armstrong's *inner cycle*), in which the electron density is evenly distributed through a π-bond above and below the ring. This model more correctly represents the location of electron density within the aromatic ring.

The single bonds are formed with electrons in line between the carbon nuclei—these are called σ-bonds. Double bonds consist of a σ-bond and a π-bond. The π-bonds are formed from overlap of atomic p-orbitals above and below the plane of the ring. The following diagram shows the positions of these p-orbitals:

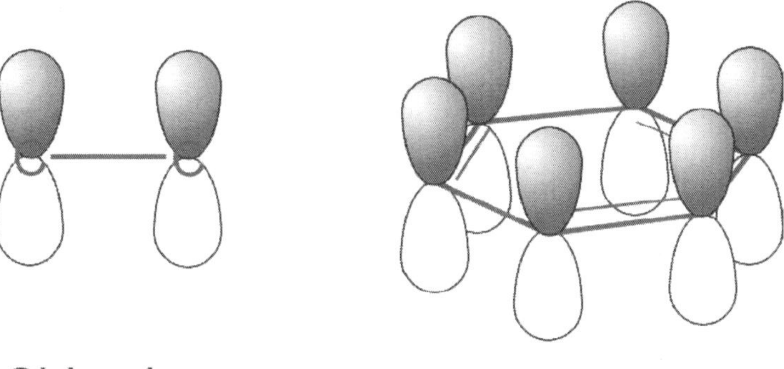

Since they are out of the plane of the atoms, these orbitals can interact with each other freely, and become delocalized. This means that instead of being tied to one atom of carbon, each electron is shared by all six in the ring. Thus, there are not enough electrons to form double bonds on all the carbon atoms, but the "extra" electrons strengthen all of the bonds on the ring equally. The resulting molecular orbital has π symmetry.

History

The first known use of the word "aromatic" as a *chemical* term—namely, to apply to compounds that contain the phenyl radical—occurs in an article by August Wilhelm Hofmann in 1855. If this is indeed the earliest introduction of the term, it is curious that Hofmann says nothing about why he introduced an adjective indicating olfactory character to apply to a group of chemical substances; only some of which have notable aromas. It is the case, however, that many of the most odoriferous organic substances known are terpenes, which are not aromatic in the chemical sense.

But terpenes and benzenoid substances do have a chemical characteristic in common, namely higher unsaturation indices than many aliphatic compounds, and Hofmann may not have been making a distinction between the two categories. The cyclohexatriene structure for benzene was first proposed by August Kekulé in 1865. Over the next few decades, most chemists readily accepted this structure, since it accounted for most of the known isomeric relationships of aromatic chemistry. However, it was always puzzling that this purportedly highly unsaturated molecule was so unreactive toward addition reactions.

The discoverer of the electron J. J. Thomson, between 1897 and 1906 placed three equivalent electrons between each carbon atom in benzene.

An explanation for the exceptional stability of benzene is conventionally attributed to Sir Robert Robinson, who was apparently the first (in 1925) to coin the term *aromatic sextet* as a group of six electrons that resists disruption.

In fact, this concept can be traced further back, via Ernest Crocker in 1922, to Henry Edward Armstrong, who in 1890, in an article entitled *The structure of cycloid hydrocarbons*, wrote *the (six) centric affinities act within a cycle...benzene may be represented by a double ring (*sic*)... and when an additive compound is formed, the inner cycle of affinity suffers disruption, the contiguous carbon-atoms to which nothing has been attached of necessity acquire the ethylenic condition.*

Here, Armstrong is describing at least four modern concepts. First, his "affinity" is better known nowadays as the electron, which was only to be discovered seven years later by J. J. Thomson. Second, he is describing electrophilic aromatic substitution, proceeding (third) through a Wheland intermediate, in which (fourth) the conjugation of the ring is broken. He introduced the symbol C centred on the ring as a shorthand for the *inner cycle*, thus anticipating Eric Clar's notation. It is argued that he also anticipated the nature of wave mechanics, since he recognized that his affinities had direction, not merely being point particles, and collectively having a distribution that could be altered by introducing substituents onto the benzene ring (*much as the distribution of the electric charge in a body is altered by bringing it near to another body*).

The quantum mechanical origins of this stability, or aromaticity, were first modelled by Hückel in 1931. He was the first to separate the bonding electrons into sigma and pi electrons.

Characteristics of Aromatic (aryl) Compounds

An aromatic (or aryl) compound contains a set of covalently bound atoms with specific characteristics:

1. A delocalized conjugated π system, most commonly an arrangement of alternating single and double bonds
2. Coplanar structure, with all the contributing atoms in the same plane
3. Contributing atoms arranged in one or more rings
4. A number of π delocalized electrons that is even, but not a multiple of 4. That is, $4n + 2$ number of π electrons, where n=0, 1, 2, 3, and so on. This is known as Hückel's Rule.

Whereas benzene is aromatic (6 electrons, from 3 double bonds), cyclobutadiene is not, since the number of π delocalized electrons is 4, which of course is a multiple of 4. The cyclobutadienide (2") ion, however, is aromatic (6 electrons). An atom in an aromatic system can have other electrons that are not part of the system, and are therefore ignored for the 4n + 2 rule. In furan, the oxygen atom is sp^2 hybridized. One lone pair is in the π system and the other in the plane of the ring (analogous to C-H bond on the other positions). There are 6 π electrons, so furan is aromatic.

Aromatic molecules typically display enhanced chemical stability, compared to similar non-aromatic molecules. A molecule that can be aromatic will tend to alter its electronic or conformational structure to be in this situation. This extra stability changes the chemistry of the molecule. Aromatic compounds undergo electrophilic aromatic substitution and nucleophilic aromatic substitution reactions, but not electrophilic addition reactions as happens with carbon-carbon double bonds. Many of the earliest-known examples of aromatic compounds, such as benzene and toluene, have distinctive pleasant smells. This property led to the term "aromatic" for this class of compounds, and hence the term "aromaticity" for the eventually discovered electronic property. The circulating π electrons in an aromatic molecule produce ring currents that oppose the applied magnetic field in NMR. The NMR signal of protons in the plane of an aromatic ring are shifted substantially further down-field than those on non-aromatic sp^2 carbons. This is an important way of detecting aromaticity. By the same mechanism, the signals of protons located near the ring axis are shifted up-field. Aromatic molecules are able to interact with each other in so-called π-π stacking: the π systems form two parallel rings overlap in a "face-to-face" orientation. Aromatic molecules are also able to interact with each other in an "edge-to-face" orientation: the slight positive charge of the substituents on the ring atoms of one molecule are attracted to the slight negative charge of the aromatic system on another molecule.

Planar monocyclic molecules containing 4n π electrons are called antiaromatic and are, in general, destabilized. Molecules that could be antiaromatic will tend to alter their electronic or conformational structure to avoid this situation, thereby becoming non-aromatic. For example, cyclooctatetraene (COT) distorts itself out of planarity, breaking π overlap between adjacent double bonds. Relatively recently, cyclobutadiene was discovered to adopt an asymmetric, rectangular configuration in which single and double bonds indeed alternate; there is no resonance and the single bonds are markedly longer than the double bonds, reducing unfavourable p-orbital overlap. Hence, cyclobutadiene is non-aromatic; the strain of the asymmetric configuration outweighs the anti-aromatic destabilization that would afflict the symmetric, square configuration.

Importance of aromatic compounds

Aromatic compounds play key roles in the biochemistry of all living things. The four aromatic amino acids histidine, phenylalanine, tryptophan, and tyrosine each serve as one of the 20 basic building blocks of proteins. Further, all 5 nucleotides (adenine, thymine, cytosine, guanine, and uracil) that make up the sequence of the genetic code in DNA and RNA are aromatic purines or pyrimidines. As well as that, the molecule heme contains an aromatic system with 22 π electrons. Chlorophyll also has a similar aromatic system.

Aromatic compounds are important in industry. Key aromatic hydrocarbons of commercial interest are benzene, toluene, *ortho*-xylene and *para*-xylene. About 35 million tonnes are produced worldwide every year. They are extracted from complex mixtures obtained by the refining of oil or by distillation of coal tar, and are used to produce a range of important chemicals and polymers, including styrene, phenol, aniline, polyester and nylon.

Types of Aromatic Compounds

The overwhelming majority of aromatic compounds are compounds of carbon, but they need not be hydrocarbons.

Heterocyclics

In heterocyclic aromatics (heteroaromats), one or more of the atoms in the aromatic ring is of an element other than carbon. This can lessen the ring's aromaticity, and thus (as in the case of furan) increase its reactivity. Other examples include pyridine, pyrazine, imidazole, pyrazole, oxazole, thiophene, and their benzannulated analogs (benzimidazole, for example).

Polycyclics

Polycyclic aromatic hydrocarbons are molecules containing two or more simple aromatic rings fused together by sharing two neighbouring carbon atoms. Examples are naphthalene, anthracene and phenanthrene.

Substituted Aromatics

Many chemical compounds are aromatic rings with other things attached. Examples include trinitrotoluene (TNT), acetylsalicylic acid (aspirin), paracetamol, and the nucleotides of DNA.

Atypical Aromatic Compounds

Aromaticity is found in ions as well: the cyclopropenyl cation (2e system), the cyclopentadienyl anion (6e system), the tropylium ion (6e) and the cyclooctatetraene dianion (10e). Aromatic properties have been attributed to non-benzenoid compounds such as tropone. Aromatic properties are tested to the limit in a class of compounds called cyclophanes.

A special case of aromaticity is found in homoaromaticity where conjugation is interrupted by a single sp^3 hybridized carbon atom.

When carbon in benzene is replaced by other elements in borabenzene, silabenzene, germanabenzene, stannabenzene, phosphorine or pyrylium salts the aromaticity is still retained. Aromaticity also occurs in compounds that are not carbon-based at all. Inorganic 6 membered ring compounds analogous to benzene have been synthesized. Hexasilabenzene (Si_6H_6) and borazine ($B_3N_3H_6$) are structurally analogous to benzene, with the carbon atoms replaced by another element or elements. In borazine, the boron and nitrogen atoms alternate around the ring.

Metal aromaticity is believed to exist in certain metal clusters of aluminium. Mobius aromaticity occurs when a cyclic system of molecular orbitals, formed from p_π atomic orbitals and populated in a closed shell by 4n (n is an integer) electrons, is given a single half-twist to correspond to a Mobius strip. Because the twist can be left-handed or right-handed, the resulting Mobius aromatics are *dissymmetric* or chiral. Up to now there is no doubtless proof that a Mobius aromatic molecule was synthesized. Aromatics with two half-twists corresponding to the paradromic topologies, first suggested by Johann Listing, have been proposed by Rzepa in 2005. In carbo-benzene the ring bonds are extended with alkyne and allene groups.

Aromatic Hydrocarbon

An aromatic hydrocarbon or arene (or sometimes aryl hydrocarbon) is a hydrocarbon with alternating double and single bonds between carbon atoms. The term 'aromatic' was assigned before the physical mechanism determining aromaticity was discovered, and was derived from the fact that many of the compounds have a sweet scent. The configuration of six carbon atoms in aromatic compounds is known as a benzene ring, after the simplest possible such hydrocarbon, benzene. Aromatic hydrocarbons can be *monocyclic* (MAH) or *polycyclic* (PAH).

Some non-benzene-based compounds called heteroarenes, which follow Hückel's rule, are also aromatic compounds. In these compounds, at least one carbon atom is replaced by one of the heteroatoms oxygen, nitrogen, or sulfur. Examples of non-benzene compounds with aromatic properties are furan, a heterocyclic compound with a five-membered ring that includes an oxygen atom, and pyridine, a heterocyclic compound with a six-membered ring containing one nitrogen atom.

Benzene Ring Model

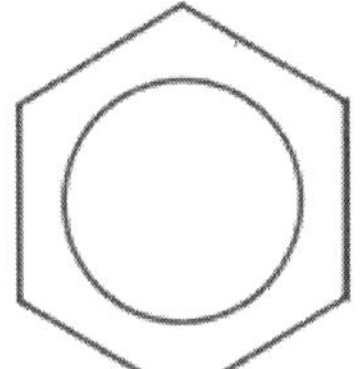

Figure: *Benzene*

Benzene, C_6H_6, is the simplest aromatic hydrocarbon and was recognized as the first aromatic hydrocarbon, with the nature of its bonding first being recognized by Friedrich August Kekulé von Stradonitz in the 19th century. Each carbon atom in the hexagonal cycle has four electrons to share. One goes to the hydrogen atom, and one each to the two neighbouring carbons. This leaves one to share with one of its two neighbouring carbon atoms, which is why the benzene molecule is drawn with alternating single and double bonds around the hexagon.

The structure is also illustrated as a circle around the inside of the ring to show six electrons floating around in delocalized molecular orbitals the size of the ring itself. This also represents the equivalent nature of the six carbon-carbon bonds all of bond order ~1.5. This equivalency is well explained by resonance forms. The electrons are visualized as floating above and below the ring with the electromagnetic fields they generate acting to keep the ring flat.

General properties:

1. Display aromaticity.
2. The carbon-hydrogen ratio is high.
3. They burn with a sooty yellow flame because of the high carbon-hydrogen ratio.
4. They undergo electrophilic substitution reactions and nucleophilic aromatic substitutions.

The circle symbol for aromaticity was introduced by Sir Robert Robinson in 1925 and popularized starting in 1959 by the Morrison & Boyd textbook on organic chemistry. The proper use of the symbol is debated, it is used to describe any cyclic pi system in some publications, or only those pi systems that obey Hückel's rule on others. Jensen argues that in line with Robinson's original proposal, the use of the circle symbol should be limited to monocyclic 6 pi-electron systems. In this way the circle symbol for a 6c–6e bond can be compared to the Y symbol for a 3c–2e bond.

Arene Synthesis

A reaction that forms an arene compound from an unsaturated or partially unsaturated cyclic precursor is simply called an aromatization. Many laboratory methods exist for the organic synthesis of arenes from non-arene precursors. Many methods rely on cycloaddition reactions. Alkyne trimerization describes the [2+2+2] cyclization of three alkynes, in the Dotz reaction an alkyne, carbon monoxide and a chromium carbene complex are the reactants. Diels-Alder reactions of alkynes with pyrone or cyclopentadienone with expulsion of carbon dioxide or carbon monoxide also form arene compounds. In Bergman cyclization the reactants are an enyne plus a hydrogen donor.

Another set of methods is the aromatization of cyclohexanes and other aliphatic rings: reagents are catalysts used in hydrogenation such as platinum, palladium and nickel (reverse hydrogenation), quinones and the elements sulfur and selenium.

Arene Reactions

Arenes are reactants in many organic reactions.

Aromatic Substitution

In aromatic substitution one substituent on the arene ring, usually hydrogen, is replaced by another substituent. The two main types are electrophilic aromatic substitution when the active reagent is an electrophile and nucleophilic aromatic substitution when the reagent is a nucleophile. In radical-nucleophilic aromatic substitution the active reagent is a radical. An example is the nitration of salicylic acid:

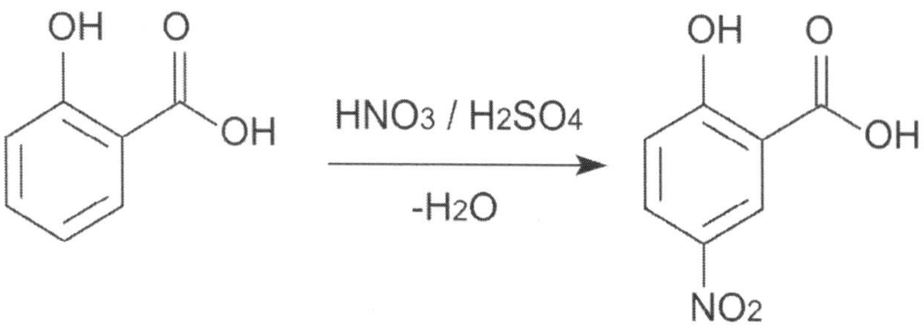

OH O OH —— HNO_3 / H_2SO_4, $-H_2O$ ——> OH O OH, NO_2

Coupling reactions

In coupling reactions a metal catalyses a coupling between two formal radical fragments. Common coupling reactions with arenes result in the formation of new carbon-carbon bonds e.g., alkylarenes, vinyl arenes, biraryls, new carbon-nitrogen bonds (anilines) or new carbon-oxygen bonds (aryloxy compounds).

An example is the direct arylation of perfluorobenzenes.

Pd(OAc)$_2$ (5 mol%)
$^tBu_2PCH_3.HBF_4$
K_2CO_3, DMA, 120 °C
98% yield

Hydrogenation

Hydrogenation of arenes create saturated rings. The compound 1-naphthol is completely reduced to a mixture of decalin-ol isomers.

H_2, 60 psi
Rh, Al_2O_3
95% EtOH, AcOH, rt, 12 hrs.
94% (isomers)

The compound resorcinol, hydrogenated with Raney nickel in presence of aqueous sodium hydroxide forms an enolate which is alkylated with methyl iodide to *2-methyl-1,3-cyclohexandione*:

H_2 900 psi
Raney nickel
50% NaOH aq. 50 °C
acidic workup, then
MeI, dioxane, reflux
54%

Cycloadditions

Cycloaddition reaction are not common. Unusual thermal Diels-Alder reactivity of arenes can be found in the Wagner-Jauregg reaction. Other photochemical cycloaddition reactions with alkenes occur through excimers.

Benzene and derivatives of benzene

Benzene derivatives have from one to six substituents attached to the central benzene core. Examples of benzene compounds with just one substituent are phenol, which carries a hydroxyl group and toluene with a methyl group.

When there is more than one substituent present on the ring, their spatial relationship becomes important for which the arene substitution patterns *ortho*, *meta*, and *para* are devised. For example, three isomers exist for cresol because the methyl group and the hydroxyl group can be placed next to each other (ortho), one position removed from each other (meta), or two positions removed from each other (para). Xylenol has two methyl groups in addition to the hydroxyl group, and, for this structure, 6 isomers exist.

- Representative arene compounds

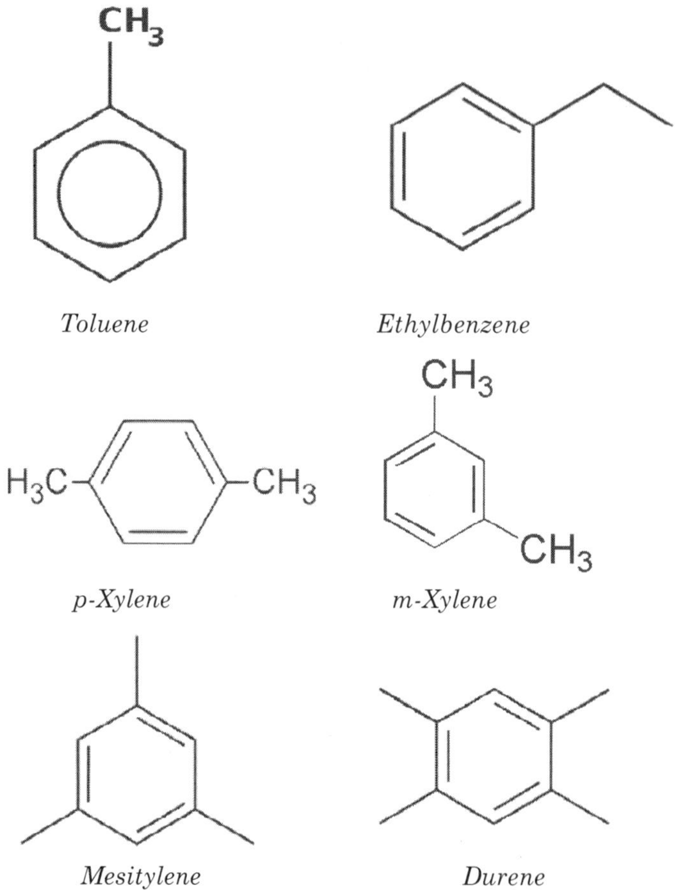

Toluene *Ethylbenzene*

p-Xylene *m-Xylene*

Mesitylene *Durene*

2-Phenylhexane

Biphenyl

OH

Phenol

NH_2

Aniline

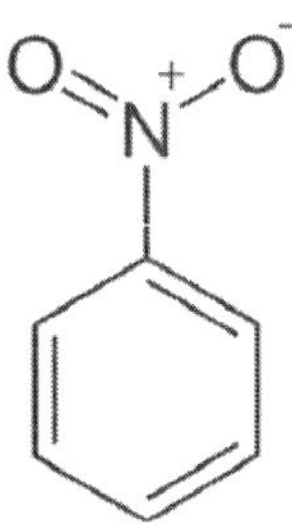

Nitrobenzene

Benzoic acid

Aspirin

Paracetamol

Picric acid

The arene ring has an ability to stabilize charges. This is seen in, for example, phenol (C_6H_5-OH), which is acidic at the hydroxyl (OH), since a charge on this oxygen (alkoxide $-O^-$) is partially delocalized into the benzene ring.

Polyaromatic Hydrocarbons

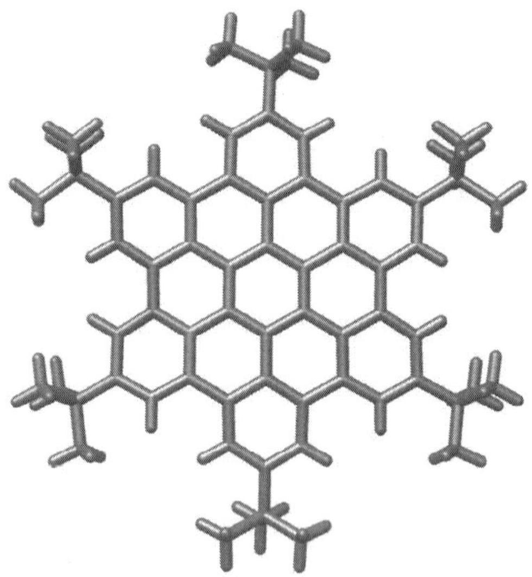

Figure: *Crystal structure of a hexa-tert-butyl derivatized hexa-peri-hexabenzo (bc,ef,hi,kl,no,qr)coronene, reported by Klaus Müllen and co-workers. The tert-butyl groups make this compound soluble in common solvents such as hexane, in which the unsubstituted PAH is insoluble.*

Poly-aromatic hydrocarbons (PAHs), also known as polycyclic aromatic hydrocarbons or polynuclear aromatic hydrocarbons, are potent atmospheric pollutants that consist of fused aromatic rings and do not contain heteroatoms or carry substituents. Naphthalene is the simplest example of a PAH.

PAHs occur in oil, coal, and tar deposits, and are produced as byproducts of fuel burning (whether fossil fuel or biomass). As a pollutant, they are of concern because some compounds have been identified as carcinogenic, mutagenic, and teratogenic.

PAHs are also found in cooked foods. Studies have shown that high levels of PAHs are found, for example, in meat cooked at high temperatures such as grilling or barbecuing, and in smoked fish.

They are also found in the interstellar medium, in comets, and in meteorites and are a candidate molecule to act as a basis for the earliest forms of life. In graphene the PAH motif is extended to large 2D sheets.

Occurrence and Pollution

Polycyclic aromatic hydrocarbons are lipophilic, meaning they mix more easily with oil than water. The larger compounds are less water-soluble and less volatile (i.e., less prone to evaporate). Because of these properties, PAHs in the environment are found primarily in soil, sediment and oily substances, as opposed to in water or air. However, they are also a component of concern in particulate matter suspended in air.

Natural crude oil and coal deposits contain significant amounts of PAHs, arising from chemical conversion of natural product molecules, such as steroids, to aromatic hydrocarbons. They are also found in processed fossil fuels, tar and various edible oils.

PAHs are one of the most widespread organic pollutants. In addition to their presence in fossil fuels they are also formed by incomplete combustion of carbon-containing fuels such as wood, coal, diesel, fat, tobacco, and incense. Different types of combustion yield different distributions of PAHs in both relative amounts of individual PAHs and in which isomers are produced. Thus, coal burning produces a different mixture than motor-fuel combustion or a forest fire, making the compounds potentially useful as indicators of the burning history. Hydrocarbon emissions from fossil fuel-burning engines are regulated in developed countries.

List of PAHs

Although the health effects of individual PAHs are not exactly alike, the following 17 PAHs are considered as a group in this profile issued by the Agency for Toxic Substances and Disease Registry (ATSDR):

- acenaphthene
- acenaphthylene
- anthracene
- benz[*a*]anthracene
- benzo[*a*]pyrene
- benzo[*e*]pyrene
- benzo[*b*]fluoranthene
- benzo[*ghi*]perylene
- benzo[*j*]fluoranthene
- benzo[*k*]fluoranthene
- chrysene

- dibenz(a,h)anthracene
- fluoranthene
- fluorene
- indeno(1,2,3-cd)pyrene
- phenanthrene
- pyrene.

Human Health

PAHs toxicity is very structurally dependent, with isomers (PAHs with the same formula and number of rings) varying from being nontoxic to being extremely toxic. Thus, highly carcinogenic PAHs may be small or large. One PAH compound, benzo[*a*]pyrene, is notable for being the first chemical carcinogen to be discovered (and is one of many carcinogens found in cigarette smoke). The EPA has classified seven PAH compounds as probable human carcinogens: benz[*a*]anthracene, benzo[*a*]pyrene, benzo[*b*]fluoranthene, benzo[*k*]fluoranthene, chrysene, dibenz(a,h)anthracene, and indeno(1,2,3-cd)pyrene.

PAHs known for their carcinogenic, mutagenic and teratogenic properties are benz[*a*]anthracene and chrysene, benzo[*b*]fluoranthene, benzo[*j*]fluoranthene, benzo[*k*]fluoranthene, benzo[*a*]pyrene, benzo[*ghi*]perylene, coronene, dibenz(a,h)anthracene ($C_{20}H_{14}$), indeno(1,2,3-cd)pyrene ($C_{22}H_{12}$) and ovalene.

High prenatal exposure to PAH is associated with lower IQ and childhood asthma. The Center for Children's Environmental Health reports studies that demonstrate that exposure to PAH pollution during pregnancy is related to adverse birth outcomes including low birth weight, premature delivery, and heart malformations.

Cord blood of exposed babies shows DNA damage that has been linked to cancer. Follow-up studies show a higher level of developmental delays at age three, lower scores on IQ tests and increased behavioural problems at ages six and eight.

Chemistry

The simplest PAHs, as defined by the International Union of Pure and Applied Chemistry (IUPAC) (G.P Moss, IUPAC nomenclature for fused-ring systems), are phenanthrene and anthracene, which both contain three fused aromatic rings. Smaller molecules, such as benzene, are not PAHs.

PAHs may contain four-, five-, six- or seven-member rings, but those with five or six are most common. PAHs composed only of six-membered rings are called alternant PAHs. Certain alternant PAHs are called benzenoid PAHs.

The name comes from benzene, an aromatic hydrocarbon with a single, six-membered ring. These can be benzene rings interconnected with each other by single carbon-carbon bonds and with no rings remaining that do not contain a complete benzene ring.

The set of alternant PAHs is closely related to a set of mathematical entities called polyhexes, which are planar figures composed by conjoining regular hexagons of identical size.

PAHs containing up to six fused aromatic rings are often known as "small" PAHs, and those containing more than six aromatic rings are called "large" PAHs. Due to the availability of samples of the various small PAHs, the bulk of research on PAHs has been of those of up to six rings. The biological activity and occurrence of the large PAHs does appear to be a continuation of the small PAHs. They are found as combustion products, but at lower levels than the small PAHs due to the kinetic limitation of their production through addition of successive rings. In addition, with many more isomers possible for larger PAHs, the occurrence of specific structures is much smaller.

PAHs possess very characteristic UV absorbance spectra. These often possess many absorbance bands and are unique for each ring structure.

Thus, for a set of isomers, each isomer has a different UV absorbance spectrum than the others. This is particularly useful in the identification of PAHs.

Most PAHs are also fluorescent, emitting characteristic wavelengths of light when they are excited (when the molecules absorb light). The extended pi-electron electronic structures of PAHs lead to these spectra, as well as to certain large PAHs also exhibiting semi-conducting and other behaviours.

Naphthalene ($C_{10}H_8$ constituent of mothballs), consisting of two coplanar six-membered rings sharing an edge, is another aromatic hydrocarbon. By formal convention, it is not a true PAH, though is referred to as a bicyclic aromatic hydrocarbon.

Aqueous solubility decreases approximately one order of magnitude for each additional ring.

PAH Compounds

The United States Environmental Protection Agency (EPA) has designated 32 PAH compounds as priority pollutants. The original 16 are listed. They are naphthalene, acenaphthylene, acenaphthene, fluorene, phenanthrene, anthracene, fluoranthene, pyrene, benzo[*a*]anthracene, chrysene, benzo[*b*]fluoranthene, benzo[*k*] flouranthene, benzo[*a*]pyrene, dibenz(ah)anthracene, benzo[*ghi*] perylene, and indeno(1,2,3-cd)pyrene. This list of the 16 EPA priority PAHs is often targeted for measurement in environmental samples.

Aromaticity

Although PAHs clearly are aromatic compounds, the degree of aromaticity can be different for each ring segment. According to Clar's rule (formulated by Erich Clar in 1964) for PAHs the resonance structure with the most disjoint aromatic ï-sextets—i.e., benzene-like moieties—is the most important for the characterization of the properties.

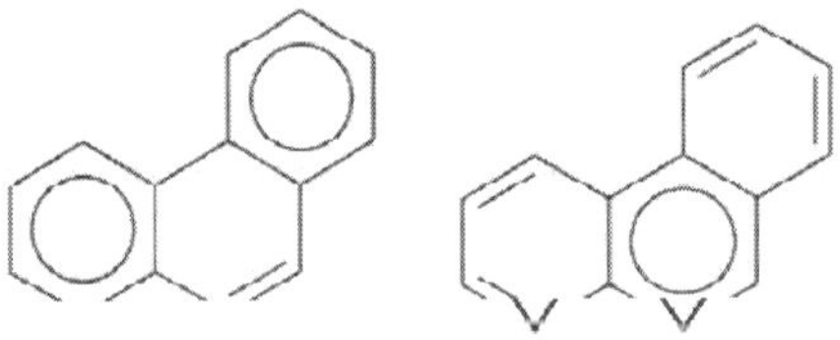

Phenanthrene (1) 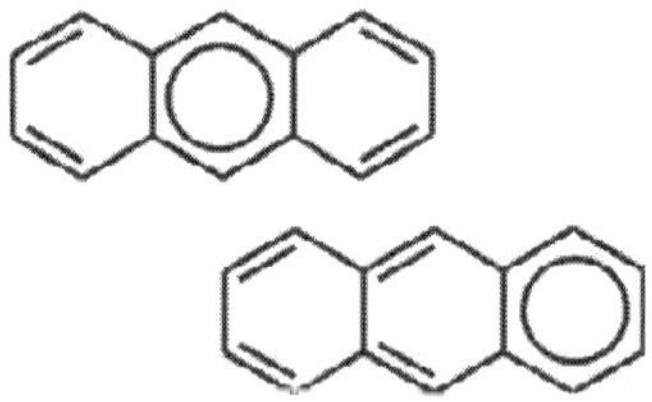 *Anthracene (2)*

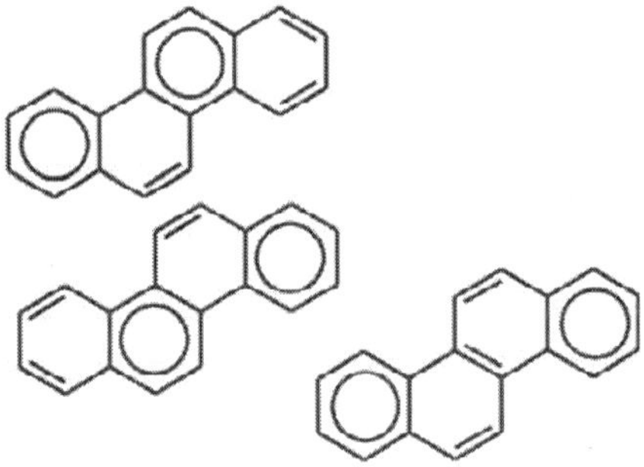

Chrysene (3) 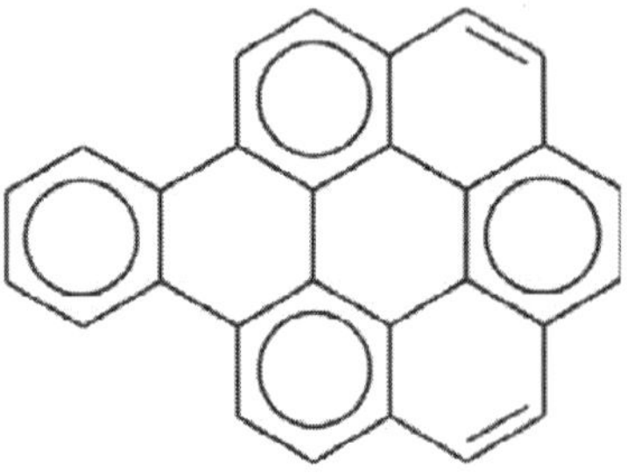 *Clar rule*

For example, in phenanthrene (1) one Clar structure has two sextets at the extremities, while the other resonance structure has just one central sextet. Therefore in this molecule the outer rings are firmly aromatic while its central ring is less aromatic and therefore more reactive. In contrast, in anthracene (2) the number of sextets is just one and aromaticity spreads out. This difference in number of sextets is reflected in the UV absorbance spectra of these two isomers. Phenanthrene has a highest wavelength absorbance around 290 nm, while anthracene has highest wavelength bands around 380 nm. Three Clar structures with two sextets are present in chrysene (3) and by superposition the aromaticity in the outer ring is larger than in the inner rings. Another relevant Clar hydrocarbon is zethrene.

Origins of Life

In January 2004 (at the 203rd Meeting of the American Astronomical Society), it was reported that a team led by A. Witt of the University of Toledo, Ohio studied ultraviolet light emitted by the Red Rectangle nebula and found the spectral signatures of anthracene and pyrene (no other such complex molecules had ever before been found in space). This discovery was considered as a controversial confirmation of a hypothesis that as nebulae of the same type as the Red Rectangle approach the ends of their lives, convection currents cause carbon and hydrogen in the nebulae's core to get caught in stellar winds, and radiate outward. As they cool, the atoms supposedly bond to each other in various ways and eventually form particles of a million or more atoms.

Figure: *Two extremely bright stars illuminate a mist of PAHs in this Spitzer image.*

Witt and his team inferred that since they discovered PAHs—which may have been vital in the formation of early life on Earth—in a nebula, by necessity they must originate in nebulae. More recently, fullerenes (or "buckyballs"), have been detected in other nebulae. Fullerenes are also implicated in the origin of life; according to astronomer Letizia Stanghellini, "It's possible that buckyballs from outer space provided seeds for life on Earth."

Detection

Detection of PAHs in materials is often done using gas chromatography-mass spectrometry or liquid chromatography with ultraviolet-visible or fluorescence spectroscopic methods or by using rapid test PAH indicator strips.

Aryl

In the context of organic molecules, aryl refers to any functional group or substituent derived from an aromatic ring, be it phenyl, naphthyl, thienyl, indolyl, etc.. “Aryl” is used for the sake of abbreviation or generalization, and “Ar” is used as a placeholder for the aryl group in chemical structure diagrams.

A simple aryl group is phenyl, C_6H_5; it is derived from benzene. The tolyl group, $CH_3C_6H_4$, is derived from toluene (methylbenzene). The xylyl group, $(CH_3)_2C_6H_3$, is derived from xylene (dimethylbenzene), while the naphthyl group, $C_{10}H_7$, is derived from naphthalene.

***Figure:** Sample aryl groups. From left to right phenyl, tolyl, o-xylyl, naphthyl.*

Arylation is simply any chemical process in which an aryl group is attached to a substrate.

Alkane

Alkanes (also known as paraffins or saturated hydrocarbons) are chemical compounds that consist only of hydrogen and carbon atoms and are bonded exclusively by single bonds (i.e., they are saturated compounds) without any cycles (or loops; i.e., cyclic structure).

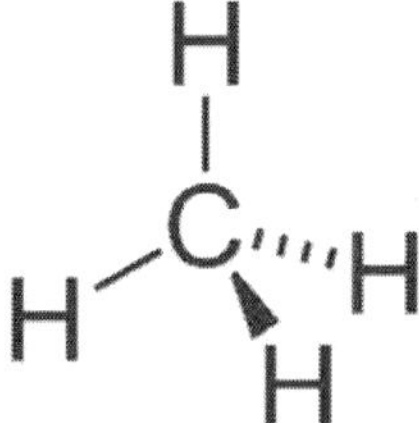

***Figure:** Chemical structure of methane, the simplest alkane*

Alkanes belong to a homologous series of organic compounds in which the members differ by a constant relative molecular mass of 14.

Each carbon atom must have 4 bonds (either C-H or C-C bonds), and each hydrogen atom must be joined to a carbon atom (H-C bonds). A series of linked carbon atoms is known as the carbon skeleton or carbon backbone. In general, the number of carbon atoms is often used to define the size of the alkane (e.g., C_2-alkane). An alkyl group, generally abbreviated with the symbol R, is a functional group or side-chain that, like an alkane, consists solely of single-bonded carbon and hydrogen atoms, for example a methyl or ethyl group.

The simplest possible alkane (the parent molecule) is methane, CH_4. There is no limit to the number of carbon atoms that can be linked together, the only limitation being that the molecule is acyclic, is saturated, and is a hydrocarbon. Saturated oils and waxes are examples of larger alkanes where the number of carbons in the carbon backbone tends to be greater than 10. Alkanes are not very reactive and have little biological activity. Alkanes can be viewed as a molecular tree upon which can be hung the interesting biologically active/reactive portions (functional groups) of the molecule.

Structure Classification

Saturated hydrocarbons can be:

- linear (general formula C_nH_{2n+2}) wherein the carbon atoms are joined in a snake-like structure
- branched (general formula C_nH_{2n+2}, $n > 3$) wherein the carbon backbone splits off in one or more directions
- cyclic (general formula C_nH_{2n}, $n > 2$) wherein the carbon backbone is linked so as to form a loop.

According to the definition by IUPAC, the former two are alkanes, whereas the third group is called cycloalkanes. Saturated hydrocarbons can also combine any of the linear, cyclic (e.g., polycyclic) and branching structures, and they are still alkanes (no general formula) as long as they are acyclic (i.e., having no loops). They also have single covalent bonds between their carbons

Isomerism

Alkanes with more than three carbon atoms can be arranged in numerous ways, forming different structural isomers. An isomer, in part, similar to a chemical anagram but unlike an anagram, may contain varying number of atoms and components, for which in a chemical compound can be structurally arranged in a multitude of different combinations and permutations.

The simplest isomer of an alkane is the one in which the carbon atoms are arranged in a single chain with no branches. This isomer is sometimes called the *n*-isomer (*n* for "normal", although it is not necessarily the most common). However the chain of carbon atoms may also be branched at one or more points. The number of possible isomers increases rapidly with the number of carbon atoms (sequence A000602 in OEIS).

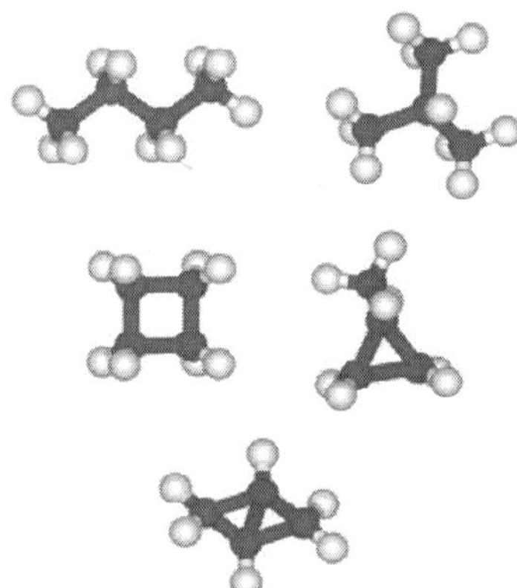

Figure: *Different C4-alkanes and -cycloalkanes (left to right): n-butane and isobutane are the two C4H10 isomers; cyclobutane and methylcyclopropane are the two C4H8 isomers. Bicyclo[1.1.0]butane is the only C4H6 compound and has no isomer; tetrahedrane (not shown) is the only C4H4 compound and has also no isomer.*

For example:

- C_1: no isomers: methane
- C_2: no isomers: ethane
- C_3: no isomers: propane
- C_4: 2 isomers: *n*-butane & isobutane
- C_5: 3 isomers: pentane, isopentane, neopentane
- C_6: 5 isomers: hexane, 2-Methylpentane, 3-Methylpentane, 2,3-Dimethylbutane & 2,2-Dimethylbutane
- C_{12}: 355 isomers
- C_{32}: 27,711,253,769 isomers
- C_{60}: 22,158,734,535,770,411,074,184 isomers, many of which are not stable.

Branched alkanes can be chiral: 3-methylhexane and its higher homologues are chiral due to their stereogenic center at carbon atom number 3. Chiral alkanes are of certain importance in biochemistry, as they occur as sidechains in chlorophyll and tocopherol (vitamin E). Chiral alkanes can be resolved into their enantiomers by enantioselective chromatography. In addition to these isomers, the chain of carbon atoms may form one or more loops. Such compounds are called cycloalkanes.

Nomenclature

The IUPAC nomenclature (systematic way of naming compounds) for alkanes is based on identifying hydrocarbon chains. Unbranched, saturated hydrocarbon chains are named systematically with a Greek numerical prefix denoting the number of carbons and the suffix "-ane". August Wilhelm von Hofmann suggested systematizing nomenclature by using the whole sequence of vowels a, e, i, o and u to create suffixes -ane, -ene, -ine (or -yne), -one, -une, for the hydrocarbons. The first three name hydrocarbons with single, double and triple bonds; "-one" represents a ketone; "-ol" represents an alcohol or OH group; "-oxy-" means an ether and refers to oxygen between two carbons, so that methoxy-methane is the IUPAC name for dimethyl ether. It is difficult or impossible to find compounds with more than one IUPAC name. This is because shorter chains attached to longer chains are prefixes and the convention includes brackets. Numbers in the name, referring to which carbon a group is attached to, should be as low as possible, so that 1- is implied and usually omitted from names of organic compounds with only one side-group; "1-" is implied in Nitro-octane. Symmetric compounds will have two ways of arriving at the same name.

Linear Alkanes

Straight-chain alkanes are sometimes indicated by the prefix *n*- (for *normal*) where a non-linear isomer exists. Although this is not strictly necessary, the usage is still common in cases where there is an important difference in properties between the straight-chain and branched-chain isomers, e.g., *n*-hexane or 2- or 3-methylpentane.

The members of the series (in terms of number of carbon atoms) are named as follows:

- methane, CH_4 - one carbon and four hydrogen
- ethane, C_2H_6 - two carbon and six hydrogen
- propane, C_3H_8 - three carbon and 8 hydrogen
- butane, C_4H_{10} - four carbon and 10 hydrogen
- pentane, C_5H_{12} - five carbon and 12 hydrogen
- hexane, C_6H_{14} - six carbon and 14 hydrogen.

These names were derived from methanol, ether, propionic acid and butyric acid, respectively. Alkanes with five or more carbon atoms are named by adding the suffix -ane to the appropriate numerical multiplier prefix with elision of any terminal vowel (*-a* or *-o*) from the basic numerical term. Hence, pentane, C_5H_{12}; hexane, C_6H_{14}; heptane, C_7H_{16}; octane, C_8H_{18}; etc. The prefix is generally Greek, with the exceptions of nonane which has a Latin prefix, and undecane and tridecane which have mixed-language prefixes.

Branched Alkanes

Simple branched alkanes often have a common name using a prefix to distinguish them from linear alkanes, for example *n*-pentane, isopentane, and neopentane.

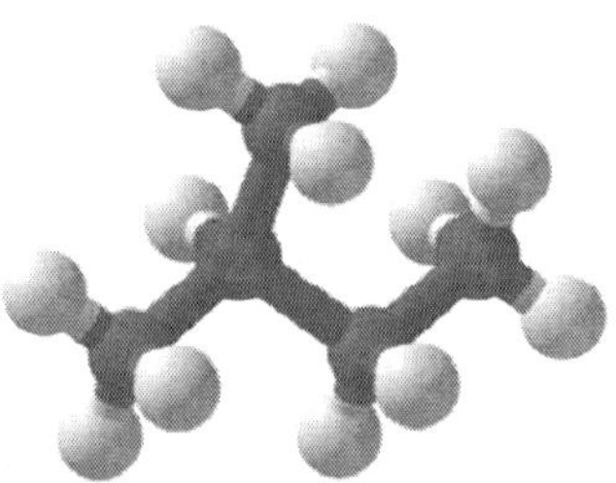

Figure 10: *Ball-and-stick model of isopentane (common name) or 2-methylbutane (IUPAC systematic name)*

IUPAC naming conventions can be used to produce a systematic name. The key steps in the naming of more complicated branched alkanes are as follows:

- Identify the longest continuous chain of carbon atoms
- Name this longest root chain using standard naming rules
- Name each side chain by changing the suffix of the name of the alkane from "-ane" to "-yl"
- Number the root chain so that sum of the numbers assigned to each side group will be as low as possible
- Number and name the side chains before the name of the root chain
- If there are multiple side chains of the same type, use prefixes such as "di-" and "tri-" to indicate it as such, and number each one.
- Add side chain names in alphabetical (disregarding "di-" etc. prefixes) order in front of the name of the root chain.

Table: *Comparison of nomenclatures for three isomers of C_5H_{12}*

Common name	***n-pentane***	***isopentane***	***neopentane***
IUPAC name	***pentane***	***2-methyl-butane***	***2,2-dimethyl-propane***
Structure		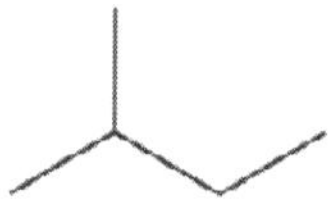	

Cyclic Alkanes

So-called cyclic alkanes are, in the technical sense, *not* alkanes, but cycloalkanes. They are hydrocarbons just like alkanes, but contain one or more rings. Simple cycloalkanes have a prefix "cyclo-" to distinguish them from alkanes. Cycloalkanes are named as per their acyclic counterparts with respect to the number of carbon atoms, e.g., cyclopentane (C_5H_{10}) is a cycloalkane with 5 carbon atoms just like pentane (C_5H_{12}), but they are joined up in a five-membered ring. In a similar manner, propane and cyclopropane, butane and cyclobutane, etc. Substituted cycloalkanes are named similar to substituted alkanes — the cycloalkane ring is stated, and the substituents are according to their position on the ring, with the numbering decided by Cahn-Ingold-Prelog rules.

Trivial Names

The trivial (non-systematic) name for alkanes is "paraffins." Together, alkanes are known as the *paraffin series*. Trivial names for compounds are usually historical artifacts. They were coined before the development of systematic names, and have been retained due to familiar usage in industry. Cycloalkanes are also called naphthenes. It is almost certain that the term paraffin stems from the petrochemical industry. Branched-chain alkanes are called *isoparaffins*. The use of the term "paraffin" is a general term and often does not distinguish between a pure compounds and mixtures of isomers with the same chemical formula (i.e., like a chemical anagram), e.g., pentane and isopentane.

Examples

The following trivial names are retained in the IUPAC system:

- isobutane for 2-methylpropane
- isopentane for 2-methylbutane
- neopentane for 2,2-dimethylpropane.

Physical Properties

Table: *Table of alkanes*

Alkane	*Formula*	*Boiling point [°C]*	*Melting point [°C]*	*Density [g·cm³] (at 20°C)*
Methane	CH_4	-162	-183	gas
Ethane	C_2H_6	-89	-172	gas
Propane	C_3H_8	-42	-188	gas
Butane	C_4H_{10}	0	-138	gas

Contd...

Alkane	*Formula*	*Boiling point [°C]*	*Melting point [°C]*	*Density [g·cm³] (at 20°C)*
Pentane	C_5H_{12}	36	-130	0.626(liquid)
Hexane	C_6H_{14}	69	-95	0.659(liquid)
Heptane	C_7H_{16}	98	-91	0.684(liquid)
Octane	C_8H_{18}	126	-57	0.703(liquid)
Nonane	C_9H_{20}	151	-54	0.718(liquid)
Decane	$C_{10}H_{22}$	174	-30	0.730(liquid)
Undecane	$C_{11}H_{24}$	196	-26	0.740(liquid)
Dodecane	$C_{12}H_{26}$	216	-10	0.749(liquid)
Icosane	$C_{20}H_{42}$	343	37	solid
Triacontane	$C_{30}H_{62}$	450	66	solid
Tetracontane	$C_{40}H_{82}$	525	82	solid
Pentacontane	$C_{50}H_{102}$	575	91	solid
Hexacontane	$C_{60}H_{122}$	625	100	solid

Boiling Point

Alkanes experience inter-molecular van der Waals forces. Stronger inter-molecular van der Waals forces give rise to greater boiling points of alkanes.

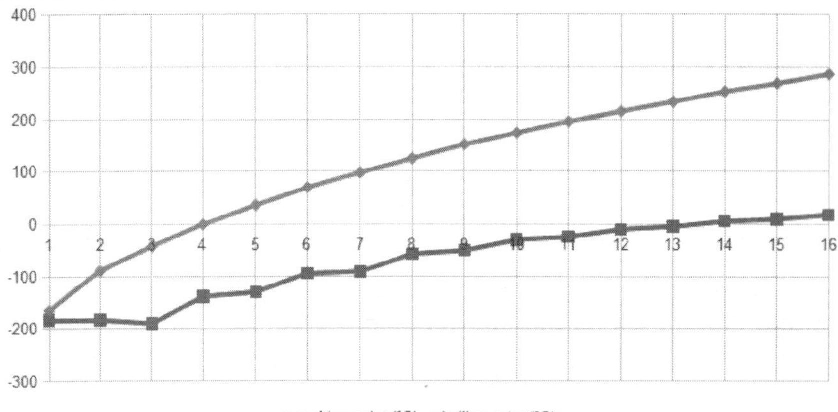

Figure : *Melting (blue) and boiling (pink) points of the first 14 n-alkanes in °C.*

There are two determinants for the strength of the van der Waals forces:

- the number of electrons surrounding the molecule, which increases with the alkane's molecular weight
- the surface area of the molecule

Under standard conditions, from CH_4 to C_4H_{10} alkanes are gaseous; from C_5H_{12} to $C_{17}H_{36}$ they are liquids; and after $C_{18}H_{38}$ they are solids. As the boiling point of alkanes is primarily determined by weight, it should not be a surprise that the boiling point has almost a linear relationship with the size (molecular weight) of the molecule. As a rule of thumb, the boiling point rises 20 - 30 °C for each carbon added to the chain; this rule applies to other homologous series.

A straight-chain alkane will have a boiling point higher than a branched-chain alkane due to the greater surface area in contact, thus the greater van der Waals forces, between adjacent molecules. For example, compare isobutane (2-methylpropane) and n-butane (butane), which boil at -12 and 0 °C, and 2,2-dimethylbutane and 2,3-dimethylbutane which boil at 50 and 58 °C, respectively. For the latter case, two molecules 2,3-dimethylbutane can "lock" into each other better than the cross-shaped 2,2-dimethylbutane, hence the greater van der Waals forces.

On the other hand, cycloalkanes tend to have higher boiling points than their linear counterparts due to the locked conformations of the molecules, which give a plane of intermolecular contact.

Melting Point

The melting points of the alkanes follow a similar trend to boiling points for the same reason as outlined above. That is, (all other things being equal) the larger the molecule the higher the melting point. There is one significant difference between boiling points and melting points. Solids have more rigid and fixed structure than liquids. This rigid structure requires energy to break down. Thus the better put together solid structures will require more energy to break apart. For alkanes, this can be seen from the graph above (i.e., the blue line). The odd-numbered alkanes have a lower trend in melting points than even numbered alkanes. This is because even numbered alkanes pack well in the solid phase, forming a well-organized structure, which requires more energy to break apart. The odd-number alkanes pack less well and so the "looser" organized solid packing structure requires less energy to break apart.

The melting points of branched-chain alkanes can be either higher or lower than those of the corresponding straight-chain alkanes, again depending on the ability of the alkane in question to packing well in the solid phase: This is particularly true for isoalkanes (2-methyl isomers), which often have melting points higher than those of the linear analogues.

Conductivity and Solubility

Alkanes do not conduct electricity, nor are they substantially polarized by an electric field. For this reason they do not form hydrogen bonds and are insoluble in polar solvents such as water. Since the hydrogen bonds between individual water molecules are aligned away from an alkane molecule, the coexistence of an alkane and water leads to an increase in molecular order (a reduction in entropy). As there is no significant bonding between water molecules and alkane molecules, the second law of thermodynamics suggests that this reduction in entropy should be minimized by minimizing the contact between alkane and water: Alkanes are said to be hydrophobic in that they repel water.

Their solubility in nonpolar solvents is relatively good, a property that is called lipophilicity. Different alkanes are, for example, miscible in all proportions among themselves.

The density of the alkanes usually increases with increasing number of carbon atoms, but remains less than that of water. Hence, alkanes form the upper layer in an alkane-water mixture.

Molecular Geometry

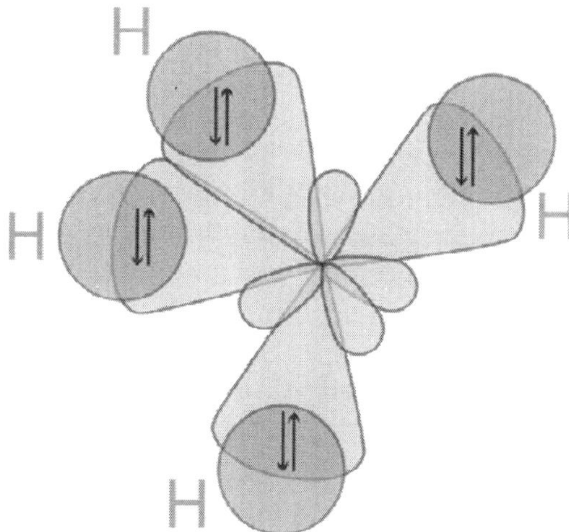

Figure : *sp^3-hybridization in methane.*

The molecular structure of the alkanes directly affects their physical and chemical characteristics. It is derived from the electron configuration of carbon, which has four valence electrons. The carbon atoms in alkanes are always sp^3 hybridized, that is to say that the valence electrons are said to be in four equivalent orbitals derived from the combination of the 2s orbital and the three 2p orbitals.

Bond Lengths and Bond Angles

An alkane molecule has only C – H and C – C single bonds. The former result from the overlap of a sp^3-orbital of carbon with the 1s-orbital of a hydrogen; the latter by the overlap of two sp^3-orbitals on different carbon atoms. The bond lengths amount to 1.09×10^{-10} m for a C – H bond and 1.54×10^{-10} m for a C – C bond.

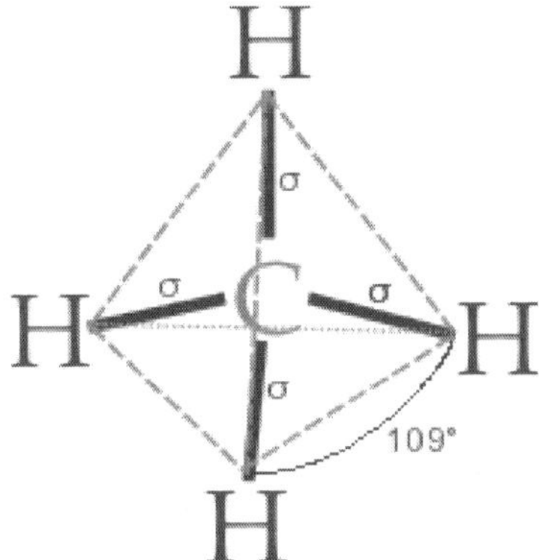

Figure: *The tetrahedral structure of methane.*

The spatial arrangement of the bonds is similar to that of the four sp^3-orbitals—they are tetrahedrally arranged, with an angle of 109.47° between them. Structural formulae that represent the bonds as being at right angles to one another, while both common and useful, do not correspond with the reality.

Conformation

The structural formula and the bond angles are not usually sufficient to completely describe the geometry of a molecule. There is a further degree of freedom for each carbon – carbon bond: the torsion angle between the atoms or groups bound to the atoms at each end of the bond. The spatial arrangement described by the torsion angles of the molecule is known as its conformation.

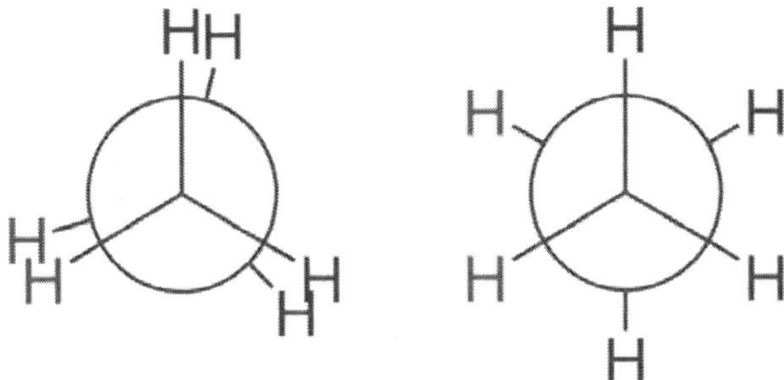

Figure: *Newman projections of the two conformations of ethane: eclipsed on the left, staggered on the right.*

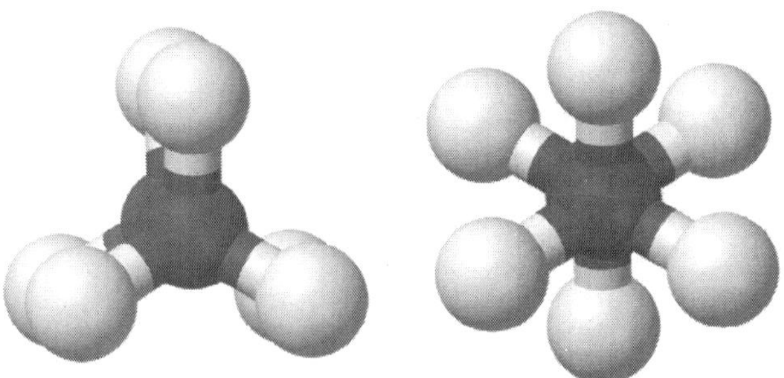

Figure: *Ball-and-stick models of the two rotamers of ethane*

Ethane forms the simplest case for studying the conformation of alkanes, as there is only one C – C bond. If one looks down the axis of the C – C bond, one will see the so-called Newman projection. The hydrogen atoms on both the front and rear carbon atoms have an angle of 120° between them, resulting from the projection of the base of the tetrahedron onto a flat plane.

However, the torsion angle between a given hydrogen atom attached to the front carbon and a given hydrogen atom attached to the rear carbon can vary freely between 0° and 360°. This is a consequence of the free rotation about a carbon – carbon single bond. Despite this apparent freedom, only two limiting conformations are important: eclipsed conformation and staggered conformation.

The two conformations, also known as rotamers, differ in energy: The staggered conformation is 12.6 kJ/mol lower in energy (more stable) than the eclipsed conformation (the least stable).

This difference in energy between the two conformations, known as the torsion energy, is low compared to the thermal energy of an ethane molecule at ambient temperature. There is constant rotation about the C-C bond.

The time taken for an ethane molecule to pass from one staggered conformation to the next, equivalent to the rotation of one CH_3-group by 120° relative to the other, is of the order of 10^{-11} seconds.

The case of higher alkanes is more complex but based on similar principles, with the antiperiplanar conformation always being the most favoured around each carbon-carbon bond. For this reason, alkanes are usually shown in a zigzag arrangement in diagrams or in models. The actual structure will always differ somewhat from these idealized forms, as the differences in energy between the conformations are small compared to the thermal energy of the molecules: Alkane molecules have no fixed structural form, whatever the models may suggest.

Spectroscopic Properties

Virtually all organic compounds contain carbon – carbon and carbon – hydrogen bonds, and so show some of the features of alkanes in their spectra. Alkanes are notable for having no other groups, and therefore for the *absence* of other characteristic spectroscopic features.

Infrared Spectroscopy

The carbon–hydrogen stretching mode gives a strong absorption between 2850 and 2960 cm^{-1}, while the carbon–carbon stretching mode absorbs between 800 and 1300 cm^{-1}. The carbon–hydrogen bending modes depend on the nature of the group: methyl groups show bands at 1450 cm^{-1} and 1375 cm^{-1}, while methylene groups show bands at 1465 cm^{-1} and 1450 cm^{-1}. Carbon chains with more than four carbon atoms show a weak absorption at around 725 cm^{-1}.

NMR Spectroscopy

The proton resonances of alkanes are usually found at $\delta_H = 0.5 – 1.5$. The carbon-13 resonances depend on the number of hydrogen atoms attached to the carbon: $\delta_C = 8 – 30$ (primary, methyl, -CH_3), 15 – 55 (secondary, methylene, -CH_2-), 20 – 60 (tertiary, methyne, C-H) and quaternary. The carbon-13 resonance of quaternary carbon atoms is characteristically weak, due to the lack of Nuclear Overhauser effect and the long relaxation time, and can be missed in weak samples, or sample that have not been run for a sufficiently long time.

Mass Spectrometry

Alkanes have a high ionization energy, and the molecular ion is usually weak. The fragmentation pattern can be difficult to interpret, but, in the case of branched chain alkanes, the carbon chain is preferentially cleaved at tertiary or quaternary carbons due to the relative stability of the resulting free radicals.

The fragment resulting from the loss of a single methyl group (M”15) is often absent, and other fragment are often spaced by intervals of fourteen mass units, corresponding to sequential loss of CH_2-groups.

Chemical Properties

In general, alkanes show a relatively low reactivity, because their C bonds are relatively stable and cannot be easily broken. Unlike most other organic compounds, they possess no functional groups.

They react only very poorly with ionic or other polar substances. The acid dissociation constant (pK_a) values of all alkanes are above 60, hence they are practically inert to acids and bases.

This inertness is the source of the term *paraffins* (with the meaning here of "lacking affinity"). In crude oil the alkane molecules have remained chemically unchanged for millions of years.

However redox reactions of alkanes, in particular with oxygen and the halogens, are possible as the carbon atoms are in a strongly reduced condition; in the case of methane, the lowest possible oxidation state for carbon (–4) is reached.

Reaction with oxygen (*if* an amount of the least is enough to meet the reaction stoichometry) leads to combustion without any smoke; with halogens, substitution.

Free radicals, molecules with unpaired electrons, play a large role in most reactions of alkanes, such as cracking and reformation where long-chain alkanes are converted into shorter-chain alkanes and straight-chain alkanes into branched-chain isomers.

In highly branched alkanes, the bond angle may differ significantly from the optimal value (109.5°) in order to allow the different groups sufficient space. This causes a tension in the molecule, known as steric hindrance, and can substantially increase the reactivity.

Reactions with Oxygen (Combustion Reaction)

All alkanes react with oxygen in a combustion reaction, although they become increasingly difficult to ignite as the number of carbon atoms increases. The general equation for complete combustion is:

$$C_nH_{2n+2} + (1.5n+0.5)O_2 \rightarrow (n+1)H_2O + nCO_2$$

In the absence of sufficient oxygen, carbon monoxide or even soot can be formed, as shown below:

$$C_nH_{(2n+2)} + nO_2 \rightarrow (n+1)H_2O + nCO$$

For example methane:

$$2CH_4 + 3O_2 \rightarrow 2CO + 4H_2O$$

$$CH_4 + 1.5O_2 \rightarrow CO + 2H_2O$$

The standard enthalpy change of combustion, $\Delta_c H^{\circ}$, for alkanes increases by about 650 kJ/mol per CH_2 group. Branched-chain alkanes have lower values of $\Delta_c H^{\circ}$ than straight-chain alkanes of the same number of carbon atoms, and so can be seen to be somewhat more stable.

Reactions with Halogens

Alkanes react with halogens in a so-called *free radical halogenation* reaction. The hydrogen atoms of the alkane are progressively replaced by halogen atoms. Free-radicals are the reactive species that participate in the reaction, which usually leads to a mixture of products. The reaction is highly exothermic, and can lead to an explosion.

These reactions are an important industrial route to halogenated hydrocarbons. There are three steps:

- Initiation the halogen radicals form by homolysis. Usually, energy in the form of heat or light is required.
- Chain reaction or Propagation then takes place—the halogen radical abstracts a hydrogen from the alkane to give an alkyl radical. This reacts further.
- Chain termination where step the radicals recombine.

Experiments have shown that all halogenation produces a mixture of all possible isomers, indicating that all hydrogen atoms are susceptible to reaction. The mixture produced, however, is not a statistical mixture: Secondary and tertiary hydrogen atoms are preferentially replaced due to the greater stability of secondary and tertiary free-radicals. An example can be seen in the monobromination of propane: [In the Figure below, the Statistical Distribution should be 25% and 75%]

H H $\xrightarrow{Br_2}$ Br + Br

Statistical distribution:	33 %	67 %
Experimental distribution:	97 %	3 %

Cracking

Cracking breaks larger molecules into smaller ones. This can be done with a thermal or catalytic method. The thermal cracking process follows a homolytic mechanism with formation of free-radicals. The catalytic cracking process involves the presence of acid catalysts (usually solid acids such as silica-alumina and zeolites), which promote a heterolytic (asymmetric) breakage of bonds yielding pairs of ions of opposite charges, usually a carbocation and the very unstable hydride anion. Carbon-localized free-radicals and cations are both highly unstable and undergo processes of chain rearrangement, C-C scission in position beta (i.e., cracking) and intra- and intermolecular hydrogen transfer or hydride transfer. In both types of processes, the corresponding reactive intermediates (radicals, ions) are permanently regenerated, and thus they proceed by a self-propagating chain mechanism. The chain of reactions is eventually terminated by radical or ion recombination.

Isomerization and Reformation

Isomerization and reformation are processes in which straight-chain alkanes are heated in the presence of a platinum catalyst. In isomerization, the alkanes become branched-chain isomers. In reformation, the alkanes become cycloalkanes or aromatic hydrocarbons, giving off hydrogen as a by-product. Both of these processes raise the octane number of the substance.

Other Reactions

Alkanes will react with steam in the presence of a nickel catalyst to give hydrogen. Alkanes can be chlorosulfonated and nitrated, although both reactions require special conditions. The fermentation of alkanes to carboxylic acids is of some technical importance. In the Reed reaction, sulfur dioxide, chlorine and light convert hydrocarbons to sulfonyl chlorides.

Occurrence

Occurrence of Alkanes in the Universe: Alkanes form a small portion of the atmospheres of the outer gas planets such as Jupiter (0.1% methane, 0.0002% ethane), Saturn (0.2% methane, 0.0005% ethane), Uranus (1.99% methane, 0.00025% ethane) and Neptune (1.5% methane, 1.5 ppm ethane). Titan (1.6% methane), a satellite of Saturn, was examined by the *Huygens* probe, which indicate that Titan's atmosphere periodically rains liquid methane onto the moon's surface.

Also on Titan, a methane-spewing volcano was spotted and this volcanism is believed to be a significant source of the methane in the atmosphere. There also appear to be Methane/Ethane lakes near the north polar regions of Titan, as discovered by Cassini's radar imaging.

Methane and ethane have also been detected in the tail of the comet Hyakutake. Chemical analysis showed that the abundances of ethane and methane were roughly equal, which is thought to imply that its ices formed in interstellar space, away from the Sun, which would have evaporated these volatile molecules. Alkanes have also been detected in meteorites such as carbonaceous chondrites.

Occurrence of Alkanes on Earth

Traces of methane gas (about 0.0001% or 1 ppm) occur in the Earth's atmosphere, produced primarily by organisms such as Archaea, found for example in the gut of cows.

The most important commercial sources for alkanes are natural gas and oil. Natural gas contains primarily methane and ethane, with some propane and butane: oil is a mixture of liquid alkanes and other hydrocarbons. These hydrocarbons were formed when dead marine animals and plants (zooplankton and phytoplankton) died and sank to the bottom of ancient seas and were covered with sediments in an anoxic environment and converted over many millions of years at high temperatures and high pressure to their current form. Natural gas resulted thereby for example from the following reaction:

$$C_6H_{12}O_6 \rightarrow 3CH_4 + 3CO_2$$

These hydrocarbons collected in porous rocks, located beneath an impermeable cap rock and so are trapped. These deposits, e.g., oil fields, have formed over millions of years and once exhausted cannot be readily replaced. The depletion of these hydrocarbons is the basis for what is known as the energy crisis.

Solid alkanes are known as tars and are formed when more volatile alkanes such as gases and oil evaporate from hydrocarbon deposits. One of the largest natural deposits of solid alkanes is in the asphalt lake known as the Pitch Lake in Trinidad and Tobago.

Methane is also present in what is called biogas, produced by animals and decaying matter, which is a possible renewable energy source. Alkanes have a low solubility in water, so the content in the oceans is negligible; however, at high pressures and low temperatures (such as at the bottom of the oceans), methane can co-crystallize with water to form a solid methane hydrate. Although this cannot be commercially exploited at the present time, the amount of combustible energy of the known methane hydrate fields exceeds the energy content of all the natural gas and oil deposits put together; methane extracted from methane hydrate is considered therefore a candidate for future fuels.

Biological Occurrence

Although alkanes occur in nature in various way, they do not rank biologically among the essential materials. Cycloalkanes with 14 to 18 carbon atoms occur in musk, extracted from deer of the family Moschidae. All further information refers to (acyclic) alkanes.

Certain types of bacteria can metabolize alkanes: they prefer even-numbered carbon chains as they are easier to degrade than odd-numbered chains. On the other hand, certain archaea, the methanogens, produce large quantities of methane by the metabolism of carbon dioxide or other oxidized organic compounds. The energy is released by the oxidation of hydrogen:

$$CO_2 + 4H_2 \rightarrow CH_4 + 2H_2O$$

Methanogens are also the producers of marsh gas in wetlands, and release about two billion tonnes of methane per year—the atmospheric content of this gas is produced nearly exclusively by them. The methane output of cattle and other herbivores, which can release up to 150 litres per day, and of termites, is also due to methanogens. They also produce this simplest of all alkanes in the intestines of humans. Methanogenic archaea are, hence, at the end of the carbon cycle, with carbon being released back into the atmosphere after having been fixed by photosynthesis. It is probable that our current deposits of natural gas were formed in a similar way.

Fungi and Plants

Alkanes also play a role, if a minor role, in the biology of the three eukaryotic groups of organisms: fungi, plants and animals. Some specialized yeasts, e.g., *Candida tropicale*, *Pichia* sp., *Rhodotorula* sp., can use alkanes as a source of carbon and/or energy. The fungus *Amorphotheca resinae* prefers the longer-chain alkanes in aviation fuel, and can cause serious problems for aircraft in tropical regions.

In plants, the solid long-chain alkanes are found in the plant cuticle and epicuticular wax of many species, but are only rarely major constituents. They protect the plant against water loss, prevent the leaching of important minerals by the rain, and protect against bacteria, fungi, and harmful insects.

The carbon chains in plant alkanes are usually odd-numbered, between twenty-seven and thirty-three carbon atoms in length and are made by the plants by decarboxylation of even-numbered fatty acids. The exact composition of the layer of wax is not only species-dependent, but changes also with the season and such environmental factors as lighting conditions, temperature or humidity.

Animals

Alkanes are found in animal products, although they are less important than unsaturated hydrocarbons. One example is the shark liver oil, which is approximately 14% pristane (2,6,10,14-tetramethylpentadecane, $C_{19}H_{40}$). Their occurrence is more important in pheromones, chemical messenger materials, on which above all insects are dependent for communication. With some kinds, as the support beetle *Xylotrechus colonus*, primarily pentacosane ($C_{25}H_{52}$), 3-methylpentaicosane ($C_{26}H_{54}$) and 9-methylpentaicosane ($C_{26}H_{54}$), they are transferred by body contact. With others like the tsetse fly *Glossina morsitans morsitans*, the pheromone contains the four alkanes 2-methylheptadecane ($C_{18}H_{38}$), 17,21-dimethylheptatriacontane ($C_{39}H_{80}$), 15,19-dimethylheptatriacontane

($C_{39}H_{80}$) and 15,19,23-trimethylheptatriacontane ($C_{40}H_{82}$), and acts by smell over longer distances, a useful characteristic for pest control. Waggle-dancing honeybees produce and release two alkanes, tricosane and pentacosane.

Ecological Relations

One example, in which both plant and animal alkanes play a role, is the ecological relationship between the sand bee (*Andrena nigroaenea*) and the early spider orchid (*Ophrys sphegodes*); the latter is dependent for pollination on the former. Sand bees use pheromones in order to identify a mate; in the case of *A. nigroaenea*, the females emit a mixture of tricosane ($C_{23}H_{48}$), pentacosane ($C_{25}H_{52}$) and heptacosane ($C_{27}H_{56}$) in the ratio 3:3:1, and males are attracted by specifically this odor.

The orchid takes advantage of this mating arrangement to get the male bee to collect and disseminate its pollen; parts of its flower not only resemble the appearance of sand bees, but also produce large quantities of the three alkanes in the same ratio as female sand bees. As a result numerous males are lured to the blooms and attempt to copulate with their imaginary partner: although this endeavour is not crowned with success for the bee, it allows the orchid to transfer its pollen, which will be dispersed after the departure of the frustrated male to different blooms.

Production

Petroleum Refining: As stated earlier, the most important source of alkanes is natural gas and crude oil. Alkanes are separated in an oil refinery by fractional distillation and processed into many different products.

Fischer-Tropsch

The Fischer-Tropsch process is a method to synthesize liquid hydrocarbons, including alkanes, from carbon monoxide and hydrogen. This method is used to produce substitutes for petroleum distillates.

Laboratory Preparation

There is usually little need for alkanes to be synthesized in the laboratory, since they are usually commercially available. Also, alkanes are generally non-reactive chemically or biologically, and do not undergo functional group interconversions cleanly. When alkanes are produced in the laboratory, it is often a side-product of a reaction. For example, the use of *n*-butyllithium as a strong base gives the conjugate acid, n-butane as a side-product:

$$C_4H_9Li + H_2O \rightarrow C_4H_{10} + LiOH$$

However, at times it may be desirable to make a portion of a molecule into an alkane like functionality (alkyl group) using the above or similar methods. For example, an ethyl group is an alkyl group; when this is attached to a hydroxy group, it gives ethanol, which is not an alkane. To do so, the best-known methods are hydrogenation of alkenes:

$$RCH{=}CH_2 + H_2 \rightarrow RCH_2CH_3 \text{ (R = alkyl)}$$

Alkanes or alkyl groups can also be prepared directly from alkyl halides in the Corey-House-Posner-Whitesides reaction. The Barton-McCombie deoxygenation removes hydroxyl groups from alcohols e.g.

$$R{-}OH \xrightarrow[\text{- HCl}]{\text{Cl}-C(=S)-R1} R{-}O{-}C(=S){-}R1 \xrightarrow[\text{AiBN}]{Bu_3SnH} R{-}H$$

And the Clemmensen reduction removes carbonyl groups from aldehydes and ketones to form alkanes or alkyl-substituted compounds e.g.:

$$R_1{-}C(=O){-}R_2 \xrightarrow[\text{HCl}]{\text{Zn(Hg)}} R_1{-}CH_2{-}R_2$$

Applications

The applications of a certain alkane can be determined quite well according to the number of carbon atoms. The first four alkanes are used mainly for heating and cooking purposes, and in some countries for electricity generation. Methane and ethane are the main components of natural gas; they are normally stored as gases under pressure. It is, however, easier to transport them as liquids: This requires both compression and cooling of the gas.

Propane and butane can be liquefied at fairly low pressures, and are well known as liquified petroleum gas (LPG). Propane, for example, is used in the propane gas burner, butane in disposable cigarette lighters. The two alkanes are used as propellants in aerosol sprays.

From pentane to octane the alkanes are reasonably volatile liquids. They are used as fuels in internal combustion engines, as they vaporise easily on entry into the combustion chamber without forming droplets, which would impair the uniformity of the combustion. Branched-chain alkanes are preferred as they are much less prone to premature ignition, which causes knocking, than their straight-chain homologues. This propensity to premature ignition is measured by the octane rating of the fuel, where 2,2,4-trimethylpentane (*isooctane*) has an arbitrary value of 100, and heptane has a value of zero. Apart from their use as fuels, the middle alkanes are also good solvents for nonpolar substances.

Alkanes from nonane to, for instance, hexadecane (an alkane with sixteen carbon atoms) are liquids of higher viscosity, less and less suitable for use in gasoline. They form instead the major part of diesel and aviation fuel. Diesel fuels are characterized by their cetane number, cetane being an old name for hexadecane. However, the higher melting points of these alkanes can cause problems at low temperatures and in polar regions, where the fuel becomes too thick to flow correctly.

Alkanes from hexadecane upwards form the most important components of fuel oil and lubricating oil. In latter function, they work at the same time as anti-corrosive agents, as their hydrophobic nature means that water cannot reach the metal surface. Many solid alkanes find use as paraffin wax, for example, in candles. This should not be confused however with true wax, which consists primarily of esters.

Alkanes with a chain length of approximately 35 or more carbon atoms are found in bitumen, used, for example, in road surfacing. However, the higher alkanes have little value and are usually split into lower alkanes by cracking.

Some synthetic polymers such as polyethylene and polypropylene are alkanes with chains containing hundreds of thousands of carbon atoms. These materials are used in innumerable applications, and billions of kilograms of these materials are made and used each year.

Environmental Transformations

When released in the environment, alkanes don't undergo rapid biodegradation, because they have no functional groups (like hydroxyl or carbonyl) that are needed by most organisms in order to metabolize the compound.

However, some bacteria can metabolize some alkanes (especially those linear and short), by oxidizing the terminal carbon atom. The product is an alcohol, that could be next oxidized to an aldehyde, and finally to a carboxylic acid. The resulting fatty acid could be metabolized through the fatty acid degradation pathway.

Hazards

Methane is explosive when mixed with air (1 – 8% CH_4). Other lower alkanes can also form explosive mixtures with air. The lighter liquid alkanes are highly flammable, although this risk decreases with the length of the carbon chain. Pentane, hexane, heptane, and octane are classed as *dangerous for the environment* and *harmful*. The straight-chain isomer of hexane is a neurotoxin. Halogen-rich alkanes, like chloroform, can be carcinogenic as well.

Alkene

In organic chemistry, an alkene, olefin, or olefine is an unsaturated chemical compound containing at least one carbon-to-carbon double bond. The simplest acyclic alkenes, with only one double bond and no other functional groups, form an homologous series of hydrocarbons with the general formula C_nH_{2n}.

The simplest alkene is ethylene (C_2H_4), which has the International Union of Pure and Applied Chemistry (IUPAC) name *ethene*. Alkenes are also called *olefins* (an archaic synonym, widely used in the petrochemical industry). For bridged alkenes, the Bredt's rule states that a double bond cannot be placed at the bridgehead of a bridged ring system, unless the rings are large enough. Aromatic compounds are often drawn as cyclic alkenes, but their structure and properties are different and they are not considered to be alkenes.

Structure

Bonding: Like single covalent bonds, double bonds can be described in terms of overlapping atomic orbitals, except that, unlike a single bond (which consists of a single sigma bond), a carbon-carbon double bond consists of one sigma bond and one pi bond. This double bond is stronger than a single covalent bond (611 kJ/mol for C=C vs. 347 kJ/mol for C—C) and also shorter with an average bond length of 1.33 Angstroms (133 pm).

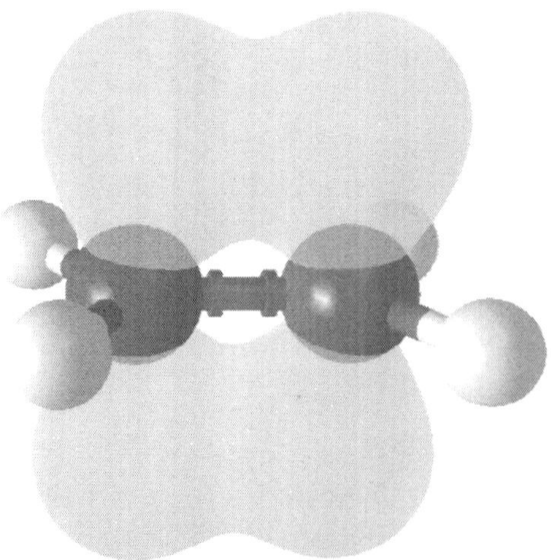

Figure: *Ethylene (ethene), showing the pi bond in green.*

Each carbon of the double bond uses its three sp^2 hybrid orbitals to form sigma bonds to three atoms. The unhybridized *2p* atomic orbitals, which lie perpendicular to the plane created by the axes of the three sp^2 hybrid orbitals, combine to form the pi bond. This bond lies outside the main C—C axis, with half of the bond on one side and half on the other.

Rotation about the carbon-carbon double bond is restricted because it involves breaking the pi bond, which requires a large amount of energy (264 kJ/mol in ethylene). As a consequence, substituted alkenes may exist as one of two isomers, called *cis* or *trans* isomers. More complex alkenes may be named using the E-Z notation, used to describe molecules having three or four different substituents (side groups). For example, of the isomers of butene, the two methyl groups of (*Z*)-but-2-ene (aka *cis*-2-butene) face the same side of the double bond, and in (*E*)-but2-ene (aka *trans*-2-butene) the methyl groups face the opposite side. These two isomers of butene are slightly different in their chemical and physical properties. It is certainly not impossible to twist a double bond. In fact, a 90° twist requires an energy approximately equal to half the strength of a pi bond. The misalignment of the p orbitals is less than expected because pyramidalization takes place. *Trans*-Cyclooctene is a stable strained alkene and the orbital misalignment is only 19° with a dihedral angle of 137° (normal 120°) and a degree of pyramidalization of 18°. This explains the dipole moment of 0.8 D for this compound (cis-isomer 0.4 D) where a value of zero is expected. The *trans* isomer of cycloheptene is only stable at low temperatures.

Shape

As predicted by the VSEPR model of electron pair repulsion, the molecular geometry of alkenes includes bond angles about each carbon in a double bond of about 120°. The angle may vary because of steric strain introduced by nonbonded interactions created by functional groups attached to the carbons of the double bond. For example, the C-C-C bond angle in propylene is 123.9°.

Physical Properties

The physical properties of alkenes are comparable with those of alkanes. The main differences between the two are that the acidity levels of alkenes are much higher than the ones in alkanes. The physical state depends on molecular mass (gases from ethene to butene - liquids from pentene onwards). The simplest alkenes, ethene, propene and butene are gases. Linear alkenes of approximately five to sixteen carbons are liquids, and higher alkenes are waxy solids.

Reactions

Alkenes are relatively stable compounds, but are more reactive than alkanes due to the presence of a carbon-carbon pi-bond. It is also attributed to the presence of pi-electrons in the molecule. The majority of the reactions of alkenes involve the rupture of this pi bond, forming new single bonds.

Alkenes serve as a feedstock for the petrochemical industry because they can participate in a wide variety of reactions.

Addition Reactions

Alkenes react in many addition reactions, which occur by opening up the double-bond. Most addition reactions to alkenes follow the mechanism of electrophilic addition. Examples of addition reactions are hydrohalogenation, halogenation, halohydrin formation, oxymercuration, hydroboration, dichlorocarbene addition, Simmons-Smith reaction, catalytic hydrogenation, epoxidation, radical polymerization and hydroxylation.

$$H_2C{=}CH_2 + X{-}Y \longrightarrow H{-}CH(X){-}CH(Y){-}H$$

Hydrogenation

Hydrogenation of alkenes produces the corresponding alkanes. The reaction is carried out under pressure at a temperature of 200 °C in the presence of a metallic catalyst. Common industrial catalysts are based on platinum, nickel or palladium. For laboratory syntheses, Raney nickel (an alloy of nickel and aluminium) is often employed. The simplest example of this reaction is the catalytic hydrogenation of ethylene to yield ethane:

$$CH_2{=}CH_2 + H_2 \rightarrow CH_3{-}CH_3$$

Halogenation

In electrophilic halogenation the addition of elemental bromine or chlorine to alkenes yields vicinal dibromo- and dichloroalkanes (1,2-dihalides or ethylene dihalides), respectively. The decoloration of a solution of bromine in water with dichloromethylene as catalyst is an analytical test for the presence of alkenes:

$$CH_2{=}CH_2 + Br_2 \rightarrow BrCH_2{-}CH_2Br$$

It is also used as a quantitive test of unsaturation, expressed as the bromine number of a single compound or mixture. The reaction works because the high electron density at the double bond causes a temporary shift of electrons in the Br-Br bond causing a temporary induced dipole. This makes the Br closest to the double bond slightly positive and therefore an electrophile.

Hydrohalogenation

Hydrohalogenation is the addition of hydrohalic acids such as HCl or HBr to alkenes to yield the corresponding haloalkanes.

$$CH_3\text{-}CH{=}CH_2 + HBr \rightarrow CH_3\text{-}CHBr\text{-}CH_2\text{-}H$$

If the two carbon atoms at the double bond are linked to a different number of hydrogen atoms, the halogen is found preferentially at the carbon with fewer hydrogen substituents (Markovnikov's rule). But terminal olefin products don't yield by this method. For bromation an alternative method denominated Kharasch-Sosnovsky Reaction is used for this purpose. It consists to add peroxides to hydrogen bromide or Copper bromide (II).

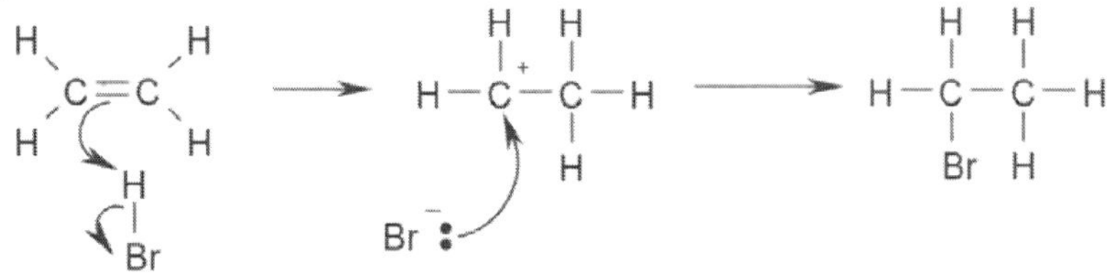

Halohydrin Formation

Alkenes react with water and halogens to form halohydrins by an addition reaction. Markovnikov regiochemistry and anti stereochemistry occur.

$$CH_2{=}CH_2 + X_2 + H_2O \rightarrow XCH_2\text{-}CH_2OH$$

Oxidation

Alkenes are oxidized with a large number of oxidizing agents. In the presence of oxygen, alkenes burn with a bright flame to produce carbon dioxide and water. Catalytic oxidation with oxygen or the reaction with percarboxylic acids yields epoxides. Reaction with ozone in ozonolysis leads to the breaking of the double bond, yielding two aldehydes or ketones. Reaction with concentrated, hot $KMnO_4$ (or other oxidizing salts) in an acidic solution will yield ketones or carboxylic acids.

$$R_1\text{-}CH{=}CH\text{-}R_2 + O_3 \rightarrow R_1\text{-}CHO + R_2\text{-}CHO + H_2O$$

This reaction can be used to determine the position of a double bond in an unknown alkene.

Oxymercuration

Hydration of alkenes via oxymercuration to produces alcohols. Reaction takes place on treatment of alkenes with strong acid as catalyst.

$$CH_2{=}CH_2 + H_2O \rightarrow CH_3\text{-}CH_2OH$$

Polymerization

Polymerization of alkenes is a reaction that yields polymers of high industrial value at great economy, such as the plastics polyethylene and polypropylene.

Polymers from alkene monomers are referred to in a general way as *polyolefins* or in rare instances as *polyalkenes*. A polymer from alpha-olefins is called a polyalphaolefin (PAO). Polymerization can proceed via either a free-radical or an ionic mechanism, converting the double to a single bond and forming single bonds to join the other monomers.

Polymerization of conjugated dienes such as buta-1,3-diene or isoprene (2-methylbuta-1,3-diene) results in largely 1,4-addition with possibly some 1,2-addition of the diene monomer to a growing polymer chain.

Reaction Overview

Reaction name	*Product*	*Comment*
Hydrogenation	alkanes	addition of hydrogen
Hydroalkenylation	alkenes	hydrometalation / insertion / beta elimination by metal catalyst
Halogen addition reaction	1,2-dihalide	electrophilic addition of halogens
Hydrohalogenation (Markovnikov)	haloalkanes	addition of hydrohalic acids
Kharasch-Sosnovsky Reaction (Antimarkovnikov Hydrohalogenation)	haloalkanes	free radicals mediated addition of hydrohalic acids
Hydroamination	amines	addition of N-H bond across C-C double bond
Hydroformylation	aldehydes	industrial process, addition of CO and H_2
Sharpless bishydroxylation	diols	oxidation, reagent: osmium tetroxide, chiral ligand
Woodward cis-hydroxylation	diols	oxidation, reagents: iodine, silver acetate

Contd...

Reaction name	***Product***	***Comment***
ozonolysis	aldehydes or ketones	reagent: ozone
Olefin metathesis	alkenes	two alkenes rearrange to form two new alkenes
Diels-Alder reaction	cyclohexenes	cycloaddition with a diene
Pauson-Khand reaction	cyclopentenones	cycloaddition with an alkyne and CO
Hydroboration–oxidation	alcohols	reagents: borane, then a peroxide
oxymercuration-reduction	alcohols	electrophilic addition of mercuric acetate, then reduction
Prins reaction	1,3-diols	electrophilic addition with aldehyde or ketone
Paterno–Büchi reaction	oxetanes	photochemical reaction with aldehyde or ketone
Epoxidation	epoxide	electrophilic addition of a peroxide
Cyclopropanation	cyclopropanes	addition of carbenes or carbenoids
Hydroacylation	ketones	oxidativc addition / reductive elimination by metal catalyst

Synthesis

Industrial Methods: Alkenes are produced by hydrocarbon cracking. Raw materials are mostly natural gas condensate components (principally ethane and propane) in the US and Mideast and naphtha in Europe and Asia. Alkanes are broken apart at high temperatures, often in the presence of a zeolite catalyst, to produce a mixture of primarily aliphatic alkenes and lower molecular weight alkanes. The mixture is feedstock dependent and separated by fractional distillation. This is mainly used for the manufacture of small alkenes (up to six carbons).

Δ (catalyst) + etc.

Related to this is catalytic dehydrogenation, where an alkane loses hydrogen at high temperatures to produce a corresponding alkene. This is the reverse of the catalytic hydrogenation of alkenes.

Δ catalyst + + +

Both of these processes are endothermic, but they are driven towards the alkene at high temperatures by entropy (the TÄS portion of the equation $\Delta G = \Delta H - T\ddot{A}S$ dominates for high T). Catalytic synthesis of higher α-alkenes (of the type $RCH{=}CH_2$) can also be achieved by a reaction of ethylene with the organometallic compound triethylaluminium in the presence of nickel, cobalt, or platinum.

Elimination Reactions

One of the principal methods for alkene synthesis in the laboratory is the elimination of alkyl halides, alcohols, and similar compounds. Most common is the β-elimination via the E2 or E1 mechanism, but α-eliminations are also known.

The E2 mechanism provides a more reliable β-elimination method than E1 for most alkene syntheses. Most E2 eliminations start with an alkyl halide or alkyl sulfonate ester (such as a tosylate or triflate). When an alkyl halide is used, the reaction is called a dehydrohalogenation.

For unsymmetrical products, the more substituted alkenes (those with fewer hydrogens attached to the C=C) tend to predominate. Two common methods of elimination reactions are dehydrohalogenation of alkyl halides and dehydration of alcohols. A typical example is shown below; note that if possible, the H is *anti* to the leaving group, even though this leads to the less stable Z-isomer.

X, CH_3, Ph, Ph, H, H $\xrightarrow[CH_3CH_2OH]{CH_3CH_2O^-,\ \Delta}$ H, CH_3, Ph, Ph (alkene) $+\ CH_3CH_2OH + X^-$

$:\ddot{O}^-{-}CH_2CH_3$

X = Cl, Br

Alkenes can be synthesized from alcohols via dehydration, in which case water is lost via the E1 mechanism. For example, the dehydration of ethanol produces ethene:

$$CH_3CH_2OH + H_2SO_4 \rightarrow H_2C{=}CH_2 + H_3O^+ + HSO_4^-$$

An alcohol may also be converted to a better leaving group (e.g., xanthate), so as to allow a milder *syn*-elimination such as the Chugaev elimination and the Grieco elimination. Related reactions include eliminations by β-haloethers (the Boord olefin synthesis) and esters (ester pyrolysis).Alkenes can be prepared indirectly from alkyl amines. The amine or ammonia is not a suitable leaving group, so the amine is first either alkylated (as in the Hofmann elimination) or oxidized to an amine oxide (the Cope reaction) to render a smooth elimination possible.

Hofmann elimination is unusual in that the *less* substituted (non-Saytseff) alkene is usually the major product. The Cope reaction is a *syn*-elimination that occurs at or below 150 °C, for example:

(i) H_2O_2
(ii) 110-120 °C

Alkenes are generated from α-halo sulfones in the Ramberg-Bäcklund reaction, via a three-membered ring sulfone intermediate.

Synthesis from Carbonyl Compounds

Another important method for alkene synthesis involves construction of a new carbon-carbon double bond by coupling of a carbonyl compound (such as an aldehyde or ketone) to a carbanion equivalent. Such reactions are sometimes called *olefinations*. The most well-known of these methods is the Wittig reaction, but other related methods are known.

The Wittig reaction involves reaction of an aldehyde or ketone with a Wittig reagent (or phosphorane) of the type $Ph_3P{=}CHR$ to produce an alkene and $Ph_3P{=}O$. The Wittig reagent is itself prepared easily from triphenylphosphine and an alkyl halide. The reaction is quite general and many functional groups are tolerated, even esters, as in this example:

$Ph_3P{=}CH{-}CHMe_2$

Related to the Wittig reaction is the Peterson olefination. This uses a less accessible silicon-based reagent in place of the phosphorane, but it allows for the selection of E or Z products. If an E-product is desired, another alternative is the Julia olefination, which uses the carbanion generated from a phenyl sulfone. The Takai olefination based on an organochromium intermediate also delivers E-products. A titanium compound, Tebbe's reagent, is useful for the synthesis of methylene compounds; in this case, even esters and amides react.

A pair of carbonyl compounds can also be reductively coupled together (with reduction) to generate an alkene. Symmetrical alkenes can be prepared from a single aldehyde or ketone coupling with itself, using Ti metal reduction (the McMurry reaction). If two different ketones are to be coupled, a more complex, indirect method such as the Barton-Kellogg reaction may be used. A single ketone can also be converted to the corresponding alkene via its tosylhydrazone, using sodium methoxide (the Bamford-Stevens reaction) or an alkyllithium (the Shapiro reaction).

Synthesis from Alkenes: Olefin Metathesis and Hydrovinylation

Alkenes can be prepared by exchange with other alkenes, in a reaction known as olefin metathesis. Frequently, loss of ethene gas is used to drive the reaction towards a desired product. In many cases, a mixture of geometric isomers is obtained, but the reaction tolerates many functional groups.

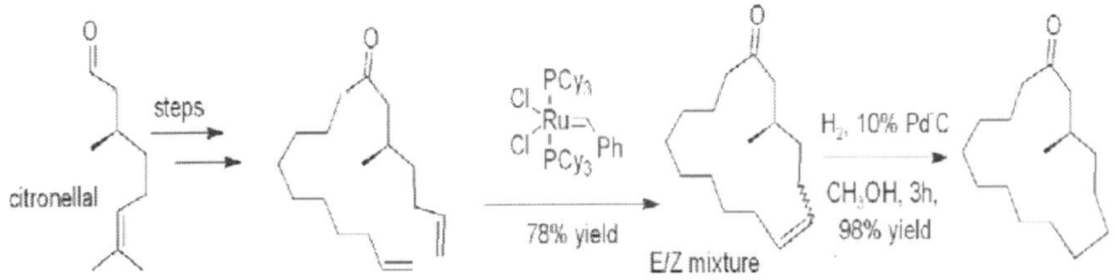

citronellal → steps → ; Ru catalyst ($Cl_2Ru(PCy_3)_2=CHPh$), 78% yield → E/Z mixture → H_2, 10% Pd/C, CH_3OH, 3h, 98% yield

Transition metal catalyzed hydrovinylation is another important alkene synthesis process starting from alkene itself. In general, it involves the addition of a hydrogen and a vinyl group (or an alkenyl group) across a double bond. The hydrovinylation reaction was first reported by Alderson, Jenner, and Lindsey by using rhodium and ruthenium salts, other metal catalysts commonly employed nowadays included iron, cobalt, nickel, and palladium. The addition can be done highly regio- and stereo-selectively, the choices of metal centres, ligands, substrates and counterions often play very important role.

From Alkynes

Reduction of alkynes is a useful method for the stereoselective synthesis of disubstituted alkenes. If the *cis*-alkene is desired, hydrogenation in the presence of Lindlar's catalyst(-heterogeneous catalyst that consists of palladium deposited on calcium carbonate and treated with various forms of lead) is commonly used, though hydroboration followed by hydrolysis provides an alternative approach.

Reduction of the alkyne by sodium metal in liquid ammonia gives the *trans*-alkene.

$R^1-\equiv-R^2 \xrightarrow[\text{or } R_2BH\text{, then } CH_3COOH]{H_2\text{, Lindlar's catalyst}}$ *cis*-alkene

$R^1-\equiv-R^2 \xrightarrow[\text{then } {}^tBuOH]{\text{Na, liq. } NH_3}$ *trans*-alkene

Rearrangements and Related Reactions

Alkenes can be synthesized from other alkenes via rearrangement reactions. Besides olefin metathesis (described above), a large number of pericyclic reactions can be used such as the ene reaction and the Cope rearrangement.

In the Diels-Alder reaction, a cyclohexene derivative is prepared from a diene and a reactive or electron-deficient alkene.

Nomenclature

IUPAC Names: To form the root of the IUPAC names for alkenes, simply change the -an- infix of the parent to -en-. For example, CH_3-CH_3 is the alkane *ethANe*. The name of CH_2=CH_2 is therefore *ethENe*.

In higher alkenes, where isomers exist that differ in location of the double bond, the following numbering system is used:

1. Number the longest carbon chain that contains the double bond in the direction that gives the carbon atoms of the double bond the lowest possible numbers.
2. Indicate the location of the double bond by the location of its first carbon.
3. Name branched or substituted alkenes in a manner similar to alkanes.
4. Number the carbon atoms, locate and name substituent groups, locate the double bond, and name the main chain.

hex-1-ene 4-methylhex-1-ene 4-ethyl-2-methylhex-1-ene

Figure: *Naming substituted hex-1-enes*

Cis-Trans notation

In the specific case of disubstituted alkenes where the two carbons have one substituent each, Cis-trans notation may be used. If both substituents are on the same side of the bond, it is defined as (cis-). If the substituents are on either side of the bond, it is defined as (trans-).

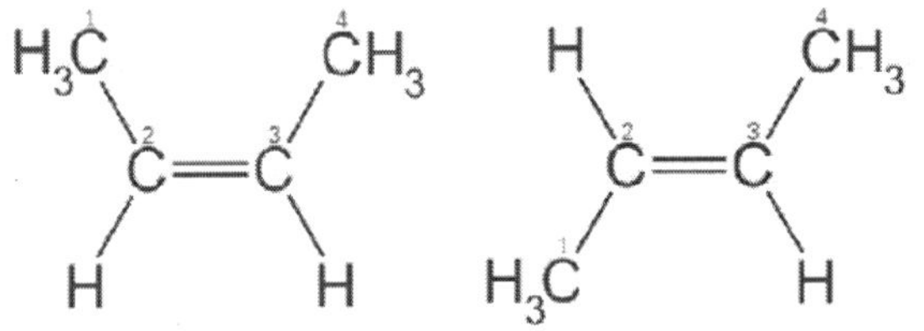

cis-but-2-ene trans-but-2-ene

Figure: *The difference between* cis- *and* trans- *isomers*

E,Z notation

When an alkene has more than one substituent (especially necessary with 3 or 4 substituents), the double bond geometry is described using the labels *E* and *Z*. These labels come from the German words "entgegen," meaning "opposite," and "zusammen," meaning "together." Alkenes with the higher priority groups (as determined by CIP rules) on the same side of the double bond have these groups together and are designated *Z*. Alkenes with the higher priority groups on opposite sides are designated *E*. A mnemonic to remember this: Z notation has the higher priority groups on "ze zame zide."

(*Z*)-2-butene
zusammen (together)

(*E*)-2-butene
entgegen (opposite)

NB. the methyl groups are on the same side

NB. the methyl groups are on opposite sides

Figure: *The difference between* E *and* Z *isomers*

Groups containing C=C double bonds

IUPAC recognizes two names for hydrocarbon groups containing carbon-carbon double bonds, the vinyl group and the allyl group. .

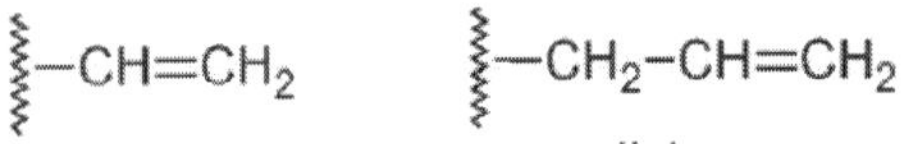

$-CH=CH_2$ $-CH_2-CH=CH_2$

vinyl allyl

Chapter 4

Cycloalkyne

In organic chemistry, a cycloalkyne is the cyclic analog of an alkyne. A cycloalkyne consists of a closed ring of carbon atoms containing one or more triple bonds. Cycloalkynes have a general formula $C_n H_{2n-4}$ Because of the linear nature of the C-CÎC-C alkyne unit, cycloalkynes are usually highly strained and can only exist when the number of carbon atoms in the ring is great enough to provide the flexibility necessary to accommodate this geometry. Consequently, cyclooctyne (C_8H_{12}) is the smallest cycloalkyne capable of being isolated and stored as a stable compound. Despite this, smaller cycloalkynes can be produced and trapped by a suitable reagent.

Cycloalkynes can be produced via β-elimination reactions with an analogous substituted cycloalkene. Alternatively, they can be produced by the ring expansion of a cyclic alkylidinecarbene.

Cycloalkanes (also called naphthenes - not to be confused with naphthalene) are types of alkanes which have one or more rings of carbon atoms in the chemical structure of their molecules. Alkanes are types of organic hydrocarbon compounds which have only single chemical bonds in their chemical structure.

Cycloalkanes consist of only carbon (C) and hydrogen (H) atoms and are saturated because there are no multiple C-C bonds to hydrogenate (add more hydrogen to). A general chemical formula for cycloalkanes would be $C_nH_{2(n+1-g)}$ where n = number of C atoms and g = number of rings in the molecule. Cycloalkanes with a single ring are named analogously to their normal alkane counterpart of the same carbon count: cyclopropane, cyclobutane, cyclopentane, cyclohexane, etc. The larger cycloalkanes, with greater than 20 carbon atoms are typically called cycloparaffins.

Cycloalkanes are classified into small, common, medium, and large cycloalkanes, where cyclopropane and cyclobutane are the small ones, cyclopentane, cyclohexane, cycloheptane are the common ones, cyclooctane through cyclotridecane are the medium ones, and the rest are the larger ones.

Nomenclature

The naming of polycyclic alkanes such as bicyclic alkanes and spiro alkanes is more complex, with the base name indicating the number of carbons in the ring system, a prefix indicating the number of rings (e.g., "bicyclo"), and a numeric prefix before that indicating the number of carbons in each part of each ring, exclusive of vertices. For instance, a bicyclooctane that consists of a six-member ring and a four-member ring, which share two adjacent carbon atoms that form a shared edge, is [4.2.0]-bicyclooctane. That part of the six-member ring, exclusive of the shared edge has 4 carbons. That part of the four-member ring, exclusive of the shared edge, has 2 carbons. The edge itself, exclusive of the two vertices that define it, has 0 carbons.

There is more than one convention (method or nomenclature) for the naming of compounds, which can be confusing for those who are just learning, and inconvenient for those who are well rehearsed in the older ways. For beginners it is best to learn IUPAC nomenclature from a source that is up to date, because this system is constantly being revised. In the above example [4.2.0]-bicyclooctane would be written bicyclo[4.2.0]octane to fit the conventions for IUPAC naming. It has then got room for an additional numerical prefix if there is the need to include details of other attachments to the molecule such as chlorine or a methyl group. Another convention for the naming of compounds is the common name, which is a shorter name and it gives less information about the compound. An example of a common name is terpineol, the name of which can only really tell us that it is an alcohol (because the suffix 'ol' is in the name) and it should then have a hydroxide (OH) group attached to it.

In this example the base name is listed first, which indicates the total number of carbons in both rings including the carbons making up the shared edge (e.g., heptane which means 'hept' or 7 carbons and 'ane' which indicates only single bonding between carbons). Then in front of the base name is the numerical prefix, which lists the number of carbons in each ring, excluding the carbons that are shared by each ring, plus the number of carbons on the bridge between the rings. In this case there are two rings with two carbons each and a single bridge with one carbon, excluding the carbons shared by it and the other two rings. There is a total of three numbers and they are listed in descending order separated by dots, thus: [2.2.1]. Before the numerical prefix is another prefix indicating the number of rings (e.g., "bicyclo"). Thus, the name is bicyclo[2.2.1]heptane. The group of cycloalkanes are also known as naphthenes, as they are compounds of petroleum or naphtha.

Properties

Cycloalkanes are similar to alkanes in their general physical properties, but they have higher boiling points, melting points, and densities than alkanes. This is due to stronger London forces because the ring shape allows for a larger area of contact. Containing only C-C and C-H bonds, unreactivity of cycloalkanes with little or no ring strain are comparable to non-cyclic alkanes.

Ring strain

The carbon atoms in cycloalkanes are sp^3 hybridized and are therefore a deviation from the ideal tetrahedral bond angles of 109°28'. This causes an increase in potential energy and an overall destabilizing effect. Eclipsing of hydrogen atoms is an important destabilizing effect, as well. The strain energy of a cycloalkane is the theoretical increase in energy caused by the compound's geometry, and is calculated by comparing the experimental standard enthalpy change of combustion of the cycloalkane with the value calculated using average bond energies.

Ring strain is highest for cyclopropane, in which the carbon atoms form a triangle and therefore have 60 degree C-C-C bond angles. There are also three pairs of eclipsed hydrogens. The ring strain is calculated to be around 120 kJ/mol. Cyclobutane has the carbon atoms in a puckered square with approximately 90-degree bond angles; "puckering" reduces the eclipsing interactions between hydrogen atoms. Its ring strain is therefore slightly less, at around 110 kJ/mol.

For a theoretical planar cyclopentane the C-C-C bond angles would be 108 degrees, very close to the measure of the tetrahedral angle. Actual cyclopentane molecules are puckered, but this changes only the bond angles slightly so that angle strain is relatively small. The eclipsing interactions are also reduced, leaving a ring strain of about 25 kJ/mol.

In cyclohexane the ring strain and eclipsing interactions are negligible because the puckering of the ring allows ideal tetrahedral bond angles to be achieved. As well, in the most stable *chair form* of cyclohexane, axial hydrogens on adjacent carbon atoms are pointed in opposite directions, virtually eliminating eclipsing strain.

After cyclohexane, the molecules are unable to take a structure with no ring strain, resulting in an increase in strain energy, which peaks at 9 carbons (around 50 kJ/mol). After that, strain energy slowly decreases until 12 carbon atoms, where it drops significantly; at 14, another significant drop occurs and the strain is on a level comparable with 10 kJ/mol. After 14 carbon atoms, sources disagree on what happens to ring strain, some indicating that it increases steadily, others saying that it disappears entirely. Generally speaking though, bond angle strain and eclipsing strain are only an issue for smaller rings.

Reactions

The simple and the bigger cycloalkanes are very stable, like alkanes, and their reactions, for example, radical chain reactions, are like alkanes. The small cycloalkanes - particularly cyclopropane - have a lower stability due to Baeyer strain and ring strain. They react similarly to alkenes, though they do not react in electrophilic addition, but in nucleophilic aliphatic substitution. These reactions are ring-opening reactions or ring-cleavage reactions of alkyl cycloalkanes. Cycloalkanes can be formed in a Diels-Alder reaction followed by a catalytic hydrogenation. A cycloalkene or cycloolefin is a type of alkene hydrocarbon which contains a closed ring of carbon atoms, but has no aromatic character. Some cycloalkenes, such as cyclobutene and cyclopentene, can be used as monomers to produce polymer chains. Unless the rings are very large, cycloalkenes are always the cis isomers, and the term cis is omitted from the names. In a large ring, a trans double bond may occur, giving a trans-cycloalekene.

Examples

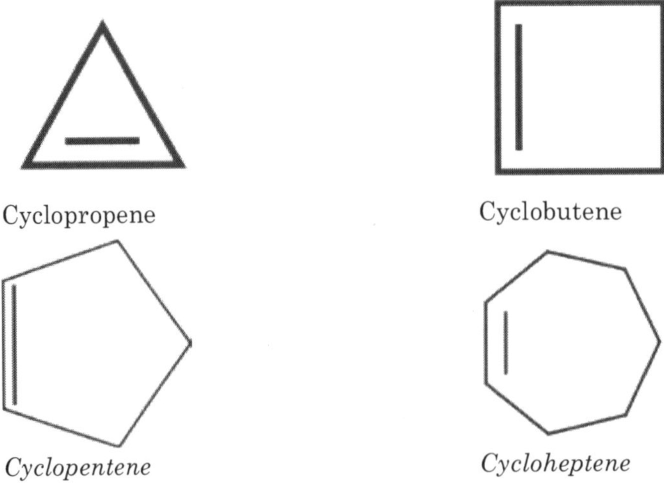

Cyclopropene

Cyclobutene

Cyclopentene

Cycloheptene

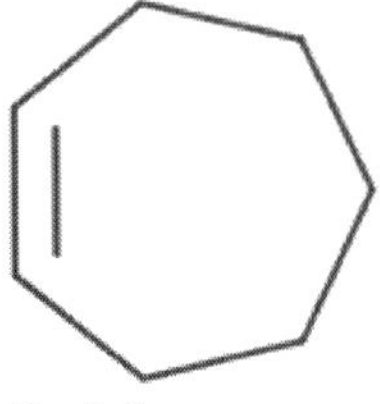

Cycloheptene

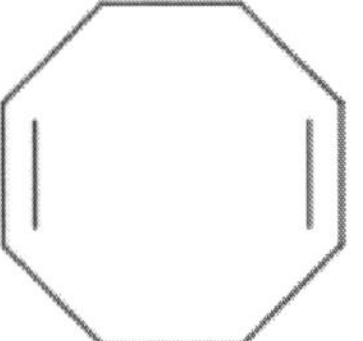

1,5-Cyclooctadiene

Basketane

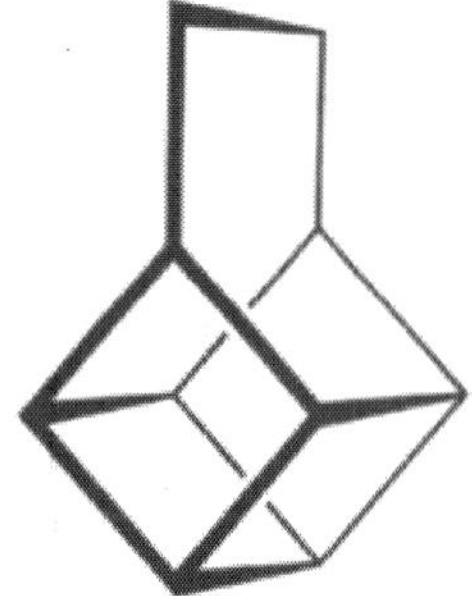

Figure: *Skeletal formula of the basketane molecule*

Basketane is a polycyclic alkane with the chemical formula $C_{10}H_{12}$. The name is taken from its structural similarity to a basket shape. Basketane was first synthesised in 1966, independently by Masamune and Dauben and Whalen.

Cubane

Cubane (C_8H_8) is a synthetic hydrocarbon molecule that consists of eight carbon atoms arranged at the corners of a cube, with one hydrogen atom attached to each carbon atom. A solid crystalline substance, cubane is one of the Platonic hydrocarbons. It was first synthesized in 1964 by Philip Eaton, a professor of chemistry at the University of Chicago.

Before Eaton and Cole's work, researchers believed that cubic carbon-based molecules could not exist, because the unusually sharp 90-degree bonding angle of the carbon atoms were expected to be too highly strained, and hence unstable. Once formed, cubane is quite kinetically stable, due to a lack of readily available decomposition paths.

The other Platonic hydrocarbons are dodecahedrane and tetrahedrane. Cubane and its derivative compounds have many important properties. The 90-degree bonding angle of the carbon atoms in cubane means that the bonds are highly strained. Therefore, cubane compounds are highly reactive, which in principle may make them useful as high-density, high-energy fuels and explosives (for example, octanitrocubane and heptanitrocubane).

Cubane also has the highest density of any hydrocarbon, further contributing to its ability to store large amounts of energy, which would reduce the size and weight of fuel tanks in aircraft and especially rocket boosters. Researchers are looking into using cubane and similar cubic molecules in medicine and nanotechnology.

Synthesis

The original 1964 cubane organic synthesis is a classic and starts from *2-cyclopentenone* (compound 1.1 in *scheme 1*):

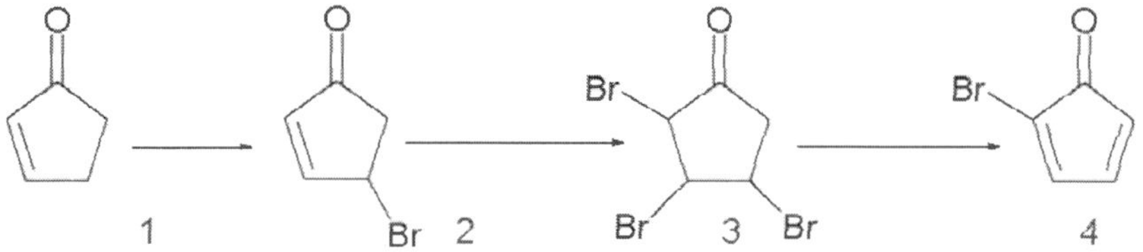

Reaction with *N*-bromosuccinimide in carbon tetrachloride places an allylic bromine atom in 1.2 and further bromination with bromine in pentane - methylene chloride gives the tribromide 1.3. Two equivalents of hydrogen bromide are eliminated from this compound with diethylamine in diethyl ether to *bromocyclopentadienone* 1.4

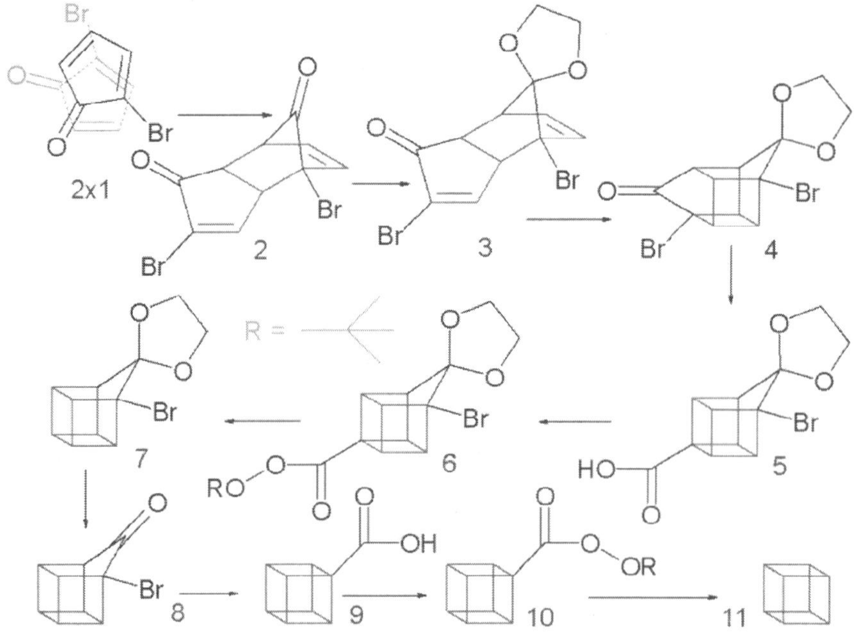

In the second part (*scheme 2*), the spontaneous Diels-Alder dimerization of 2.1 to 2.2 is analogous to the dimerization of cyclopentadiene to dicyclopentadiene. For the next steps to succeed, only the endo isomer should form; this happens because the bromine atoms, on their approach, take up positions as far away from each other, and from the carbonyl group, as possible.

In this way the like-dipole interactions are minimized in the transition state for this reaction step. Both carbonyl groups are protected as acetals with ethylene glycol and *p*-toluenesulfonic acid in benzene; one acetal is then selectively deprotected with aqueous hydrochloric acid to 2.3

In the next step, the endo isomer 2.3 (with both alkene groups in close proximity) forms the cage-like isomer 2.4 in a photochemical [2+2] cycloaddition. The bromoketone group is converted to ring-contracted carboxylic acid 2.5 in a Favorskii rearrangement with potassium hydroxide. Next, the thermal decarboxylation takes place through the acid chloride (with thionyl chloride) and the tert-butyl perester 2.6 (with t-butyl hydroperoxide and pyridine) to 2.7; afterward, the acetal is once more removed in 2.8. A second Favorskii rearrangement gives 2.9, and finally another decarboxylation gives 2.10 and 2.11.

Inorganic Cubes and Related Derivatives

The cube motif occurs outside of the area of organic chemistry. Prevalent non-organic cubes are the [Fe_4-S_4] clusters found pervasively iron-sulfur proteins. Such species contain sulfur and Fe at alternating corners. Alternatively such inorganic cube clusters can often be viewed as interpenetrated S_4 and Fe_4 tetrahedra. Many organometallic compounds adopt cube structures, examples being $(CpFe)_4(CO)_4$, $(Cp^*Ru)_4Cl_4$, and $(Ph_3PAg)_4I_4$.

Reactions

Cuneane may be produced from cubane by a metal-ion-catalyzed σ-bond rearrangement.

Aryl Hydrocarbon Receptor

The Aryl hydrocarbon receptor (AhR or AHR) is a member of the family of basic-helix-loop-helix transcription factors. AhR is a cytosolic transcription factor that is normally inactive, bound to several co-chaperones. Upon ligand binding to chemicals such as 2,3,7,8-tetrachlorodibenzo-*p*-dioxin (TCDD), the chaperones dissociate resulting in AhR translocating into the nucleus and dimerizing with ARNT (*AhR nuclear translocator*), leading to changes in gene transcription.

Protein Functional Domains

The AhR protein contains several domains critical for function and is classified as a member of the basic helix-loop-helix/Per-Arnt-Sim (bHLH/PAS) family of transcription factors. The bHLH motif is located in the N-terminal of the protein and is a common entity in a variety of transcription factors.

Members of the bHLH superfamily have two functionally distinctive and highly conserved domains. The first is the basic-region (b), which is involved in the binding of the transcription factor to DNA. The second is the helix-loop-helix (HLH) region, which facilitates protein-protein interactions.

Also contained with the AhR are two PAS domains, PAS-A and PAS-B, which are stretches of 200-350 amino acids that exhibit a high sequence homology to the protein domains that were originally found in the Drosophila genes period (Per) and single-minded (Sim) and in AhR's dimerization partner the aryl hydrocarbon receptor nuclear translocator (ARNT). The PAS domains support specific secondary interactions with other PAS domain containing proteins, as is the case with AhR and ARNT, so that heterozygous and homozygous protein complexes can form. The ligand binding site of AhR is contained within the PAS-B domain and contains several conserved residues critical for ligand binding. Finally, a Q-rich domain is located in the C-terminal region of the protein and is involved in co-activator recruitment and transactivation.

Ligands

Ahr ligands have been generally classified into two categories, synthetic or naturally occurring. The first ligands to be discovered were synthetic and members of the halogenated aromatic hydrocarbons (polychlorinated dibenzodioxins, dibenzofurans and biphenyls) and polycyclic aromatic hydrocarbons (3-methylcholanthrene, benzo(a)pyrene, benzanthracenes and benzoflavones). However, recent work has focused on naturally occurring compounds with the hope of identifying an endogenous ligand.

Naturally occurring compounds that have been identified as ligands of Ahr include derivatives of tryptophan such as indigo dye and indirubin, tetrapyroles such as bilirubin, the arachidonic acid metabolites lipoxin A4 and prostaglandin G, modified low-density lipoprotein and several dietary carotinoids. One assumption made in the search for an endogenous ligand is that the ligand will be a receptor agonist. However, work by Savouret *et al.* has shown this may not be the case since their findings demonstrate that 7-ketocholesterol competitively inhibits Ahr signal transduction.

Signalling Pathway

Cytosolic Complex: Non-ligand bound Ahr is retained in the cytoplasm as an inactive protein complex consisting of a dimer of Hsp90, prostaglandin E synthase 3 (Ptges3, p23) and a single molecule of the immunophilin-like protein hepatitis B virus X-associated protein 2 (XAP2), which was previously identified as AhR interacting protein (AIP) and AhR-activated 9 (ARA9). The dimer of Hsp90, along with p23, has a multifunctional role in the protection of the receptor from proteolysis, constraining the receptor in a conformation receptive to ligand binding and preventing the premature binding of ARNT. XAP2 interacts with carboxyl-terminal of Hsp90 and binds to the AhR nuclear localization sequence (NLS) preventing the inappropriate trafficking of the receptor into the nucleus.

Receptor Activation

Upon ligand binding to AhR, XAP2 is released resulting in exposure of the NLS, which is located in the bHLH region, leading to importation into the nucleus. It is presumed that once in the nucleus, Hsp90 dissociates exposing the two PAS domains allowing the binding of ARNT. The activated AhR/ARNT heterodimer complex is then capable of either directly and indirectly interacting with DNA by binding to recognition sequences located in the 5'- regulatory region of dioxin-responsive genes.

DNA Binding (Xenobiotic Response Element - XRE)

The classical recognition motif of the AhR/ARNT complex, referred to as either the AhR-, dioxin- or xenobiotic- responsive element (AHRE, DRE or XRE), contains the core sequence 5'-GCGTG-3' within the consensus sequence 5'-T/GNGCGTGA/CG/CA-3' in the promoter region of AhR responsive genes. The AhR/ARNT heterodimer directly binds the AHRE/DRE/XRE core sequence in an asymmetric manner such that ARNT binds to 5'-GTG-3' and AhR binding 5'-TC/TGC-3'. Recent research suggests that a second type of element termed AHRE-II, 5'-CATG(N6)C[T/A]TG-3', is capable of indirectly acting with the AhR/ARNT complex. Regardless of the response element, the end result is a variety of differential changes in gene expression.

Functional Role in Physiology and Toxicology

Role in Development: In terms of evolution, the oldest physiological role of Ahr is in development. Ahr is presumed to have evolved from invertebrates where it served a ligand-independent role in normal development processes. The Ahr homologue in *Drosophila, spineless* (ss) is necessary for development of the distal segments of the antenna and leg. *Ss* dimerizes with *tango* (tgo), which is the homologue to the mammalian Arnt, to initiate gene transcription. Evolution of the receptor in vertebrates resulted in the ability to bind ligand. In developing vertebrates, Ahr seemingly plays a role in cellular proliferation and differentiation. Despite lacking a clear endogenous ligand, AHR appears to play a role in the differentiation of many developmental pathways, including T-cells, neurons, and hepatocytes. AhR has also been found to have an important function in hematopoietic stem cells: AhR antagonism promotes their self-renewal and ex-vivo expansion and is involved in megakaryocyte differentiation.

Adaptive Response

The adaptive response is manifested as the induction of xenobiotic metabolizing enzymes. Evidence of this response was first observed from the induction of cytochrome P450, family 1, subfamily A, polypeptide 1 (Cyp1a1) resultant from TCDD exposure, which was determined to be directly related to activation of the Ahr signalling pathway.

The search for other metabolizing genes induced by Ahr ligands, due to the presence of DREs, has led to the identification of an "Ahr gene battery" of Phase I and Phase II metabolizing enzymes consisting of CYP1A1, CYP1A2, CYP1B1, NQO1, ALDH3A1, UGT1A2 and GSTA1. Presumably, vertebrates have this function to be able to detect a wide range of chemicals, indicated by the wide range of substrates Ahr is able to bind and facilitate their biotransformation and elimination.

Toxic Response

Extensions of the adaptive response are the toxic responses elicited by Ahr activation. Toxicity results from two different ways of Ahr signalling. The first is a function of the adaptive response in which the induction of metabolizing enzymes results in the production of toxic metabolites. For example, the polycyclic aromatic hydrocarbon benzo(a)pyrene (BaP), a ligand for Ahr, induces its own metabolism and bioactivation to a toxic metabolite via the induction of CYP1A1 and CYP1B1 in several tissues. The second approach to toxicity is the result of aberrant changes in global gene transcription beyond those observed in the "Ahr gene battery." These global changes in gene expression lead to adverse changes in cellular processes and function. Microarray analysis has proved most beneficial in understanding and characterizing this response.

Protein-protein Interactions

In addition to the protein interactions mentioned above, AhR has also been shown to interact with RELA, cyclin T1,SRC-1, retinoblastoma protein, NRIP1, estrogen receptor alpha, NEDD8 and ARNTL.

Dioxins and Dioxin-like Compounds

Dioxins and dioxin-like compounds (DLC) are by-products of various industrial processes, and are commonly regarded as highly toxic compounds that are environmental pollutants and persistent organic pollutants (POPs). They include:

- Polychlorinated dibenzo-*p*-dioxins (PCDDs), or simply, but inaccurately, dioxins. Technically PCDDs are derivatives of dibenzo-*p*-dioxin. There are 75 PCDDs, and seven of them are specifically toxic.
- Polychlorinated dibenzofurans (PCDFs), or simply furans. Technically PCDFs are derivatives of dibenzofuran. There are 135 congeners (derivatives differing only in the number and location of chlorine atoms). Whilst they strictly speaking are not dioxins, ten of them have "dioxin-like" properties.
- Polychlorinated biphenyls (PCBs), which also are not dioxins, but twelve of them have "dioxin-like" properties. Under certain conditions PCBs may form more toxic dibenzofurans through partial oxidation.
- Finally, dioxin may refer to dioxin proper, the basic chemical unit of the more complex dioxins. This simple compound is not persistent and has no PCDD-like toxicity.

Because dioxins refer to such a broad class of compounds that vary widely in toxicity, the concept of toxic equivalence (TEQ) has been developed to facilitate risk assessment and regulatory control. Toxic equivalence factors (TEFs) exist for seven congeners of dioxins, ten furans and twelve PCBs. The reference congener is the most toxic dioxin 2,3,7,8-tetrachlorodibenzo-*p*-dioxin (TCDD) which per definition has a TEF of one.

In reference to their importance as environmental toxicants the term dioxins is used almost exclusively to refer to the sum of compounds (as TEQ) from the above groups which demonstrate the same specific toxic mode of action associated with TCDD. These include 17 PCDD/Fs and 12 PCBs. Incidents of contamination with PCBs are also often reported as dioxin contamination incidents since it is this toxic characteristic which is of most public and regulatory concern.

Toxicity

Mechanism of Toxicity: The toxic effects of dioxins are measured in fractional equivalencies of TCDD (2,3,7,8-tetrachlorodibenzo-*p*-dioxin), the most toxic and best studied member of its class. The toxicity is mediated through the interaction with a specific intracellular protein, the aryl hydrocarbon (AH) receptor, a transcriptional enhancer, affecting a number of other regulatory proteins. This receptor is a transcription factor which is involved in expression of many genes. TCDD binding to the AH receptor induces the cytochrome P450 1A class of enzymes which function to break down toxic compounds, e.g., carcinogenic polycyclic hydrocarbons such as benzo*(a)*pyrene.

While the affinity of dioxins and related industrial toxicants to this receptor may not fully explain all their toxic effects including immunotoxicity, endocrine effects and tumour promotion, toxic responses appear to be typically dose-dependent within certain concentration ranges. A multiphasic dose-response relationship has also been reported, leading to uncertainty and debate about the true role of dioxins in cancer rates.

The endocrine disrupting activity of dioxins is thought to occur as a down-stream function of AH receptor activation, with thyroid status in particular being a sensitive marker of exposure. It is important to note that TCDD, along with the other PCDDs, PCDFs and dioxin-like coplanar PCBs are not direct agonists or antagonists of hormones, and are not active in assays which directly screen for these activities such as ER-CALUX and AR-CALUX. These compounds have also not been shown to have any direct mutagenic or genotoxic activity. Their main action in causing cancer is cancer promotion. A mixture of PCBs such as Aroclor may contain PCB compounds which are known estrogen agonists, but on the other hand are not classified as dioxin-like in terms of toxicity. Mutagenic effects have been established for some lower chlorinated chemicals such as 3-chlorodibenzofuran, which is neither persistent nor an AH receptor agonist.

Toxicity in Animals

The symptoms reported to be associated with dioxin toxicity in animal studies are incredibly wide ranging, both in the scope of the biological systems affected and in the range of dosage needed to bring these about. Acute effects of single high dose dioxin exposure include wasting syndrome, and typically a delayed death of the animal in 1 to 6 weeks. By far most toxicity studies have been performed using 2,3,7,8-tetrachlorodibenzo-*p*-dioxin.

The LD_{50} of TCDD varies wildly between species and even strains of the same species, with the most notable disparity being between the seemingly similar species of hamster and guinea pig. The oral LD_{50} for guinea pigs is as low as 0.5 to 2 μg/Kg body weight, whereas the oral LD_{50} for hamsters can be as high as 1 to 5 mg/Kg body weight. Even between different mouse or rat strains there may be tenfold to thousandfold differences in acute toxicity. Many pathological findings are seen in the liver, thymus and other organs.

Some chronic and sub-chronic exposures can be harmful at much lower levels, especially at particular developmental stages including foetal, neonatal and pubescent stages. Well established developmental effects are cleft palate, hydronephrosis, disturbances in tooth development and sexual development as well as endocrine effects.

Human Toxicity

"Dioxins are highly toxic and can cause reproductive and developmental problems, damage the immune system, interfere with hormones and also cause cancer." Studies of the toxic effects on dioxins on human beings have been surprisingly few. The best proven is chloracne. Even in poisonings with huge doses of TCDD, the only persistent effects after the initial malaise have been chloracne and amenorrhea. In occupational settings many symptoms have been seen, but exposures have always been to a multitude of chemicals including chlorophenols, chlorophenoxy acid herbicides, and solvents. Therefore proof of dioxins as causative factors has been difficult. The suspected effects in adults are liver damage, and alterations in heme metabolism, serum lipid levels, thyroid functions, as well as diabetes and immunological effects.

In line with animal studies, developmental effects may be much more important than effects in adults. These include disturbances of tooth development, and of sexual development. An example of the variation in responses is clearly seen in a study following the Seveso disaster indicating that sperm count and motility were affected in different ways in exposed males, depending on whether they were exposed before, during or after puberty.

Carcinogenicity

Dioxins are associated strongly with carcinogenesis in animal studies, although the precise mechanistic role is not clear. Dioxins are not mutagenic or genotoxic. The United States Environmental Protection Agency has categorised dioxin, and the mixture of substances associated with sources of dioxin toxicity as a "likely human carcinogen".

The International Agency for Research on Cancer has classified TCDD as a human carcinogen (class 1) on the basis of clear animal carcinogenicity and limited human data, but was not able to classify other dioxins. It is thought that the presence of dioxin can accelerate the formation of tumours and adversely affect the normal mechanisms for inhibiting tumour growth, without actually instigating the carcinogenic event.

As with all toxic endpoints of dioxin, a clear dose-response relationship is very difficult to establish. After accidental or high occupational exposures there is evidence on human carcinogenicity. There is much controversy especially on cancer risk at low population levels of dioxins. Among fishermen with high dioxin concentrations in their bodies, cancer deaths were decreased rather than increased. Some researchers have also proposed that dioxin induces cancer progression through a very different mitochondrial pathway.

Risk Assessment

The uncertainty and variability in the dose-response relationship of dioxins in terms of their toxicity, as well as the ability of dioxins to bioaccumulate mean that the tolerable daily intake (TDI) of dioxin has been set very low, 1-4 pg/Kg body weight per day, i.e. 0.000 000 000 07 to 0.000 000 000 28 g per 70-kg person per day, to allow for this uncertainty and ensure public safety in all instances. Specifically, the TDI has been assessed based on the safety of children born to mothers exposed all their lifetime prior to pregnancy to such a daily intake of dioxins. It is likely that the TDI for other population groups could be somewhat higher. The most important cause for differences in different assessments is carcinogenicity. If the dose-response of TCDD in causing cancer is linear, it might be a true risk. If the dose-response is of a threshold-type or J-shape, there is little or no risk at the present concentrations. Understanding the mechanisms of toxicity better is hoped to increase the reliability of risk assessment.

Human Intake and Levels

Most intake of dioxin-like chemicals is from food of animal origin: meat, dairy products or fish predominate depending on the country. The daily intake of dioxins and dioxin-like PCBs as TEQ is of the order of 100 pg/day, i.e. 1-2 pg/kg/day. In many countries both the absolute and relative significance of dairy products and meat have decreased due to strict emission controls, and brought about the decrease of total intake. E.g. in the United Kingdom the total intake of PCDD/F in 1982 was 239 pg/day and in 2001 only 21 pg/day (WHO-TEQ). Since the half-lives are very long (for e.g. TCDD 7–8 years), the body burden will increase almost over the whole lifetime. Therefore the concentrations may increase five- to tenfold from age 20 to age 60. For the same reason, short term higher intake such as after food contamination incidents, is not crucial unless it is extremely high or lasts for several months or years.

The highest body burdens were found in Western Europe in the 1970s and early 1980s, and the trends have been similar in the U.S. The most useful measure of time trends is concentration in breast milk measured over decades. In many countries the concentrations have decreased to about one tenth of those in 1970s, and the total TEQ concentrations are now of the order of 10-30 pg/g fat (please note the units, pg/g is the same as ng/kg, or the non-standard expression ppt used sometimes in America). The decrease is due to strict emission controls and also to the control of concentrations in food. In the U.S. young adult female population (age group 20-39), the concentration was 9.7 pg/g lipid in 2001-2002 (geometric mean).

Certain professions such as subsistence fishermen in some areas are exposed to exceptionally high amounts of dioxins and related substances. This along with high industrial exposures may be the most valuable source of information on the health risks of dioxins.

Common Uses

According to the Agency of Toxic Substances & Disease Registry:

Dioxins are not intentionally produced and have no known use. They are the by-products of various industrial processes (i.e., bleaching paper pulp, and chemical and pesticide manufacture) and combustion activities (i.e., burning household trash, forest fires, and waste incineration). The defoliant Agent Orange, used during the Vietnam War, contained dioxins. Dioxins are found at low levels throughout the world in air, soil, water, sediment, and in foods such as meats, dairy, fish, and shellfish. The highest levels of dioxins are usually found in soil, sediment, and in the fatty tissues of animals. Much lower levels are found in air and water.

CDDs are not manufactured commercially in the United States except on a small scale for use in chemical and toxicological research. They are unique among the large number of organochlorine compounds of environmental interest in that they were never produced intentionally as desired commercial products.

Dioxins belong to the persistent organic chemicals (“dirty dozen”) the production and use of which was banned by the Stockholm Commission in 2001.

Sources

Environmental Sources: PCB-compounds, always containing low concentrations dioxin-like PCBs and PCDFs, were synthesized at large amounts for various technical purposes. They have entered the environment through accidents such as fires or leaks from transformers or heat exchangers, or from PCB-containing products in waste landfills or during incineration. Because PCBs are somewhat volatile, they have also been transported long distances by air leading to global distribution including the Arctic.

PCDD/F-compounds were never synthesized for any purpose, except for small quantities for scientific research. Small amounts of PCDD/Fs are formed whenever carbon, oxygen and chlorine are available at suitable temperatures.

This is augmented by metal catalysts such as copper. The optimal temperatures are 400 to 700 °C. This means that formation is highest when organic material is burned at poor burning conditions. The most important sources of PCDD/Fs have been incineration of mixed municipal or hospital waste at too low temperatures as well as metal smelting and refining. Chlorine bleaching of pulp has been an important source of PCDD/Fs to waterways. PCDD/Fs are also formed as synthesis side products of several chemicals, especially PCBs, chlorophenols, chlorophenoxy acid herbicides and hexachlorophene. A poorly appreciated but important production of dioxins is "backyard barrel burning" of waste.

When the problems were realized, both industries and municipal authorities have worked to decrease the emissions. In waste incineration this is based on basically two improvements, increase of burning temperatures to over 1000 °C at a long enough residence time for burning gases, and flue gas cleaning and filtering techniques. This has been very important in Europe, because most of municipal mixed waste is incinerated. Then dioxins spread around the incinerator, and the fallout on pastures caused accumulation in farm animals, and high concentrations in dairy and meat products. Therefore European Union set in 2000 very strict exhaust limit values for dioxins, 0.1 ng/Nm3 (TEQ in exhaust gases).Both in Europe and in U.S.A. the emissions have decreased dramatically since the 1980s, up to 90 %. This has also led to decreases in human body burdens, which is neatly demonstrated by the decrease of dioxin concentrations in breast milk.

Unfortunately the private small-scale burning ("backyard barrel burning") has not decreased effectively, and in the U.S. it is now the most important source of dioxins. Total annual emissions decreased from 14 kg in 1987 to 1.4 kg in 2000. However, backyard barrel burning decreased only modestly from 0.6 kg to 0.5 kg, resulting in over one third of all dioxins coming in the year 2000 from backyard burning alone.

Low concentrations of dioxins have been found in some soils without any anthropogenic contamination. A puzzling case of milk contamination was detected in Germany. The reason was found out to be kaolin added to animal feed. Dioxins have been repeatedly detected in clays from Europe and USA since 1996. Contamination of clay is assumed to be the result of ancient forest fires or similar natural reasons, and enrichment during clay sedimentation.

Environmental Persistence and Bioaccumulation

All groups of dioxin-like compounds are persistent in the environment. Neither soil microbes nor animals are able to break down effectively the PCDD/Fs with lateral chlorines (positions 2,3,7, and 8). This causes very slow elimination. Ultraviolet light is able to slowly break down these compounds. Lipophilicity (tendency to seek for fat-like environments) and very poor water solubility make these compounds move from water environment to living organisms having lipid cell structures.

This is called bioaccumulation. Increase in chlorination increases both stability and lipophilicity. The compounds with the very highest chlorine numbers (e.g. octachlorodibenzo-p-dioxin) are, however, so poorly soluble that this hinders their bioaccumulation. Bioaccumulation is followed by biomagnification. Lipid soluble compounds are first accumulated to microscopic organisms such as phytoplankton (plankton of plant character, e.g. algae). Phytoplankton is consumed by animal plankton, this by invertebrates such as insects, these by small fish, and further by large fish and seals. At every stage or trophic level the concentration is higher, because the persistent chemicals are not "burned off" when the higher organism uses the fat of the prey organism to produce energy.

Due to bioaccumulation and biomagnification the species at the top of the trophic pyramid are most vulnerable to dioxin-like compounds. In Europe the white-tailed eagle and some species of seals have been close to extinct due to poisoning of persistent organic pollutants. Likewise in America the populations of bald eagle declined because of POPs causing thinning of eggs and other reproductive problems. Usually the failure has been attributed mostly to DDT, but dioxins are also a possible cause of reproductive effects. Both in America and in Europe many waterfowl have high concentrations of dioxins, but usually not high enough to disturb their reproductive success Due to supplementary winter feeding and other measures also the white-tailed eagle is recovering. Also ringed seals in the Baltic Sea are recovering.

Human being is also at the top of the trophic pyramid, but due to variable food sources as compared with seals and eagles feeding almost exclusively on fish, human concentrations are much less, 10-100 pg/g, compared with 9000 to 340,000 pg/g (TEQ in lipid) in eagles.

Because of different physicochemical properties, not all congeners of dioxin-like compounds find their routes to human beings equally well. Measured as TEQs, the dominant congeners in human tissues are 2,3,7,8-TCDD, 1,2,3,7,8-PeCDD, 1,2,3,6,7,8-HxCDD and 2,3,4,7,8-PeCDF. This is very different from most sources where hepta- and octa-congeners may predominate. The WHO panel re-evaluating the TEF values in 2005 expressed their concern that emissions should not be uncritically measured as TEQs, because all congeners are not equally important. They stated that "when a human risk assessment is to be done from abiotic matrices, factors such as fate, transport, and bioavailability from each matrix be specifically considered".

All POPs are poorly water soluble, especially dioxins. Therefore ground water contamination has not been a problem even in cases of severe contamination due to the main chemicals such as chlorophenols. In surface waters dioxins are bound to organic and inorganic particles.

Fate of Dioxins in Human Body

The same features causing persistence of dioxins in the environment, also cause very slow elimination in man and other animals. Because of low water solubility, kidneys are not able to secrete them in urine as such. They should be metabolised to more water-soluble metabolites, but also metabolism especially in humans is extremely slow. This results in biological half-lives of several years for all dioxins. That of TCDD is estimated to be 7 to 8 years, and for other PCDD/Fs from 1.4 to 13 years, PCDFs on average slightly shorter than PCDDs.

Dioxins are absorbed well from the digestive tract, if they are dissolved in fats or oils (e.g. in fish or meat). On the other hand, dioxins tend to adsorb tightly to soil particles, and absorption may be quite low: 13.8 % of the given dose of TEQs in contaminated soil was absorbed.

In mammalian organisms dioxins are found mostly in fat. Concentrations in fat seem to be relatively similar, be it serum fat, adipose tissue fat or milk fat. This gives the possibility to measure body burden of dioxins by analysing breast milk. Initially, however, at least in laboratory animals, after a single dose high concentrations are found in the liver, but in a few days adipose tissue will predominate. In rat liver, however, high doses cause induction of CYP1A2 enzyme, and this binds dioxins. Thus depending on the dose, the ratio of fat and liver tissue concentrations may vary considerably in rodents.

Sources of Human Exposure

The most important source of human exposure is fatty food of animal origin. There is quite a lot of variation between different countries as to the most important items. In U.S. and Central Europe milk and dairy products and meat have been by far the most important sources. In some countries, notably in Finland and to some extent in Sweden, fish is important due to contaminated Baltic fish and very low intake from any other sources. In most countries a significant decrease of dioxin intake has occurred due to stricter controls during the last 20 years.

Historically occupational exposure to dioxins has been a major problem. Dioxins are formed as important toxic side products in the production of PCBs, chlorophenols, chlorophenoxy acid herbicides, and other chlorinated organic chemicals. This caused very high exposures to workers in poorly controlled hygienic conditions. Many workers had chloracne. In a NIOSH study in the U.S., the average concentration of TCDD in exposed persons was 233 ng/kg (in serum lipid) while it was 7 ng/kg in unexposed workers, even though the exposure had been 15-37 years earlier. This indicates a huge previous exposure, in fact the exact back-calculation is debated, and the concentrations may have been even several times higher than originally estimated.

Handling and spraying of chlorophenoxy acid herbicides may also cause quite high exposures, as clearly demonstrated by the users of agent orange in Vietnam War. The highest concentrations were detected in nonflying enlisted personnel (e.g. filling the tanks of planes), although the variation was huge, 0 to 618 ng/kg TCDD (mean 23.6 ng/kg). Other occupational exposures (working at paper and pulp mills, steel mills and incinerators) have been remarkably lower.

Accidental exposures have been huge in some cases. The highest concentrations in people after the Seveso accident were 56,000 ng/kg, and the highest exposure ever recorded was found in Austria in 1998, 144,000 ng/kg. This is equivalent to a dose of 20 to 30 μg/kg TCDD, a dose that would be lethal to guinea pigs and some rat strains.

Exposure from contaminated soil is possible, if children eat dirt or if dioxins are blown up in dust. This was clearly demonstrated in Missouri, when waste oils were used as dust suppressant in horse arenas. Many horses and other animals were killed due to poisoning. Dioxins are neither volatile nor water soluble, and therefore exposure of human beings depends on direct eating of soil or production of dust which carries the chemical. Contamination of ground water or breathing vapour of the chemical are not likely to cause a significant exposure. Currently, in the United States, there are 126 Superfund sites (with a completed exposure pathway) contaminated with dioxins.

TEF Values

All dioxin-like compounds share a common mechanism of action via the aryl hydrocarbon receptor (AHR), but their potencies are very different. This means that similar effects are caused by all of them, but much larger doses of some of them are needed than of TCDD. Binding to the AHR as well as persistence in the environment and in the organism depends on the presence of so called "lateral chlorines", in case of dioxins and furans, chlorine substitutes in positions 2,3,7, and 8. Each additional chlorine decreases the potency, but qualitatively the effects remain similar. Therefore a simple sum of different dioxin congeners is not a meaningful measure of toxicity. To compare the toxicities of various congeners and to render it possible to make a toxicologically meaningful sum of a mixture, a toxicity equivalency (TEQ) concept was created.

Each congener has been given a toxicity equivalence factor (TEF). This indicates its relative toxicity as compared with TCDD. Most TEFs have been extracted from *in vivo* toxicity data on animals, but if these are missing (e.g. in case of some PCBs), less reliable *in vitro* data have been used. After multiplying the actual amount or concentration of a congener by its TEF, the product is the virtual amount or concentration of TCDD having effects of the same magnitude as the compound in question. This multiplication is done for all compounds in a mixture, and these "equivalents of TCDD" can then simply be added, resulting in TEQ, the amount or concentration of TCDD toxicologically equivalent to the mixture.

The TEQ conversion makes it possible to use all studies on the best studied TCDD to assess the toxicity of a mixture. This resembles the common measure of all alcoholic drinks: beer, wine and whiskey can be added together as absolute alcohol, and this sum gives the toxicologically meaningful measure of the total impact.

The TEQ only applies to dioxin-like effects mediated by the AHR. Some toxic effects (especially of PCBs) may be independent of the AHR, and those are not taken into account by using TEQs.

TEFs are also approximations with certain amount of scientific judgement rather than scientific facts. Therefore they may be re-evaluated from time to time.

There have been several TEF versions since the 1980s. The most recent re-assessment was by an expert group of the World Health organization in 2005.

WHO Toxic Equivalence Factors (WHO-TEF) for the dioxin-like congeners of concern

Polychlorinated Dioxins	
2,3,7,8-TCDD	1
1,2,3,7,8-PeCDD	1
1,2,3,4,7,8-HxCDD	0.1
1,2,3,6,7,8-HxCDD	0.1
1,2,3,7,8,9-HxCDD	0.1
1,2,3,4,6,7,8-HpCDD	0.01
OCDD	0.0003
Polychlorinated Dibenzofurans	
2,3,7,8-TCDF	0.1
1,2,3,7,8-PeCDF	0.03
2,3,4,7,8-PeCDF	0.3
1,2,3,4,7,8-HxCDF	0.1
1,2,3,6,7,8-HxCDF	0.1
1,2,3,7,8,9-HxCDF	0.1
2,3,4,6,7,8-HxCDF	0.1
1,2,3,4,6,7,8-HpCDF	0.01
1,2,3,4,7,8,9-HpCDF	0.01
OCDF	0.0003
Non-ortho-substituted PCBs	
3,3',4,4'-TCB (PCB77)	0.0001
3,4,4',5-TCB (PCB81)	0.0003
3,3',4,4',5-PeCB (PCB126)	0.1
3,3',4,4',5,5'-HxCB (PCB169)	0.03

Mono-ortho-substituted PCBs	
2,3,3',4,4'-PeCB (PCB105)	0.00003
2,3,4,4',5-PeCB (PCB114)	0.00003
2,3',4,4',5-PeCB (PCB118)	0.00003
2',3,4,4',5-PeCB (PCB123)	0.00003
2,3,3',4,4',5-HxCB (PCB156)	0.00003
2,3,3',4,4',5'-HxCB (PCB157)	0.00003
2,3',4,4',5,5'-HxCB (PCB167)	0.00003
2,3,3',4,4',5,5'-HpCB (PCB189)	0.00003

(T = tetra, Pe = penta, Hx = hexa, Hp = hepta, O = octa)

Asphaltene

Asphaltenes are molecular substances that are found in crude oil, along with resins, aromatic hydrocarbons, and alkanes (i.e., saturated hydrocarbons). The word "asphaltene" was coined by Boussingault in 1837 when he noticed that the distillation residue of some bitumens had asphalt-like properties. Asphaltenes in the form of distillation products from oil refineries are used as "tar-mats" on roads.

Composition

Asphaltenes consist primarily of carbon, hydrogen, nitrogen, oxygen, and sulfur, as well as trace amounts of vanadium and nickel. The C:H ratio is approximately 1:1.2, depending on the asphaltene source. Asphaltenes are defined operationally as the n-heptane (C_7H_{16})-insoluble, toluene ($C_6H_5CH_3$)-soluble component of a carbonaceous material such as crude oil, bitumen, or coal. Asphaltenes have been shown to have a distribution of molecular masses in the range of 400 u to 1500 u, with an average being around 750 u.

Analysis

All techniques are now roughly in accord, including many different mass spectral methods (ESI FT-ICR MS, APPI, APCI FIMS, LDI) and many different diffusion techniques (time-resolved fluorescence depolarization (TRFD), fluorescence correlation spectroscopy (FCS), Taylor dispersion).

This result deviates substantially from previous conventional wisdom, the old and incorrect view is promulgated on many popular web sites putatively illustrating asphaltene molecular structures. Asphaltene molecular weight is ~750 Da with 500 - 1000 FWHM. Aggregation of asphaltenes at very low concentrations (in toluene) led to aggregate weights being misinterpreted as molecular weights with techniques such as VPO or GPC.

The chemical structure is difficult to ascertain, due to the complex nature of the asphaltenes, but has been studied by all available techniques including X-ray, elemental, and pyrolysis GC-FID-GC-MS. However, it is undisputed that the asphaltenes are composed mainly of polyaromatic carbon i.e. polycondensed aromatic benzene units with oxygen, nitrogen, and sulfur, (NSO-compounds) combined with minor amounts of a series of heavy metals, particularly vanadium and nickel which occur in porphyrin structures.

Asphaltene molecular architecture has also been controversial. The TRFD rotation diffusion measurements have indicated there is predominantly one PAH per asphaltene molecule. Asphaltene rotational diffusion measurements show that small PAH chromophores (blue fluorescing) are in small asphaltene molecules while big PAH chromophores (red fluorescing) are in big molecules. This implies that there is only one fused polycyclic aromatic hydrocarbon (PAH) ring system per molecule. Very recent fragmentation studies by FT ICR-MS and by L2MS (two-colour laser mass spectrometry) strongly support this 'island' molecular architecture as shown by TRFD and refuting the 'archipelago' molecular architecture.

Nanocolloidal structure: it has been shown that asphaltenes have two distinct nanocolloidal structures. The island molecule architecture, with attractive forces in the molecule interior (PAH) and steric repulsion from alkane peripheral groups gives rise to nanocolloids with aggregation numbers less than 10. Methods to show these structures in clude SANS, SAXS, high-Q ultrasonics, NMR, AC-conductivity, DC-conductivity, centrifugation. These structures are also observed in oil reservoirs with extensive vertical offset (where gravitational effects are evident). In addition to these 'primary' nanoaggrgeates, CLUSTERS of nanoaggregates can also form as seen by a variety of techniques.

Geology

Asphaltenes are today widely recognised as soluble, chemically altered fragments of kerogen, which migrated out of the source rock for the oil, during oil catagenesis. Asphaltenes had been thought to be held in solution in oil by resins (similar structure and chemistry, but smaller), but recent data shows that this is incorrect. Indeed, it has recently been suggested that asphaltenes are nanocolloidally suspended in crude oil and in toluene solutions of sufficient concentrations. In any event, for low surface tension liquids, such as alkanes and toluene, surfactants are not necessary to maintain nanocolloidal suspensions of asphaltenes.

The nickel to vanadium contents of asphaltenes reflect the pH and Eh conditions of the paleo-depositional environment of the source rock for oil (Lewan, 1980;1984), and this ratio is, therefore, in use in the petroleum industry for oil-oil correlation and for identification of potential source rocks for oil (oil exploration).

Occurrence

Heavy oils, tar sands, and biodegraded oils (as bacteria can not assimilate asphaltenes, but readily consume saturated hydrocarbons and certain aromatic hydrocarbon isomers - enzymatically controlled) contain much higher proportions of asphaltenes than do medium-API oils or light oils. Condensates are virtually devoid of asphaltenes.

Production Problems

They are of particular interest to the petroleum industry because of their depositional effect in production equipment, such as tubulars in oil wells. In addition, asphaltenes impart high viscosity to crude oils, negatively impacting production. The variable asphaltene concentration in crude oils within individual reservoirs creates a myriad of production problems.

Asphaltene Removal

Chemical treatments for removing asphaltene include:

1. solvents
2. dispersant/ solvents
3. oil/dispersants/solvents.

The dispersant/solvent approach is used for removing asphaltenes from formation minerals. Continuous treating may be required to inhibit asphaltene deposition in the tubing. Batch treatments are common for dehydration equipment and tank bottoms. There are also asphaltene precipitation inhibitors that can be used by continuous treatment or squeeze treatments.

Chapter 5

Stereochemistry

Stereochemistry, a subdiscipline of chemistry, involves the study of the relative spatial arrangement of atoms within molecules. An important branch of stereochemistry is the study of chiral molecules.

Stereochemistry is also known as 3D chemistry because the prefix "stereo-" means "three-dimensionality".

The study of stereochemical problems spans the entire range of organic, inorganic, biological, physical and supramolecular chemistries. Stereochemistry includes methods for determining and describing these relationships; the effect on the physical or biological properties these relationships impart upon the molecules in question, and the manner in which these relationships influence the reactivity of the molecules in question (dynamic stereochemistry).

History and Significance

Louis Pasteur could rightly be described as the first stereochemist, having observed in 1849 that salts of tartaric acid collected from wine production vessels could rotate plane polarized light, but that salts from other sources did not. This property, the only physical property in which the two types of tartrate salts differed, is due to optical isomerism. In 1874, Jacobus Henricus van 't Hoff and Joseph Le Bel explained optical activity in terms of the tetrahedral arrangement of the atoms bound to carbon.

An infamous demonstration of the significance of stereochemistry was the thalidomide disaster. Thalidomide is a drug, first prepared in 1957 in Germany, prescribed for treating morning sickness in pregnant women. The drug however was discovered to cause deformation in babies. It was discovered that one optical isomer of the drug was safe while the other had teratogenic effects, causing serious genetic damage to early embryonic growth and development.

In the human body, thalidomide undergoes racemization: even if only one of the two stereoisomers is ingested, the other one is produced. Thalidomide is currently used as a treatment for leprosy and must be used with contraceptives in women to prevent pregnancy-related deformations. This disaster was a driving force behind requiring strict testing of drugs before making them available to the public.

Cahn-Ingold-Prelog priority rules are part of a system for describing a molecule's stereochemistry. They rank the atoms around a stereocenter in a standard way, allowing the relative position of these atoms in the molecule to be described unambiguously. A Fischer projection is a simplified way to depict the stereochemistry around a stereocenter.

Definitions

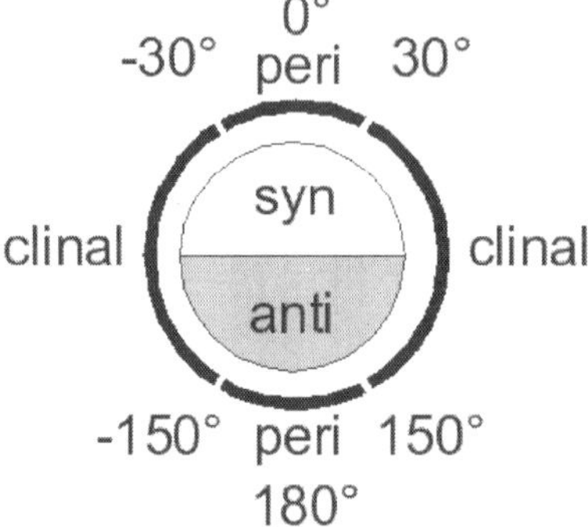

Figure 1: *Syn/anti peri/clinal*

Many definitions that describe a specific conformation (IUPAC Gold Book) exist:

- a torsion angle of ±60° is called gauche
- a torsion angle between 0° and ± 90° is called syn (s)
- a torsion angle between ± 90° and 180° is called anti (a)
- a torsion angle between 30° and 150° or between –30° and –150° is called clinal
- a torsion angle between 0° and 30° or 150° and 180° is called periplanar (p)
- a torsion angle between 0° to 30° is called synperiplanar or syn- or cis-conformation (sp)

- a torsion angle between 30° to 90° and –30° to –90° is called synclinal or gauche or skew (sc)
- a torsion angle between 90° to 150°, and –90° to –150° is called anticlinal (ac)
- a torsion angle between ± 150° to 180° is called antiperiplanar or anti or trans (ap).

Torsional strain results from resistance to twisting about a bond.

Types

- Atropisomerism
- *Cis-trans* isomerism
- Conformational isomerism
- Diastereomers
- Enantiomers
- Rotamers.

Alkane Stereochemistry

Alkane stereochemistry concerns the stereochemistry of alkanes. Alkane conformers are one of the subjects of alkane stereochemistry.

Conformations

Alkane conformers arise from rotation around sp^3 hybridised carbon carbon sigma bonds. The smallest alkane with such a chemical bond, ethane, exists as an infinite number of conformations with respect to rotation around the C–C bond. Two of these are recognised as energy minimum (staggered conformation) and energy maximum (eclipsed conformation) forms. The existence of specific conformations is due to hindered rotation around sigma bonds, although a role for hyperconjugation is proposed by a competing theory.

The importance of energy minimum and energy maximum is seen by extension of these concepts to more complex molecules for which stable conformations may be predicted as minimum energy forms. The determination of stable conformations has also played a large role in the establishment of the concept of asymmetric induction and the ability to predict the stereochemistry of reactions controlled by steric effects.

In the example of staggered ethane in Newman projection, a hydrogen atom on one carbon atom has a 60° torsional angle or torsion angle with respect to the nearest hydrogen atom on the other carbon so that steric hindrance is minimised. The staggered conformation is more stable by 12.5 kJ/mol than the eclipsed conformation, which is the energy maximum for ethane. In the eclipsed conformation the torsional angle is minimized.

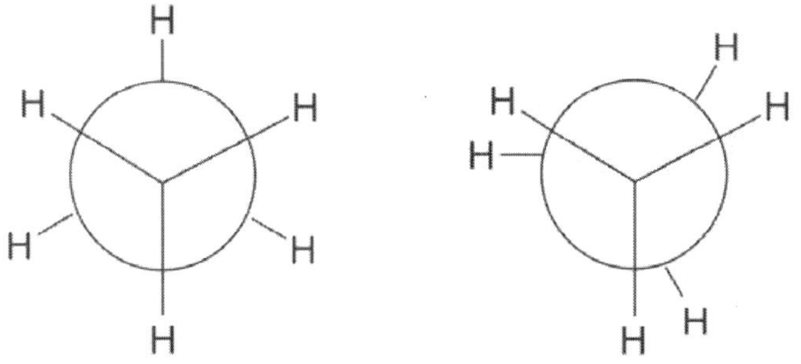

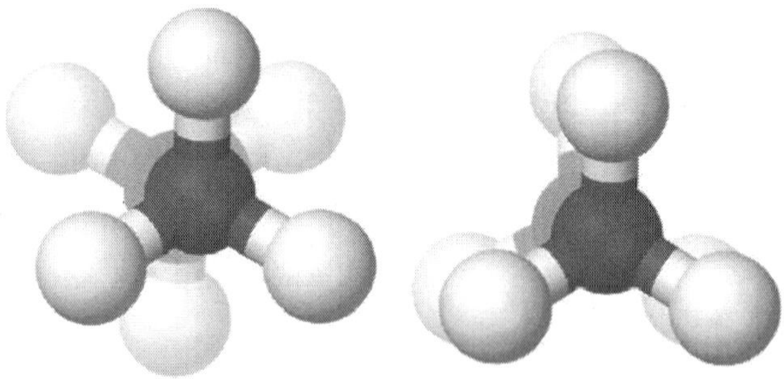

In butane, the two staggered conformations are no longer equivalent and represent two distinct conformers: the anti-conformation (left-most, below) and the gauche conformation (right-most, below).

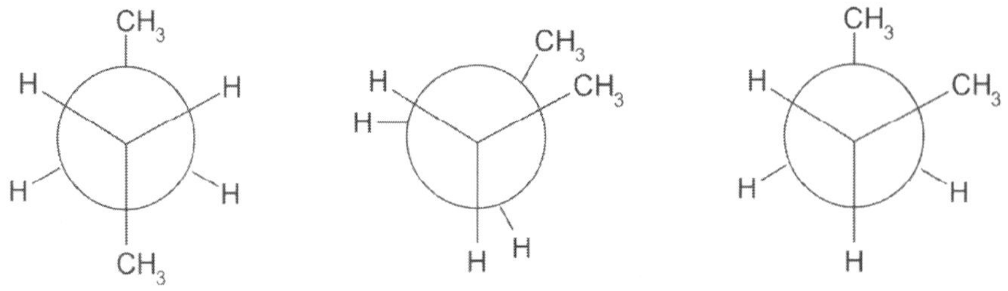

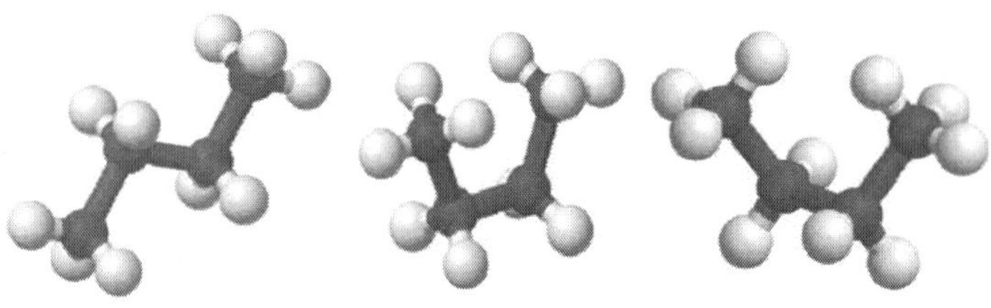

Both conformations are free of torsional strain, but, in the gauche conformation, the two methyl groups are in closer proximity than the sum of their van der Waals radii. The interaction between the two methyl groups is repulsive (van der Waals strain), and an energy barrier results. A measure of the potential energy stored in butane conformers with greater steric hindrance than the 'anti'-conformer ground state is given by these values:

- Gauche, conformer - 3.8 kJ/mol
- Eclipsed H and CH_3 - 16 kJ/mol
- Eclipsed CH_3 and CH_3 - 19 kJ/mol.

The eclipsed methyl groups exert a greater steric strain because of their greater electron density compared to lone hydrogen atoms.

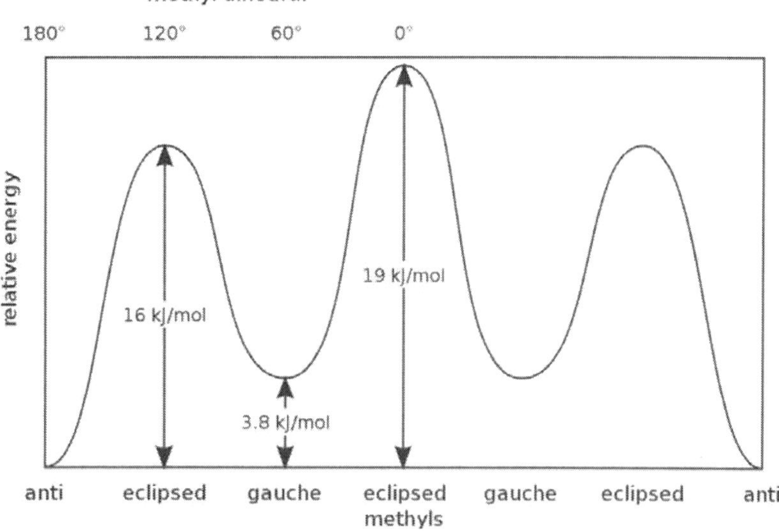

The textbook explanation for the existence of the energy maximum for an eclipsed conformation in ethane is steric hindrance, but, with a C-C bond length of 154 pm and a Van der Waals radius for hydrogen of 120 pm, the hydrogen atoms in ethane are never in each other's way. The question of whether steric hindrance is responsible for the eclipsed energy maximum is a topic of debate to this day. One alternative to the steric hindrance explanation is based on hyperconjugation as analyzed within the Natural Bond Orbital framework.

In the staggered conformation, one C-H sigma bonding orbital donates electron density to the antibonding orbital of the other C-H bond. The energetic stabilization of this effect is maximized when the two orbitals have maximal overlap, occurring in the staggered conformation. There is no overlap in the eclipsed conformation, leading to a disfavoured energy maximum. On the other hand, an analysis within quantitative molecular orbital theory shows that 2-orbital-4-

electron (steric) repulsions are dominant over hyperconjugation. A valence bond theory study also emphasizes the importance of steric effects.

Definitions

Many definitions that describe a specific conformation (IUPAC Gold Book) exist:

- a torsion angle of ±60° is called gauche
- a torsion angle between 0° and ± 90° is called syn (s)
- a torsion angle between ± 90° and 180° is called anti (a)
- a torsion angle between 30° and 150° or between –30° and –150° is called clinal
- a torsion angle between 0° and 30° or 150° and 180° is called periplanar (p)
- a torsion angle between 0° to 30° is called synperiplanar or syn- or cis-conformation (sp)
- a torsion angle between 30° to 90° and –30° to –90° is called synclinal or gauche or skew (sc)
- a torsion angle between 90° to 150°, and –90° to –150° is called anticlinal (ac)
- a torsion angle between ± 150° to 180° is called antiperiplanar or anti or trans (ap).

Torsional strain results from resistance to twisting about a bond.

Special Cases

In n-pentane, the terminal methyl groups experience additional pentane interference.

Replacing hydrogen by fluorine in polytetrafluoroethylene changes the stereochemistry from the zigzag geometry to that of a helix due to electrostatic repulsion of the fluorine atoms in the 1,3 positions. Evidence for the helix structure in the crystalline state is derived from X-ray crystallography and from NMR spectroscopy and circular dichroism in solution.

Cyclohexane Conformation

A cyclohexane conformation is any of several three-dimensional shapes that a cyclohexane molecule can assume while maintaining the integrity of its chemical bonds. The internal angles of a flat regular hexagon are 120 degrees, while the preferred angle between successive bonds in a carbon chain is about 109 degrees. Therefore the cyclohexane ring tends to assume certain non-planar (warped) conformations, which have all angles closer to 109 degrees and therefore a lower strain energy than the flat hexagonal shape. The most important shapes are called *chair*, *half-chair*, *boat*, and *twist-boat*. The molecule can easily switch between these conformations, and only two of them — *chair* and *twist-boat* — can be isolated in pure form.

Cyclohexane conformations have been extensively studied in organic chemistry because they are the classical example of conformational isomerism and have noticeable influence on the physical and chemical properties of cyclohexane.

Historical Background

The very first suggestion that cyclohexane may not be a flat molecule goes back a surprisingly long time. In 1890, Hermann Sachse, a 28-year-old assistant in Berlin, published instructions for folding a piece of paper to represent two forms of cyclohexane he called *symmetrical* and *unsymmetrical* (what we would now call *chair* and *boat*). He clearly understood that these forms had two positions for the hydrogens (again, to use modern terminology, *axial* and *equatorial*), that two chairs would probably interconvert, and even how certain substituents might favour one of the chair forms. Because he expressed all this in mathematical language, few chemists of the time understood his arguments. He had several attempts at publishing these ideas, but none succeeded in capturing the imagination of chemists. His death in 1893 at the age of 31 meant his ideas sank into obscurity. It was only in 1918 when Ernst Mohr, using the then very new technique of x-ray crystallography, was able to determine the molecular structure of diamond, that it became recognised that Sachse's chair was the pivotal motif. Derek Barton and Odd Hassel shared the 1969 Nobel Prize for work on the conformations of cyclohexane and various other molecules.

General

The carbon-carbon bonds along the cyclohexane ring are sp^3 hybrid orbitals, which have tetrahedral symmetry. Therefore, the angles between bonds of a tetravalent carbon atom have a preferred value θ H– 109.5°. The bonds also have a fairly fixed bond length λ. On the other hand, adjacent carbon atoms are free to rotate about the axis of the bond. Therefore, a ring that is warped so that the bond lengths and angles are close to those ideal values will have less strain energy than a flat ring with 120° angles. For each particular conformation of the carbon ring, the directions of the 12 carbon-hydrogen bonds (and therefore the positions of the hydrogen atoms) are fixed.

There are exactly eight warped polygons with six distinguished corners that have all internal angles equal to θ and all sides equal to λ. They comprise two ideal *chair conformations*, where the carbons alternately lie above and below the mean ring plane; and six ideal *boat conformations*, where two opposite carbons lie above the mean plane, and the other four lie below it. In theory, a molecule with any of those ring conformations would be free of angle strain. However, due to interactions between the hydrogen atoms, the angles and bond lengths of the actual *chair* forms are slightly different from the nominal values. For the same reasons, the actual *boat* forms have slightly higher energy than the chair forms. Indeed, the *boat* forms are unstable, and deform spontaneously to *twist-boat* conformations that are local minima of the total energy, and therefore stable.

Each of the stable ring conformations can be transformed into any other without breaking the ring. However, such transformations must go through other states with stressed rings. In particular, they must go through unstable states where four successive carbon atoms lie on the same plane. These shapes are called *half-chair* conformations.

Chair Conformation

The two chair conformations have the lowest total energy, and are therefore the most stable. In the basic chair conformation, the carbons C1 through C6 alternate between two parallel planes, one with C1, C3 and C5, the other with C2, C4, and C6. The molecule has a symmetry axis perpendicular to these two planes, and is congruent to itself after a rotation of 120° about that axis. The two chair conformations have the same shape; one is congruent to the other after 60° rotation about that axis, or after being mirrored across the mean plane. The perpendicular projection of the ring onto its mean plane is a regular hexagon. All C-C bonds are tilted relative to the mean plane, but opposite bonds (such as C1-C_2 and C_4-C_5) are parallel to each other.

As a consequence of the ring warping, six of the 12 carbon-hydrogen bonds end up almost perpendicular to the mean plane and almost parallel to the symmetry axis, with alternating directions, and are said to be axial. The other six C-H bonds lie almost parallel to the mean plane, and are said to be equatorial. The conversion from one chair shape to the other is called ring flipping or chair-flipping. Carbon-hydrogen bonds that are axial in one configuration become equatorial in the other, and vice-versa; but their "up" or "down" character remains the same. In cyclohexane, the two chair conformations have the same energy. At 25°C, 99.99% of all molecules in a cyclohexane solution will be in a chair conformation.

In cyclohexane derivatives, the two chair conformations may have different energies, depending upon the identity and location of the substituents. For example, in methylcyclohexane the lowest energy conformation is a chair one where the methyl group is in equatorial position. This configuration reduces interaction between the methyl group (on carbon number 1) and the hydrogens at carbons 3 and 5; more importantly, it avoids two gauche butane interactions (of the C1-CH3 bond with the C2-C3 and C5-C6 ring bonds). Similarly, *cis*-1,3-dimethylcyclohexane usually has both methyls in the equatorial position so as to avoid interaction between them. In six-membered heterocyles such as pyran, a substituent next to an heteroatom may prefer the axial position due to the anomeric effect.

The preference of a substituent towards the equatorial conformation is measured in terms of its *A value*, which is the Gibbs free energy difference between the two chair conformations, with the substituent in equatorial or in axial position. A positive A value indicates preference towards the equatorial position. The magnitude of the A values ranges from nearly zero for very small substituents such as deuterium, to about 5 kcal/mol for very bulky substituents such as the tert-butyl group.

Boat Conformation

In the basic boat conformation (C_{2v} symmetry), carbons C_2, C_3, C_5 and C_6 are coplanar, while C_1 and C_4 are displaced away from that plane in the same direction. Bonds C2-C3 and C5-C6 are therefore parallel. In this form, the molecule has two perpendicular planes of symmetry as well as a C_2 axis. The boat conformations have higher energy than the chair conformations. The interaction between the two flagpole hydrogens, in particular, generates steric strain. There is also torsional strain involving the C_2-C_3 and C_5-C_6 bonds, which are eclipsed. Because of this strain, the boat configuration is unstable (not a local minimum of the energy function).

Twist-boat Conformation

The *twist-boat* conformation, sometimes called twist (D_2 symmetry) can be derived from the boat conformation by applying a slight twist to the molecule about the axes connecting the two unique carbons. The result is a structure that has three C_2 axes and no plane of symmetry.

The concentration of the twist-boat conformation at room temperature is very low (less than 0.1%) but at 1073 Kelvin it can reach 30%. Rapid cooling from 1073 K to 40 K will freeze in a large concentration of twist-boat conformation, which will then slowly convert to the chair conformation upon heating .

Half-chair Conformation

The half-chair conformation is a transition state with C_2 symmetry generally considered to be on the pathway between chair and twist-boat. It involves rotating one of the dihedrals to zero such that four adjacent atoms are coplanar and the other two atoms are out of plane (one above and one below).

Interconversions between Conformations

At room temperature there is a rapid equilibrium between the two chair conformations of cyclohexane. The interconversion of these two conformations has been much debated and still lacks consensus. What is known is that the twist-boat and chair are both energy minima—the twist-boat being a local minimum; the chair being a global minimum (ground state).The half-chair state (2, below) is the transition state in the interconversion between the chair and twist-boat conformations. Due to the D2 symmetry of the twist-boat, there are two energy-equivalent pathways that it can take to two *different* half-chair conformations, leading to the two different chair conformations of cyclochexane. Thus, at a minimum, the interconversion between the two chair conformations involves the following sequence: chair - half-chair - twist-boat - half-chair' - chair'.

The boat conformation (4, below) is also a transition state, allowing the interconversion between two different twist-boat conformations. While the boat conformation is not *necessary* for interconversion between the two chair conformations of cyclohexane, it is often included in the reaction coordinate diagram used to describe this interconversion because its energy is considerably lower than that of the half-chair, so any molecule with enough energy to go from twist-boat to chair also has enough energy to go from twist-boat to boat. Thus, there are multiple pathways by which a molecule of cyclohexane in the twist-boat conformation can achieve the chair conformation again.

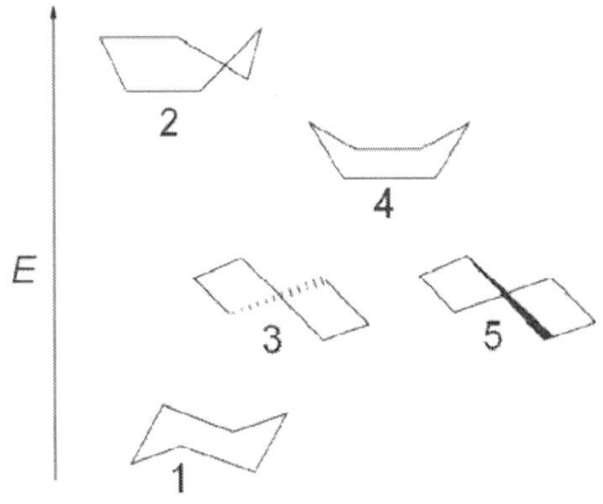

E

2

4

3

5

1

Forced Conformations

[6.6]Chiralane is a point group T molecule wholly composed of identical fused twist-boat cyclohexanes. Twistane is another compound with a forced twist-boat conformation.

Cyclohexane Derivatives

Substituents found on cyclohexane adopt cis and trans formations and cannot be easily switched by simple single sigma bond rotation as with linear molecules. Cis formation means that both substituents are found on the upper side of the 2 substituent placements on the carbon, while trans would mean that they were on opposing sides. Despite the fact that carbons on cyclohexane are linked by a single bond, the ring remains rigid, in that switching from cis to trans would require breaking the ring. The nomenclature for cis is dubbed (Z) while the name for trans is (E) to be placed in front of the IUPAC name.

For di-substituted cyclohexane rings (i.e. two groups on the ring), the relative orientation of the two substituents affect the energy of the possible conformations. For 1,2- and 1,4-di-substituted cyclohexane, a *cis* configuration leads to one axial and one equatorial group. This configuration can undergo chair flipping. For 1,2- and 1,4-di-substituted cyclohexane, a *trans* configuration leads to either both groups axial or both equatorial. In this case, the diaxial conformation is effectively prevented by its high steric strain (four gauche interactions more than the diequatorial).

For 1,3-di-substituted cyclohexanes, the *cis* form is diequatorial and the flipped conformation suffers additional steric interaction between the two axial groups. *Trans*-1,3-di-substituted cyclohexanes are like *cis*-1,2- and *cis*-1,4- and can flip between the two equivalent axial/equatorial forms. Derivatives of cyclohexane do exist that have a more stable twist-boat conformation. An example is 1,2,4,5-tetrathiane, an organosulfur compound with 4 methylene groups replaced by a sulfide group thus removing unfavourable 1,3-diaxial interactions. In the tetramethyl analogue 3,3,6,6-tetramethyl-1,2,4,5-tetrathiane the twist-boat conformation actually dominates. Also in cyclohexane-1,4-dione with the steric 1,4-hydrogen interaction removed, the actual stable conformation is the twist-boat.

Cis-1,4-di-tert-butylcyclohexane has an axial tert-butyl group in the chair conformation and conversion to the twist-boat conformation places both groups in more favourable equatorial positions. As a result the twist-boat conformation is more stable by 0.47 kcal/mol (1.96 kJ/mol) at 125 K as measured by NMR spectroscopy.

Effect of polar substituent:

- cis-cyclohexane-1,3-diol prefers diaxial conformation "formation of intrahydrogen bond".
- 2,5-di-tert-butyl-1,4-cyclohexanediol present in boat or twist-boat form "also intra-H-bond"
- 2-bromocyclohexanone prefers a-Br "min.dipolar repulsion"
- 2-bromo-4,4-dimethylcyclohexanone prefers e-Br –1,3 diaxial interaction(-ve in e-Br)more than dipolar repulsion:
- trans-1,2-dibromocyclohexane present in axial form in non-polar solvents "dipoles cancel" while present in equtorial form in polar solvents "dipoles reinforce".

Gauche Effect

The term "gauche" refers to conformational isomers (conformers) where two vicinal groups are separated by a 60° torsion angle. IUPAC defines groups as gauche if they have a "synclinal alignment of groups attached to adjacent atoms".

In stereochemistry, gauche interactions hinder bond rotation. For example, sighting along the C_2-C_3 bond in staggered butane, there are two possible relative potential energies. The two methyl groups can be in an anti-bonding relationship, or offset at sixty degree dihedral angles. In the latter configuration, the two methyls are said to be in a gauche relationship, and the relative potential energy of each methyl-methyl gauche interaction is 0.9 kilocalories per mole (4 kJ/mol). In general a gauche rotamer is less stable than an anti-rotamer.

The Gauche effect characterizes any gauche rotamer which is actually more stable than the anti rotamer. This effect is present in 1,2-difluoroethane (H_2FCCFH_2) for which the gauche conformation is more stable by 2.4 to 3.4 kJ/mole in the gas phase. Another example is 1,2-dimethoxyethane. There are two main explanations for the gauche effect: hyperconjugation and bent bonds. In the hyperconjugation model, the donation of electron density from the C–H σ bonding orbital to the C–F σ^* antibonding orbital is considered the source of stabilization in the gauche isomer. Due to the greater electronegativity of fluorine, the C–H σ orbital is a better electron donor than the C–F σ orbital, while the C–F σ^* orbital is a better electron acceptor than the C–H σ^* orbital. Only the gauche conformation allows good overlap between the better donor and the better acceptor.

Key in the bent bond explanation of the gauche effect in difluoroethane is the increased p orbital character of both C-F bonds due to the large electronegativity of fluorine. As a result electron density builds up above and below to the left and right of the central C-C bond.

The resulting reduced orbital overlap can be partially compensated when a gauche conformation is assumed, forming a bent bond. Of these two models, hyperconjugation is generally considered the principal cause behind the gauche effect in difluoroethane.

The molecular geometry of both rotamers can be obtained experimentally by high resolution infrared spectroscopy augmented with in silico work. In accordance with the model described above, the carbon - carbon bond length is higher for the anti-rotamer (151.4 pm vs. 150).

The steric repulsion between the fluorine atoms in the gauche rotamer causes increased CCF bond angles (by 3.2°) and increased FCCF dihedral angles (from the default 60° to 71°).

In the related *1,2-difluorodiphenylethane* (two protons replaced by phenyl) the threo isomer is found (by X-ray diffraction and from NMR coupling constants) to have an anti conformation between the two phenyl groups and the two fluorine groups and a gauche conformation is found for both groups for the erythro isomer.

According to in silico results this conformation is more stable by 0.21 kcal/mol (880 J/mol).

The gauche effect is very sensitive to solvent effects, due to the large difference in polarity of the two conformers. For example, 2,3-dinitro-2,3-dimetylbutane, which in the solid state exists only in the gauche conformation, prefers the gauche conformer in benzene solution by a ratio of 79:21, but in carbon tetrachloride it prefers the anti conformer by a ratio of 58:42.

Another case is trans-1,2 difluorocyclohexane, which has a larger preference for the eq/eq conformer in more polar solvents.

A related effect is the cis effect in alkenes.

Structural Isomer

Skeletal Isomerism

In skeletal isomerism, or chain isomerism, components of the (usually carbon) skeleton are distinctly re-ordered to create different structures. Pentane exists as three isomers: *n*-pentane (often called simply "pentane"), isopentane (methylbutane) and neopentane (dimethylpropane).

Skeletal isomerism of pentane:

n-Pentane 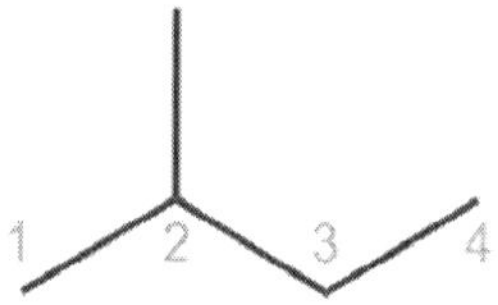 *Isopentane* 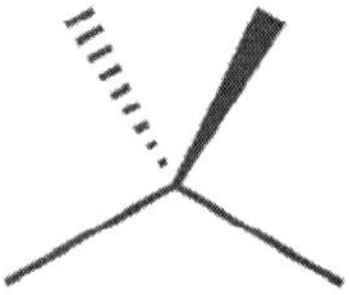 *Neopentane*

Position Isomerism

In position isomerism a functional group or other substituent changes position on a parent structure. In the table below, the hydroxyl group can occupy three different positions on an n-pentane chain forming three different compounds.

Many aromatic isomers exist because substituents can be positioned on different parts of the benzene ring. Only one isomer of phenol or hydroxybenzene exists but cresol or methylphenol has three isomers where the additional methyl group can be placed on three different positions on the ring. Xylenol has one hydroxyl group and two methyl groups and a total of 6 isomers exist.

Positional isomers of xylenol:

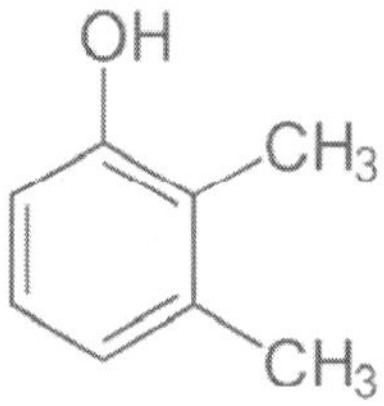

2,3-Xylenol 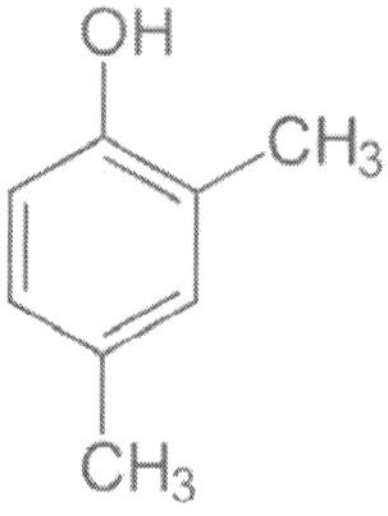 *2,4-Xylenol* 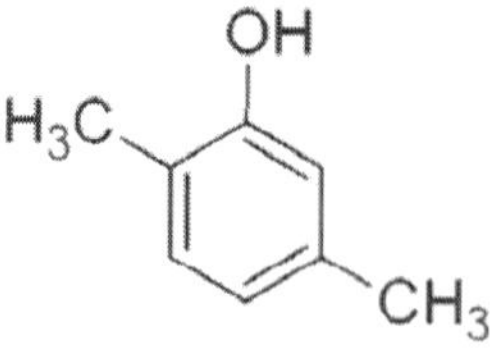 *2,5-Xylenol*

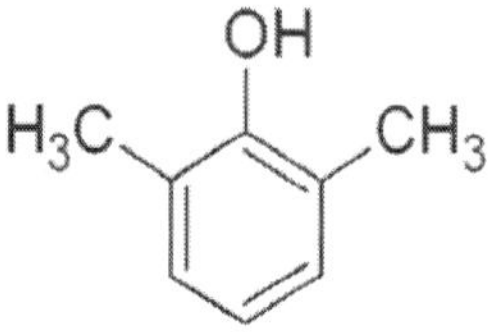

2,6-Xylenol 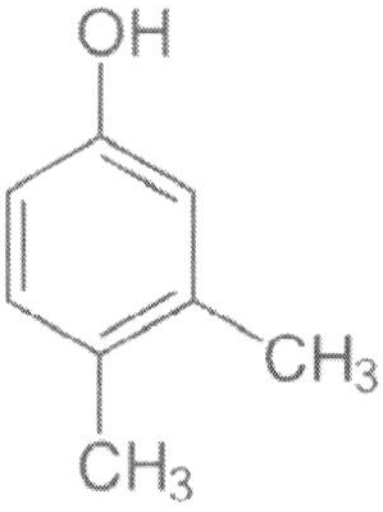 *3,4-Xylenol* 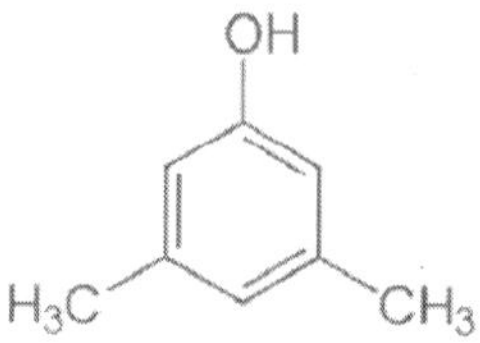 *3,5-xylenol*

Functional Group Isomerism

Functional isomers are structural isomers that have the same molecular formula (that is, the same number of atoms of the same elements), but the atoms are connected together in different ways so that the groupings are dissimilar. These groups of atoms are called functional groups, functionalities, or moieties. Another way to say this is that two compounds with the same molecular formula, but different functional groups, are functional isomers.

For two molecules to be functional isomers, they must contain key groups of atoms arranged in particular ways. Some of the best examples come from organic chemistry. C_2H_6O is a molecular formula. Depending on how the atoms are arranged, it can represent two different compounds dimethyl ether CH_3-O-CH_3 or ethanol CH_3CH_2-O-H. Dimethyl ether and ethanol are functional isomers. The first is an ether. The carbon chain-oxygen-carbon chain functionality is called an ether. The second is an alcohol. The carbon chain-oxygen-hydrogen functionality is called an alcohol.

If the functionalities stay the same, but their locations change, the structural isomers are not functional isomers. 1-Propanol and 2-propanol are structural isomers, but they are not functional isomers. Both of them are alcohols. The functional group (carbon chain-O-H) is present in both of these compounds, but they are not the same.

While some chemists use the terms structural isomer and functional isomer interchangeably, not all structural isomers are functional isomers. Functional isomers are most often identified in chemistry using infrared spectroscopy. Infrared radiation corresponds to the energies associated primarily with molecular vibration. The alcohol functionality has a very distinct vibration called OH-stretch that is due to hydrogen bonding. All alcohols in liquid and solid form absorb infrared radiation at certain wavelengths.

Compounds with the same functional groups will all absorb certain wavelengths of infrared light because of the vibrations associated with those groups. In fact, the infrared spectrum is divided into two regions. The first part is called the functional group region. Dimethyl ether and ethanol would have dissimilar infrared spectra in the functional group region.

The second part of the infrared spectrum is called the fingerprint region; it is associated with types of motion allowed by the symmetry of the molecule and influenced by the bond energies. The fingerprint region is more specific to an individual compound. Even though 1-

propanol and 2-propanol have similar infrared spectra in the functional group region, they differ in the fingerprint region.

In simple terms, functional isomers are structural isomers that have different functional groups like alcohol and ether.

Pentane

Pentane is an organic compound with the formula C_5H_{12} — that is, an alkane with five carbon atoms. The term may refer to any of three structural isomers, or to a mixture of them: in the IUPAC nomenclature, however, pentane means exclusively the *n*-pentane isomer; the other two being called "methylbutane" and "dimethylpropane". Cyclopentane is not an isomer of pentane.

Pentanes are components of some fuels and are employed as specialty solvents in the laboratory. Their properties are very similar to those of butanes and hexanes.

Industrial Uses

Pentane is one of the primary blowing agents used in the production of polystyrene foam. Because of its low boiling point, low cost, and relative safety, pentane is used as a working medium in geothermal power stations. It is added into some refrigerant blends as well.

Laboratory Use

Pentanes are relatively inexpensive and are the most volatile alkanes that are liquid at room temperature, so they are often used in the laboratory as solvents that can be conveniently evaporated. However, because of their nonpolarity and lack of functionality, they can only dissolve non-polar and alkyl-rich compounds. Pentanes are miscible with most common nonpolar solvents such as chlorocarbons, aromatics, and ethers. They are also often used in liquid chromatography.

Physical Properties

Table: *Melting point (MP), boiling point (BP) and density of pentanes*

Isomer	*MP (°C)*	*BP (°C)*	*Density (g/L)*
n-pentane	–129.8	36.0	621
isopentane	–159.9	27.7	616
neopentane	–16.6	9.5	586

The boiling points of the pentane isomers range from about 9 to 36 °C. As is the case for other alkanes, the more branched isomers tend to have lower boiling points.

The same trend normally holds for the melting points of alkane isomers, and indeed that of isopentane is 30 °C lower than that of *n*-pentane. However, the melting point of neopentane, the most heavily branched of the three, is 100 °C *higher* that of isopentane.

The anomalously high melting point of neopentane has been attributed to the better solid-state packing assumed to be possible with its tetrahedral molecule; but this explanation has been challenged on account of it having a lower density than the other two isomers.

The branched isomers are more stable (have lower heat of formation and heat of combustion) than normal pentane. The difference is 1.8 kcal/mol for isopentane, and 5 kcal/mol for neopentane.

Rotation about two central single C-C bonds of *n*-pentane produces four different conformations.

Reactions

All pentane isomers burn with oxygen to form carbon dioxide and water:

$$C_5H_{12} + 8\ O_2 \rightarrow 5\ CO_2 + 6\ H_2O$$

As with other hydrocarbons, pentanes undergo free radical chlorination:

$$C_5H_{12} + Cl_2 \rightarrow C_5H_{11}Cl + HCl$$

Such reactions are unselective; with *n*-pentane, the result is a mixture of the 1-, 2-, and 3-chloropentanes, as well as more highly chlorinated derivatives. Other radical halogenations can also occur.

While *n*-butane is the conventional feedstock in the production of maleic anhydride, *n*-pentane is also a substrate:

$$CH_3CH_2CH_2CH_2CH_3 + 5\ O_2 \rightarrow C_2H_2(CO)_2O + 5\ H_2O + CO_2$$

Isopentane

Isopentane, C_5H_{12}, also called methylbutane or 2-methylbutane, is a branched-chain alkane with five carbon atoms. Isopentane is an extremely volatile and extremely flammable liquid at room temperature and pressure.

The normal boiling point is just a few degrees above room temperature and isopentane will readily boil and evaporate away on a warm day. Isopentane is commonly used in conjunction with liquid nitrogen to achieve a liquid bath temperature of -160 °C.

An isopentyl group is a subset of the generic pentyl group. It has the chemical structure $-CH_3CH_2CH(CH_3)_2$.

Nomenclature

Isopentane is the name recommended by the International Union of Pure and Applied Chemistry (IUPAC) in its *1993 Recommendations for the Nomenclature of Organic Chemistry*. It is one of only four acyclic hydrocarbons to retain its pre-IUPAC name. An isopentyl group is a subset of the generic pentyl group. It has the chemical structure -$CH_3CH_2CH(CH_3)_2$.

Isomers

Isopentane is one of three structural isomers with the molecular formula C_5H_{12}, the others being pentane (*n*-pentane) and dimethyl propane (neopentane).

Uses

Isopentane is one of the ingredients in both Aquafresh® and Sensodyne®.Neopentane

Neopentane, also called dimethylpropane, is a double-branched-chain alkane with five carbon atoms. Neopentane is an extremely flammable gas at room temperature and pressure which can condense into a highly volatile liquid on a cold day, in an ice bath, or when compressed to a higher pressure.

Neopentane is the simplest alkane with a quaternary carbon. It is one of the three structural isomers with the molecular formula C_5H_{12} (pentanes), the other two being *n*-pentane and isopentane.

Nomenclature

IUPAC nomenclature retains the trivial name neopentane. The systematic name is 2,2-dimethylpropane, but the substituent numbers are unnecessary because it is the only possible "dimethylpropane".

Figure: *A neopentyl group attached to a generic group R.*

A neopentyl substituent, often symbolized by Np, has the structure Me_3C-CH_2- for instance neopentyl alcohol (Me_3CCH_2OH or NpOH).

Physical Properties

Boiling and Melting Points

The boiling point of neopentane is only 9.5°C, significantly lower than those of isopentane (27.7°C) and normal pentane (36.0°C).

Therefore, neopentane is a gas at room temperature and atmospheric pressure, while the other two isomers are (barely) liquids.

The melting point of neopentane (-16.6°C), on the other hand, is some 140 degrees higher than that of isopentane (-159.9°C) and some 110 degrees higher than that of *n*-pentane (-129.8°C). This anomaly has been attributed to the better solid-state packing assumed to be possible with the tetrahedral neopentane molecule; but this explanation has been challenged on account of it having a lower density than the other two isomers.

Moreover, its enthalpy of fusion is lower than the enthalpies of fusion of both n-pentane and isopentane, thus indicating that its high melting point is due to an entropy effect. Indeed, the entropy of fusion of neopentane is about 4 times lower than that of n-pentane and isopentane.

NMR Spectrum

Neopentane has T_d symmetry. As a result, all protons are chemically equivalent leading to a single NMR chemical shift $\delta = 0.902$ when dissolved in carbon tetrachloride. In this respect, neopentane is similar to its silane analog, tetramethylsilane, whose single chemical shift is zero by convention.

The symmetry of the neopentane molecule can be broken if some hydrogen atoms are replaced by deuterium atoms. In particular, if each methyl group has a different number of substituted atoms (0, 1, 2, and 3), one obtains a chiral molecule. The chirality in this case arises solely by the mass distribution of its nuclei, while the electron distribution is still essentially achiral.

1-Pentanol

1-Pentanol, (or n-pentanol, pentan-1-ol), is an alcohol with five carbon atoms and the molecular formula $C_5H_{12}O$. 1-Pentanol is a colourless liquid with an unpleasant aroma. There are 8 alcohols with this molecular formula. The ester formed from butanoic acid and 1-pentanol, pentyl butyrate, smells like apricot. The ester formed from acetic acid and 1-pentanol, amyl acetate (pentyl acetate), smells like banana.

Pentanol can be prepared by fractional distillation of fusel oil. To reduce the use of fossil fuels, research is underway to discover cost effective methods of utilizing fermentation to produce Bio-Pentanol. Pentanol can be used as a solvent for coating CDs and DVDs. Another use is a replacement for gasoline.

2-Pentanol

2-Pentanol (IUPAC name, also called *sec*-amyl alcohol) is an organic chemical compound. It is used as a solvent and an intermediate in the manufacture of other chemicals. 2-Pentanol is a component of many mixtures of amyl alcohols sold industrially.

Reactions

2-Pentanol can be manufactured by halogenation of pentane.

3-Pentanol

3-Pentanol is one of the isomers of amyl alcohol.

Cis–trans Isomerism

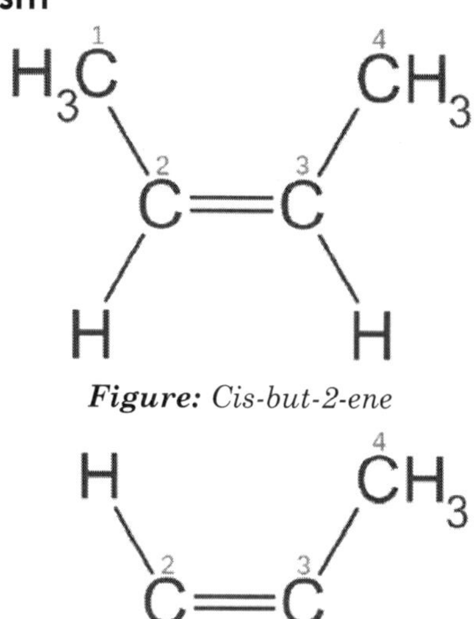

Figure: *Cis-but-2-ene*

Figure: *Trans-but-2-ene*

In organic chemistry, *cis/trans* isomerism or geometric isomerism or configuration isomerism or *E/Z* isomerism is a form of stereoisomerism describing the orientation of functional groups within a molecule. In general, such isomers contain double bonds, which cannot rotate, but they can also arise from ring structures, wherein the rotation of bonds is greatly restricted. *Cis* and *trans* isomers occur both in organic molecules and in inorganic coordination complexes.

The terms *cis* and *trans* are from Latin, in which *cis* means "on the same side" and *trans* means "on the other side" or "across". The term "geometric isomerism" is considered an obsolete synonym of "*cis/trans* isomerism" by IUPAC. It is sometimes used as a synonym for general stereoisomerism (e.g., optical isomerism being called geometric isomerism); the correct term for non-optical stereoisomerism is diastereomerism.

In Organic Chemistry

When the substituent groups are oriented in the same direction, the diastereomer is referred to as *cis*, whereas, when the substituents are oriented in opposing directions, the diastereomer is referred to as *trans*. An example of a small hydrocarbon displaying *cis/trans* isomerism is 2-butene.

Alicyclic compounds can also display *cis/trans* isomerism. As an example of a geometric isomer due to a ring structure, consider 1,2-dichlorocyclohexane:

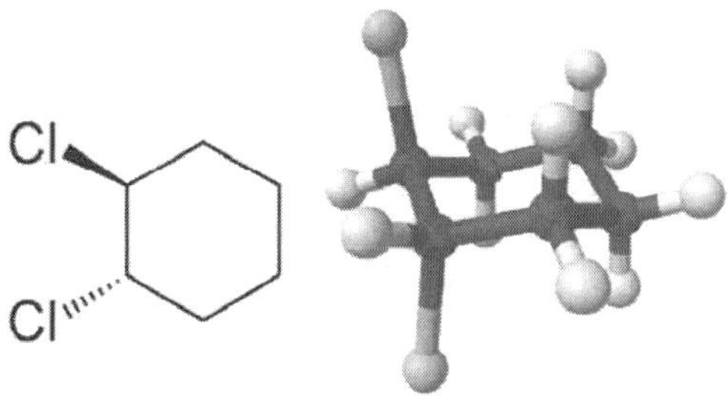

Figure: trans-*1,2-dichlorocyclohexane*

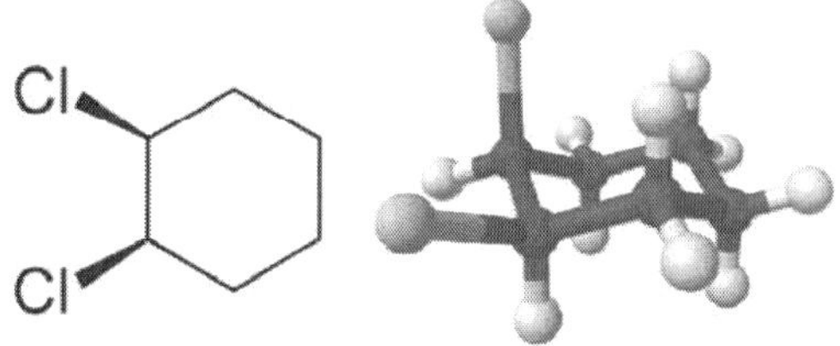

Figure: cis-*1,2-dichlorocyclohexane*

Comparison of Physical Properties

Cis and *trans* isomers often have different physical properties. Differences between isomers, in general, arise from the differences in the shape of the molecule or the overall dipole moment.

These differences can be very small, as in the case of the boiling point of straight-chain alkenes, such as 2-pentene, which is 37°C in the *cis* isomer and 36°C in the *trans* isomer. The differences between *cis* and *trans* isomers can be larger if polar bonds are present, as in the

1,2-dichloroethenes. The *cis* isomer in this case has a boiling point of 60.3°C, while the *trans* isomer has a boiling point of 47.5°C. In the *cis* isomer the two polar C-Cl bond dipole moments combine to give an overall molecular dipole, so that there are intermolecular dipole–dipole forces (or Keesom forces) which add to the London dispersion forces and raise the boiling point. In the *trans* isomer on the other hand, this does not occur because the two C-Cl bond moments cancel and the molecule is non-polar.

The two isomers of butenedioic acid have such large differences in properties and reactivities that they were actually given completely different names.

The *cis* isomer is called maleic acid and the *trans* isomer fumaric acid. Polarity is key in determining relative boiling point as it causes increased intermolecular forces, thereby raising the boiling point. In the same manner, symmetry is key in determining relative melting point as it allows for better packing in the solid state, even if it does not alter the polarity of the molecule.

One example of this is the relationship between oleic acid and elaidic acid; oleic acid, the *cis* isomer, has a melting point of 13.4 degrees Celsius, making it a liquid at room temperature, while the *trans* isomer, elaidic acid, has the much higher melting point of 43 degrees Celsius, due to the straighter *trans* isomer being able to pack more tightly, and is solid at room temperature.

Thus, *trans*-alkenes, which are less polar and more symmetrical, have lower boiling points and higher melting points, and *cis*-alkenes, which are generally more polar and less symmetrical, have higher boiling points and lower melting points.

In the case of geometric isomers that are a consequence of double bonds, and, in particular, when both substituents are the same, some general trends usually hold. These trends can be attributed to the fact that the dipoles of the substituents in a *cis* isomer will add up to give an overall molecular dipole. In a *trans* isomer, the dipoles of the substituents will cancel out due to their being on opposite site of the molecule. *Trans* isomers also tend to have lower densities than their *cis* counterparts.

March observes that *trans* alkenes tend to have higher melting points and lower solubility in inert solvents, as *trans* alkenes, in general, are more symmetrical than *cis* alkenes. Vicinal coupling constants ($^3J_{HH}$), measured by NMR spectroscopy, are larger for *trans*– (range: 12–18 Hz; typical: 15 Hz) than for *cis*– (range: 0–12 Hz; typical: 8 Hz) isomers.

Stability

Usually, *trans* isomers are more stable than *cis* isomers. This is due partly to their shape; the straighter shape of *trans* isomers leads to hydrogen intermolecular forces that make them more stable. According to Jerry March, *trans* isomers also have a lower heat of combustion, indicating higher thermochemical stability. In the Benson heat of formation group additivity dataset, *cis* isomers suffer a 1.10 kcal/mol stability penalty. Exceptions to this rule exist, such as 1,2-difluoroethylene, 1,2-difluorodiazene (FN=NF), and several other halogen- and oxygen-substituted ethylenes. In these cases, the *cis* isomer is more stable than the *trans* isomer. This phenomenon is called the *cis* effect.

E/Z Notation

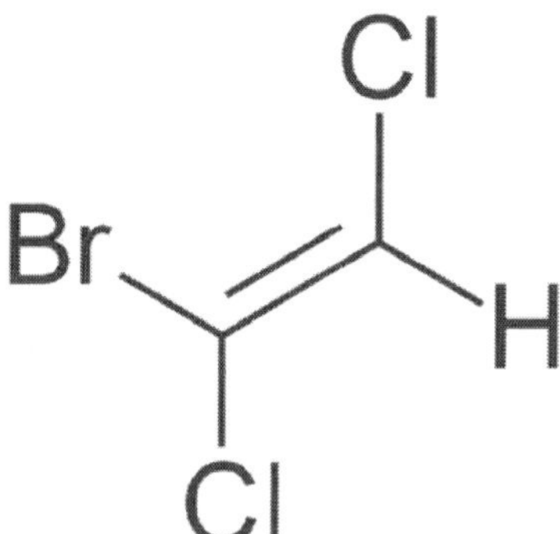

Figure: *Bromine has a higher CIP priority than chlorine, so this alkene is the Z isomer:*

The *cis/trans* system for naming isomers is not effective when there are more than two different substituents on a double bond. The *E/Z* notation should then be used. *Z* (from the German *zusammen*) means "together" and corresponds to the term *cis*; *E* (from the German *entgegen*) means "opposite" and corresponds to *trans*.

Whether a molecular configuration is designated *E* or *Z* is determined by the Cahn-Ingold-Prelog priority rules; higher atomic numbers are given higher priority. For each of the two atoms in the double bond, it is necessary to determine the priority of each substituent. If both the higher-priority substituents are on the same side, the arrangement is *Z*; if on opposite sides, the arrangement is *E*.

Inorganic Coordination Complexes

In inorganic coordination complexes with octahedral or square planar geometries, there are also *cis* isomers in which similar ligands are closer together and *trans* isomers in which they are further apart.

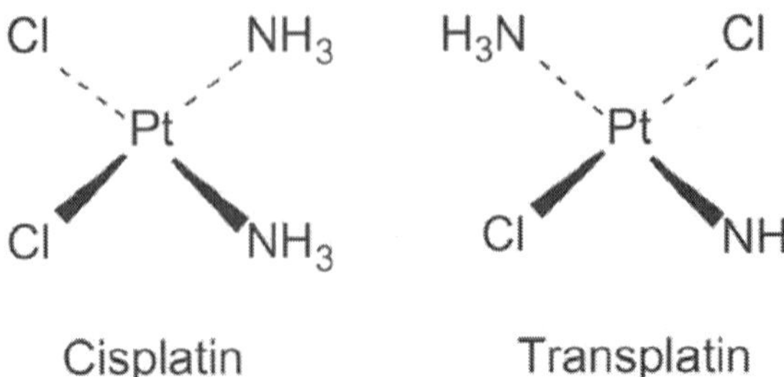

Figure: *The two isomeric complexes, cisplatin and transplatin*

For example, there are two isomers of square planar $Pt(NH_3)_2Cl_2$, as explained by Alfred Werner in 1893. The *cis* isomer, whose full name is *cis*-diamminedichloroplatinum(II), was shown in 1969 by Barnett Rosenberg to have antitumor activity, and is now a chemotherapy drug known by the short name cisplatin. In contrast, the *trans* isomer (transplatin) has no useful anticancer activity. Each isomer can be synthesized using the trans effect to control which isomer is produced.

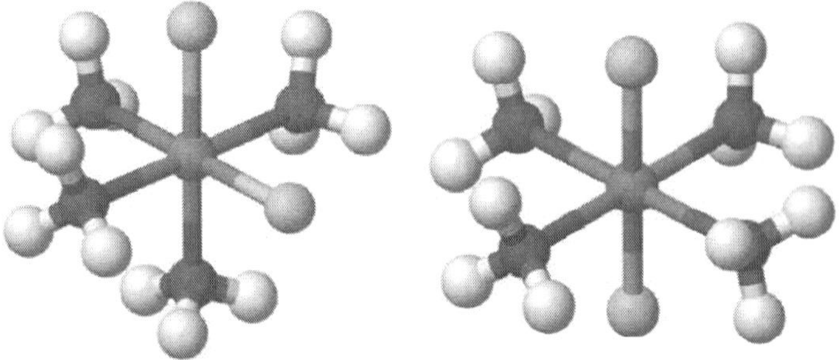

Figure: cis-$[Co(NH_3)_4 Cl_2]^+$ *and* trans-$[Co(NH_3)_4 Cl_2]^+$

For octahedral complexes of formula MX_4Y_2, two isomers also exist. (Here M is a metal atom, and X and Y are two different types of ligands.) In the *cis* isomer, the two Y ligands are adjacent to each other at 90°, as is true for the two chlorine atoms shown in green in *cis*-$[Co(NH_3)_4Cl_2]^+$, at left. In the *trans* isomer shown at right, the two Cl atoms are on opposite sides of the central Co atom. A related type of isomerism in octahedral MX_3Y_3 complexes is facial-meridional (or *fac/mer*) isomerism, in which different numbers of ligands are *cis* or *trans* to each other.

E-Z Notation

E-Z notation, or the *E-Z* convention, is the IUPAC preferred method of describing the stereochemistry of double bonds in organic chemistry. It is an extension of *cis/trans* notation that can be used to describe double bonds having three or four substituents.

Following a set of defined rules (Cahn-Ingold-Prelog priority rules), each substituent on a double bond is assigned a priority.

If the two groups of higher priority are on opposite sides of the double bond, the bond is assigned the configuration *E*.

If the two groups of higher priority are on the same side of the double bond, the bond is assigned the configuration *Z* (from *zusammen*, German: [tsuÈzamYn], the German word for "together").

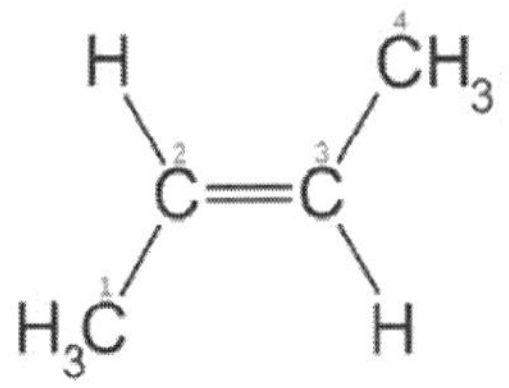

(E)-But-2-ene 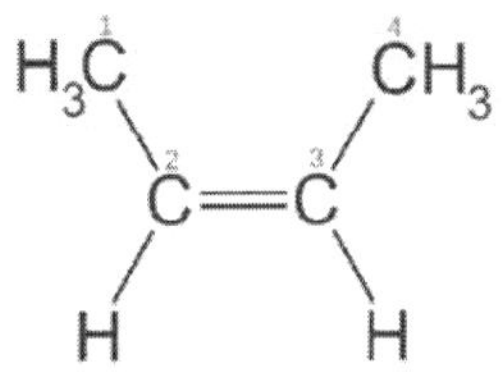 *(Z)-But-2-ene*

The letters *E* and *Z* are conventionally printed in italic type, within parentheses, and separated from the rest of the name with a hyphen. They are always printed as full capitals (not lower case or small capitals), but do not constitute the first letter of the name for English capitalization rules (as in the example above).

Chapter 6

Hydrodealkylation

Hydrodealkylation is a chemical reaction that often involves reacting an aromatic hydrocarbon, such as toluene, in the presence of hydrogen gas to form a simpler aromatic hydrocarbon devoid of functional groups.

An example is the conversion of 1,2,4-trimethylbenzene to xylene. This chemical process usually occurs at high temperature, at high pressure, or in the presence of a catalyst. These are predominantly transition metals, such as chromium or molybdenum.

- Toluene hydrodealkylation to benzene
- Transalkylation.

1,2,4-Trimethylbenzene

1,2,4-Trimethylbenzene is a colourless liquid with chemical formula C_9H_{12}. It is a flammable aromatic hydrocarbon with a strong odor. It occurs naturally in coal tar and petroleum (about 3%). It is nearly insoluble in water, but well-soluble in ethanol, diethyl ether, and benzene.

Industrially, it is isolated from the C_9 aromatic hydrocarbon fraction during petroleum distillation. Approximately 40% of this fraction is 1,2,4-trimethylbenzene.

Uses

1,2,4-Trimethylbenzene dissolved in mineral oil is used as a liquid scintillator. It is also used as a sterilizing agent and in the manufacture of dyes, perfumes, and resins. Another major use is as a gasoline additive.

Benzene

Benzene is a natural constituent of crude oil, and is one of the most basic petrochemicals. Benzene is an aromatic hydrocarbon and the second [*n*]-annulene ([6]-annulene), a cyclic hydrocarbon with a continuous pi bond. It is sometimes abbreviated Ph–H. Benzene is a colourless and highly flammable liquid with a sweet smell. Because it is a known carcinogen, its use as an additive in gasoline is now limited, but it is an important industrial solvent and precursor to basic industrial chemicals including drugs, plastics, synthetic rubber, and dyes.

History

Discovery: The word "benzene" derives historically from "gum benzoin", sometimes called "benjamin" (i.e., benzoin resin), an aromatic resin known to European pharmacists and perfumers since the 15th century as a product of southeast Asia. "Benzoin" is itself a corruption of the Arabic expression "*luban jawi*", or "frankincense of Java". An acidic material was derived from benzoin by sublimation, and named "flowers of benzoin", or benzoic acid. The hydrocarbon derived from benzoic acid thus acquired the name benzin, benzol, or benzene.

Michael Faraday first isolated and identified benzene in 1825 from the oily residue derived from the production of illuminating gas, giving it the name *bicarburet of hydrogen.*

In 1833, Eilhard Mitscherlich produced it via the distillation of benzoic acid (from gum benzoin) and lime. He gave the compound the name *benzin.*

In 1836, the French chemist Auguste Laurent named the substance "phène"; this is the root of the word phenol, which is hydroxylated benzene, and phenyl, which is the radical formed by abstraction of a hydrogen atom (free radical H*) from benzene.

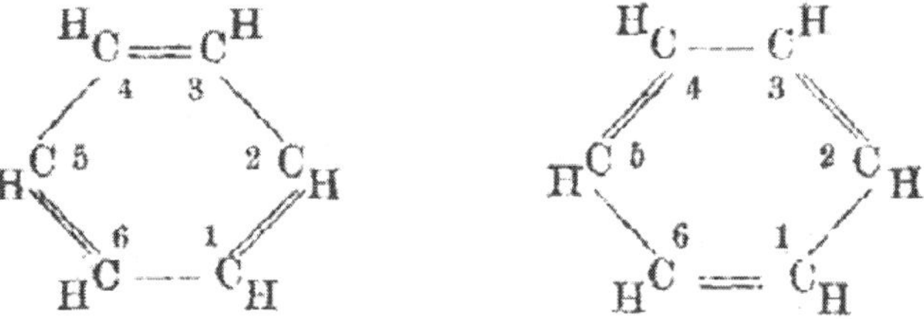

In 1845, Charles Mansfield, working under August Wilhelm von Hofmann, isolated benzene from coal tar. Four years later, Mansfield began the first industrial-scale production of benzene, based on the coal-tar method. Gradually the sense developed among chemists that substances related to benzene represent a diverse chemical family. In 1855 August Wilhelm Hofmann used the word "aromatic" to designate this family relationship, after a characteristic property of many of its members.

Ring formula

The empirical formula for benzene was long known, but its highly polyunsaturated structure, with just one hydrogen atom for each carbon atom, was challenging to determine. Archibald Scott Couper in 1858 and Joseph Loschmidt in 1861 suggested possible structures that contained multiple double bonds or multiple rings, but too little evidence was then available to help chemists decide on any particular structure.

In 1865, the German chemist Friedrich August Kekulé published a paper in French (for he was then teaching in Francophone Belgium) suggesting that the structure contained a six-membered ring of carbon atoms with alternating single and double bonds. The next year he published a much longer paper in German on the same subject. Kekulé used evidence that had accumulated in the intervening years—namely, that there always appeared to be only one isomer of any monoderivative of benzene, and that there always appeared to be exactly three isomers of every diderivative—now understood to correspond to the ortho, meta, and para patterns of arene substitution—to argue in support of his proposed structure. Kekulé's symmetrical ring could explain these curious facts, as well as benzene's 1:1 carbon-hydrogen ratio.

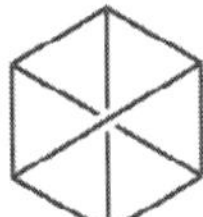     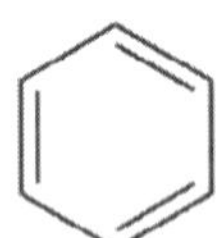

Historic benzene formulae (from left to right) by Claus (1867), Dewar (1867), Ladenburg (1869), Armstrong (1887), Thiele (1899) and Kekulé (1865). Dewar benzene and prismane are different chemicals that have Dewar's and Ladenburg's structures. Thiele and Kekulé's structures are used today.

The new understanding of benzene, and hence of all aromatic compounds, proved to be so important for both pure and applied chemistry that in 1890 the German Chemical Society organized an elaborate appreciation in Kekulé's honour, celebrating the twenty-fifth anniversary of his first benzene paper. Here Kekulé spoke of the creation of the theory. He said that he had discovered the ring shape of the benzene molecule after having a reverie or day-dream of a snake seizing its own tail (this is a common symbol in many ancient cultures known as the Ouroboros or Endless knot). This vision, he said, came to him after years of studying the nature of carbon-carbon bonds. This was 7 years after he had solved the problem of how carbon atoms could bond to up to four other atoms at the same time.

It is curious that a similar, humorous depiction of benzene had appeared in 1886 in the *Berichte der Durstigen Chemischen Gesellschaft* (Journal of the Thirsty Chemical Society), a parody of the *Berichte der Deutschen Chemischen Gesellschaft*, only the parody had monkeys seizing each other in a circle, rather than snakes as in Kekulé's anecdote. Some historians have suggested that the parody was a lampoon of the snake anecdote, possibly already well-known through oral transmission even if it had not yet appeared in print.

Others have speculated that Kekulé's story in 1890 was a re-parody of the monkey spoof, and was a mere invention rather than a recollection of an event in his life. Kekulé's 1890 speech in which these anecdotes appeared has been translated into English. If one takes the anecdote as the memory of a real event, circumstances mentioned in the story suggest that it must have happened early in 1862.

The cyclic nature of benzene was finally confirmed by the crystallographer Kathleen Lonsdale in 1929.

Structure

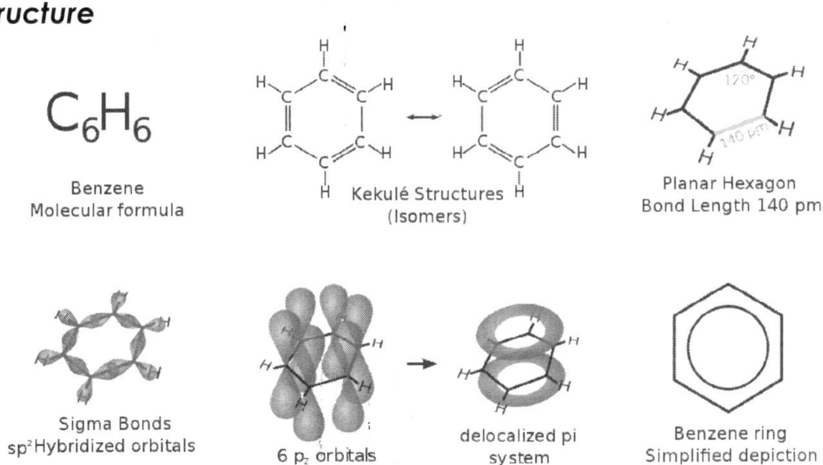

Figure: *The various representations of benzene*

Benzene represents a special problem in that, to account for all the bonds, there must be alternating double carbon bonds:

X-ray diffraction shows that all of six carbon-carbon bonds in benzene are of the same length of 140 picometres (pm). The C–C bond lengths are greater than a double bond (135 pm) but shorter than a single bond (147 pm). This intermediate distance is consistent with electron delocalization: the electrons for C–C bonding are distributed equally between each of the six carbon atoms. The molecule is planar. One representation is that the structure exists as a superposition of so-called resonance structures, rather than either form individually. The delocalization of electrons is one explanation for the thermodynamic stability of benzene and related aromatic compounds. It is likely that this stability contributes to the peculiar molecular and chemical properties known as aromaticity. To indicate the delocalized nature of the bonding, benzene is often depicted with a circle inside a hexagonal arrangement of carbon atoms:

The delocalized picture of benzene has been contested by Cooper, Gerratt and Raimondi in their article published in 1986 in the journal Nature. They showed that the electrons in benzene are almost certainly localized, and the aromatic properties of benzene originate from spin coupling rather than electron delocalization. This view has been supported in the next-year Nature issue, but it has been slow to permeate the general chemistry community.

As is common in organic chemistry, the carbon atoms in the diagram above have been left unlabelled. Realizing each carbon has 2p electrons, each carbon donates an electron into the delocalized ring above and below the benzene ring. It is the side-on overlap of p-orbitals that produces the pi clouds.

Derivatives of benzene occur sufficiently often as a component of organic molecules that there is a Unicode symbol in the Miscellaneous Technical block with the code U+232C (,#) to represent it with three double bonds, and U+23E3 (γ#) for a delocalized version.

Substituted Benzene Derivatives

Many important chemical compounds are derived from benzene by replacing one or more of its hydrogen atoms with another functional group. Examples of simple benzene derivatives are phenol, toluene, and aniline, abbreviated PhOH, PhMe, and $PhNH_2$, respectively. Linking benzene rings gives biphenyl, C_6H_5–C_6H_5. Further loss of hydrogen gives "fused" aromatic hydrocarbons, such as naphthalene and anthracene. The limit of the fusion process is the hydrogen-free allotrope of carbon, graphite.

In heterocycles, carbon atoms in the benzene ring are replaced with other elements. The most important derivatives are the rings containing nitrogen. Replacing one CH with N gives the compound pyridine, C_5H_5N. Although benzene and pyridine are *structurally* related, benzene cannot be converted into pyridine. Replacement of a second CH bond with N gives, depending on the location of the second N, pyridazine, pyrimidine, and pyrazine.

Production

Four chemical processes contribute to industrial benzene production: catalytic reforming, toluene hydrodealkylation, toluene disproportionation, and steam cracking. According to the ATSDR Toxicological Profile for benzene, between 1978 and 1981, catalytic reformats accounted for approximately 44–50% of the total U.S benzene production.

Until World War II, most benzene was produced as a by-product of coke production (or "coke-oven light oil") in the steel industry. However, in the 1950s, increased demand for benzene, especially from the growing polymers industry, necessitated the production of benzene from petroleum. Today, most benzene comes from the petrochemical industry, with only a small fraction being produced from coal.

Catalytic Reforming

In catalytic reforming, a mixture of hydrocarbons with boiling points between 60–200 °C is blended with hydrogen gas and then exposed to a bifunctional platinum chloride or rhenium chloride catalyst at 500–525 °C and pressures ranging from 8–50 atm. Under these conditions, aliphatic hydrocarbons form rings and lose hydrogen to become aromatic hydrocarbons. The aromatic products of the reaction are then separated from the reaction mixture (or reformate) by extraction with any one of a number of solvents, including diethylene glycol or sulfolane, and benzene is then separated from the other aromatics by distillation. The extraction step of aromatics from the reformate is designed to produce aromatics with lowest non-aromatic components. So-called BTX (benzene-toluene-xylene) process consists of such extraction and distillation steps. One such widely used process from UOP was licensed to producers and called the Udex process.

In similar fashion to this catalytic reforming, UOP and BP commercialized a method from LPG (mainly propane and butane) to aromatics.

Toluene Hydrodealkylation

Toluene hydrodealkylation converts toluene to benzene. In this hydrogen-intensive process, toluene is mixed with hydrogen, then passed over a chromium, molybdenum, or platinum oxide catalyst at 500–600 °C and 40–60 atm pressure. Sometimes, higher temperatures are used instead of a catalyst (at the similar reaction condition). Under these conditions, toluene undergoes dealkylation to benzene and methane:

$$C_6H_5CH_3 + H_2 \rightarrow C_6H_6 + CH_4$$

This irreversible reaction is accompanied by an equilibrium side reaction that produces biphenyl (aka diphenyl) at higher temperature:

$$2\ C_6H_6 \rightleftharpoons H_2 + C_6H_5\text{–}C_6H_5$$

If the raw material stream contains much non-aromatic components (paraffins or naphthenes), those are likely decomposed to lower hydrocarbons such as methane, which increases the consumption of hydrogen.

A typical reaction yield exceeds 95%. Sometimes, xylenes and heavier aromatics are used in place of toluene, with similar efficiency.

This is often called "on-purpose" methodology to produce benzene, compared to conventional BTX (benzene-toluene-xylene) processes.

Toluene Disproportionation

Where a chemical complex has similar demands for both benzene and xylene, then toluene disproportionation (TDP) may be an attractive alternative to the toluene hydrodealkylation. In the broad sense, 2 toluene molecules are reacted and the methyl groups rearranged from one toluene molecule to the other, yielding one benzene molecule and one xylene molecule.

Given that demand for *para*-xylene (*p*-xylene) substantially exceeds demand for other xylene isomers, a refinement of the TDP process called Selective TDP (STDP) may be used. In this process, the xylene stream exiting the TDP unit is approximately 90% paraxylene. In some current catalytic systems, even the benzene-to-xylenes ratio is decreased (more xylenes) when the demand of xylenes is higher.

Steam Cracking

Steam cracking is the process for producing ethylene and other alkenes from aliphatic hydrocarbons. Depending on the feedstock used to produce the olefins, steam cracking can produce a benzene-rich liquid by-product called *pyrolysis gasoline*. Pyrolysis gasoline can be blended with other hydrocarbons as a gasoline additive, or distilled (in BTX process) to separate it into its components, including benzene.

Other Sources

Trace amounts of benzene may result whenever carbon-rich materials undergo incomplete combustion. It is produced in volcanoes and forest fires, and is also a component of cigarette smoke. Benzene is a principal product from the combustion of PVC (polyvinyl chloride).

Uses

Early Uses: In the 19th and early-20th centuries, benzene was used as an after-shave lotion because of its pleasant smell. Prior to the 1920s, benzene was frequently used as an industrial solvent, especially for degreasing metal. As its toxicity became obvious, benzene was supplanted by other solvents, especially toluene (methyl benzene), which has similar physical properties but is not as carcinogenic.

In 1903, Ludwig Roselius popularized the use of benzene to decaffeinate coffee. This discovery led to the production of Sanka (the letters "ka" in the brand name stand for *kaffein*). This process was later discontinued.

Benzene was historically used as a significant component in many consumer products such as Liquid Wrench, several paint strippers, rubber cements, spot removers and other hydrocarbon-containing products. Some ceased manufacture of their benzene-containing formulations in about 1950, while others continued to use benzene as a component or significant contaminant until the late 1970s when leukemia deaths were found associated with Goodyear's Pliofilm production operations in Ohio.

Until the late 1970s, many hardware stores, paint stores, and other retail outlets sold benzene in small cans, such as quart size, for general-purpose use. Many students were exposed to benzene in school and university courses while performing laboratory experiments with little or no ventilation in many cases. This very dangerous practice has been almost totally eliminated.

As a gasoline (petrol) additive, benzene increases the octane rating and reduces knocking. As a consequence, gasoline often contained several percent benzene before the 1950s, when tetraethyl lead replaced it as the most widely-used antiknock additive.

With the global phaseout of leaded gasoline, benzene has made a comeback as a gasoline additive in some nations. In the United States, concern over its negative health effects and the possibility of benzene's entering the groundwater have led to stringent regulation of gasoline's benzene content, with limits typically around 1%. European petrol specifications now contain the same 1% limit on benzene content. The United States Environmental Protection Agencyý has new regulations that will lower the benzene content in gasoline to 0.62% in 2011.

Current Uses

Today, benzene is used mainly as an intermediate to make other chemicals. Its most widely-produced derivatives include styrene, which is used to make polymers and plastics, phenol for resins and adhesives (via cumene), and cyclohexane, which is used in the manufacture of Nylon. Smaller amounts of benzene are used to make some types of rubbers, lubricants, dyes, detergents, drugs, explosives, napalm, and pesticides.

In both the US and Europe, 50% of benzene is used in the production of ethylbenzene/styrene, 20% is used in the production of cumene, and about 15% of benzene is used in the production of cyclohexane (eventually to nylon).

In laboratory research, toluene is now often used as a substitute for benzene. The solvent-properties of the two are similar, but toluene is less toxic and has a wider liquid range.

Benzene has been used as a basic research tool in a variety of experiments including analysis of a two-dimensional gas.

Reactions

The most common reactions that benzene undergoes are substitution reactions.

- Electrophilic aromatic substitution is a general method of derivatizing benzene. Benzene is sufficiently nucleophilic that it undergoes substitution by acylium ions or alkyl carbocations to give substituted derivatives.

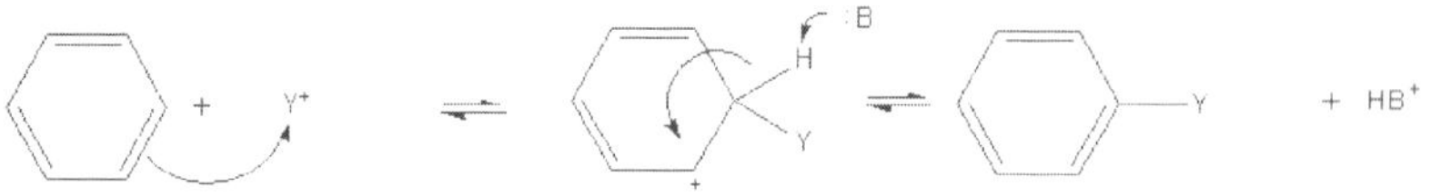

Figure 2: *Electrophilic aromatic substitution of benzene*

- The Friedel-Crafts alkylation involves the alkylation of benzene (and many other aromatic rings) using an alkyl halide in the presence of a strong Lewis acid catalyst.

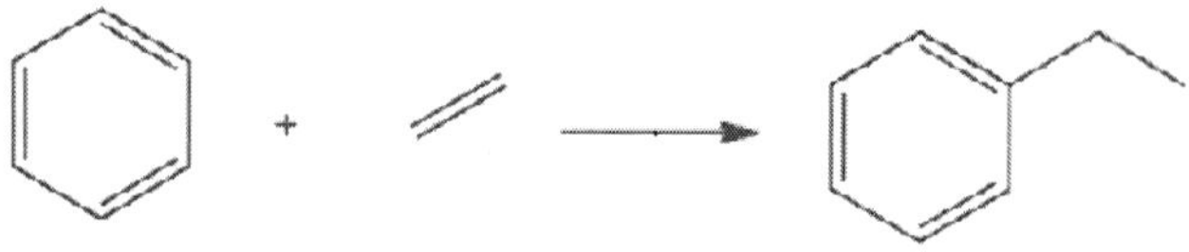

- The Friedel-Crafts acylation is a specific example of electrophilic aromatic substitution. The reaction involves the acylation of benzene (or many other aromatic rings) with an acyl chloride using a strong Lewis acid catalyst such as aluminium chloride or Iron(III) chloride, which act as a halogen carrier.

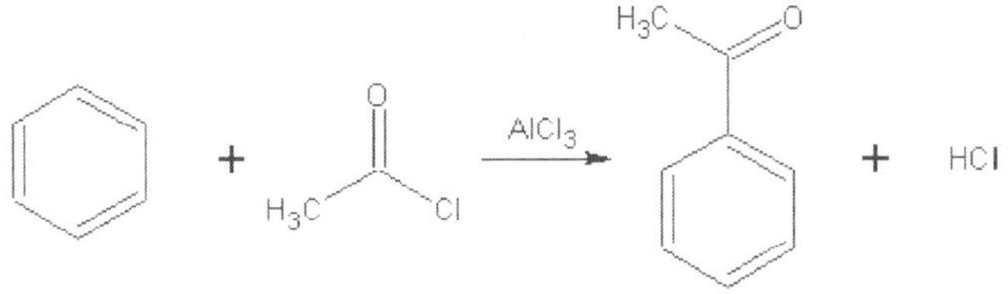

Figure 3: *Friedel-Crafts acylation of benzene by acetyl chloride*

- Sulfonation. The most common method involves mixing sulfuric acid with sulfate, a mixture called fuming sulfuric acid. The sulfuric acid protonates the sulfate, giving the sulfur atom a permanent, rather than resonance stabilized positive formal charge. This molecule is very electrophillic and Electrophillic Aromatic Substitution then occurs.
- Nitration: Benzene undergoes nitration with nitronium ions (NO_2^+) as the electrophile. Thus, warming benzene at 50–55 °C, with a combination of concentrated sulfuric and nitric acid to produce the electrophile, gives nitrobenzene.
- Hydrogenation (reduction): Benzene and derivatives convert to cyclohexane and derivatives when treated with hydrogen at 450 K and 10 atm of pressure with a finely divided nickel catalyst.
- Benzene is an excellent ligand in the organometallic chemistry of low-valent metals. Important examples include the sandwich and half-sandwich complexes, respectively, $Cr(C_6H_6)_2$ and $[RuCl_2(C_6H_6)]_2$.

Environmental Transformation

Even if it is not a common substrate for the metabolism of organisms, benzene could be oxidized by both bacteria and eukaryotes.

In bacteria, dioxygenase enzyme can add an oxygen molecule to the ring, and the unstable product is immediately reduced (by NADH) to a cyclic diol with two double bonds, breaking the aromaticity. Next, the diol is newly reduced by NADH to catechol.

The catechol is then metabolized to acetyl CoA and succinyl CoA, used by organisms mainly in the Krebs Cycle for energy production.

Health Effects

Benzene causes cancer and other illnesses. Benzene is a "notorious cause" of bone marrow failure. "Vast quantities of epidemiologic, clinical, and laboratory data" link benzene to aplastic anemia, acute leukemia, and bone marrow abnormalities.

The American Petroleum Institute (API) stated in 1948 that "it is generally considered that the only absolutely safe concentration for benzene is zero." The US Department of Health and Human Services (DHHS) classifies benzene as a human carcinogen. Long-term exposure to excessive levels of benzene in the air causes leukemia, a potentially fatal cancer of the blood-forming organs, in susceptible individuals. In particular, Acute myeloid leukemia or acute non-lymphocytic leukaemia (AML & ANLL) is not disputed to be caused by benzene. IARC rated benzene as "known to be carcinogenic to humans" (Group 1).

Outdoor air may contain low levels of benzene from automobile service stations, wood smoke, tobacco smoke, the transfer of gasoline, exhaust from motor vehicles, and industrial emissions. About 50% of the entire nationwide (United States) exposure to benzene results from smoking tobacco or from exposure to tobacco smoke.

Vapors from products that contain benzene, such as glues, paints, furniture wax, and detergents, can also be a source of exposure, although many of these have been modified or reformulated since the late 1970s to eliminate or reduce the benzene content. Air around hazardous waste sites or gas stations may contain higher levels of benzene. Because petroleum hydrocarbon products are complex mixtures of chemicals, risk assessments for these products, in general, focus on specific toxic constituents. The petroleum constituents of primary interest to human health have been the aromatic hydrocarbons (i.e., benzene, ethylbenzene, toluene, and xylenes). OSHA requires that a mixture "shall be assumed to present a carcinogenic hazard if it contains a component in concentrations of 0.1% or greater, which is considered to be a carcinogen.

The short-term breathing of high levels of benzene can result in death; low levels can cause drowsiness, dizziness, rapid heart rate, headaches, tremors, confusion, and unconsciousness. Eating or drinking foods containing high levels of benzene can cause vomiting, irritation of the stomach, dizziness, sleepiness, convulsions, and death.

The major effects of benzene are manifested via chronic (long-term) exposure through the blood. Benzene damages the bone marrow and can cause a decrease in red blood cells, leading to anemia. It can also cause excessive bleeding and depress the immune system, increasing the chance of infection. Benzene causes leukemia and is associated with other blood cancers and pre-cancers of the blood.

Human exposure to benzene is a global health problem. Benzene targets liver, kidney, lung, heart and the brain and can cause DNA strand breaks, chromosomal damage, etc. Benzene causes cancer in both animals and humans. Benzene was first reported to induce cancer in humans in the 1920s. The chemical industry claims it was not until 1979 that the cancer-inducing properties were determined "conclusively" in humans, despite many references to this fact in the medical literature. Industry exploited this "discrepancy" and tried to discredit animal studies that showed that benzene causes cancer, saying that they are not relevant to humans. Benzene has been shown to cause cancer in both sexes of multiple species of laboratory animals exposed via various routes.

Some women having breathed high levels of benzene for many months had irregular menstrual periods and a decrease in the size of their ovaries. Benzene exposure has been linked directly to the neural birth defects spina bifida and anencephaly. Men exposed to high levels of benzene are more likely to have an abnormal amount of chromosomes in their sperm, which impacts fertility and fetal development.

Animal studies have shown low birth weights, delayed bone formation, and bone marrow damage when pregnant animals breathed benzene. Benzene has been connected to a rare form of kidney cancer in two separate studies, one involving tank truck drivers, and the other involving seamen on tanker vessels, both carrying benzene-laden chemicals.

Exposure to Benzene

Workers in various industries that make or use benzene may be at risk for being exposed to high levels of this carcinogenic chemical. Industries that involve the use of benzene include the rubber industry, oil refineries, coke and chemical plants, shoe manufacturers, and gasoline-related industries. Downstream petroleum industry operations include the following categories: refinery, pipeline, marine, rail, bulk terminals and trucks, service stations, underground storage tanks, tank cleaning, and site characterization and remediation.

Exposure of the general population to benzene mainly occurs through breathing, the major sources of benzene being tobacco smoke (about 50%) as well as automobile service stations, exhaust from motor vehicles and industrial emissions (about 20% altogether). Vapors (or gases) from products that contain benzene, such as glues, paints, furniture wax, and detergents, can also be a source of exposure. The average smoker (32 cigarettes per day) takes in about 1.8 milligrams (mg) of benzene per day. This amount is about 10 times the average daily intake of benzene by nonsmokers.

Figure: *Light refraction of benzene (above) and water (below)*

In 1987, OSHA estimated that about 237,000 workers in the United States were potentially exposed to benzene, but it is not known if this number has substantially changed since then.

Water and soil contamination are important pathways of concern for transmission of benzene contact. In the US alone, there are approximately 100,000 different sites that have benzene soil or groundwater contamination.

In 2005, the water supply to the city of Harbin in China with a population of almost nine million people, was cut off because of a major benzene exposure. Benzene leaked into the Songhua River, which supplies drinking water to the city, after an explosion at a China National Petroleum Corporation (CNPC) factory in the city of Jilin on 13 November.

In March 2006, the official Food Standards Agency in Britain conducted a survey of 150 brands of soft drinks. It found that four contained benzene levels above World Health Organization limits. The affected batches were removed from sale.

Benzene Exposure Limits

The United States Environmental Protection Agency has set a maximum contaminant level (MCL) for benzene in drinking water at 0.005 mg/L (5 ppb), as promulgated via the National Primary Drinking Water Regulations. This regulation is based on preventing benzene leukemogenesis. The maximum contaminant level goal (MCLG), a nonenforceable health goal that would allow an adequate margin of safety for the prevention of adverse effects, is zero benzene concentration in drinking water. The EPA requires that spills or accidental releases into the environment of 10 pounds (4.5 kg) or more of benzene be reported to the EPA.

The US Occupational Safety and Health Administration (OSHA) has set a permissible exposure limit of 1 part of benzene per million parts of air (1 ppm) in the workplace during an 8-hour workday, 40-hour workweek. The short term exposure limit for airborne benzene is 5 ppm for 15 minutes. These legal limits were based on studies demonstrating compelling evidence of health risk to workers exposed to benzene. The risk from exposure to 1 ppm for a working lifetime has been estimated as 5 excess leukemia deaths per 1,000 employees exposed. (This estimate assumes no threshold for benzene's carcinogenic effects.) OSHA has also established an action level of 0.5 ppm to encourage even lower exposures in the workplace.

The National Institute for Occupational Safety and Health (NIOSH) revised the Immediately Dangerous to Life or Health (IDLH) concentration for benzene to 500 ppm. The current NIOSH definition for an IDLH condition, as given in the NIOSH Respirator Selection Logic, is one that poses a threat of exposure to airborne contaminants when that exposure is likely to cause death or immediate or delayed permanent adverse health effects or prevent escape from such an environment [NIOSH 2004].

The purpose of establishing an IDLH value is (1) to ensure that the worker can escape from a given contaminated environment in the event of failure of the respiratory protection equipment and (2) is considered a maximum level above which only a highly reliable breathing apparatus providing maximum worker protection is permitted [NIOSH 2004]. In September 1995, NIOSH issued a new policy for developing recommended exposure limits (RELs) for substances, including carcinogens. Because benzene can cause cancer, NIOSH recommends that all workers wear special breathing equipment when they are likely to be exposed to benzene at levels exceeding the REL (10-hour) of 0.1 ppm. The NIOSH STEL (15 min) is 1 ppm.

American Conference of Governmental Industrial Hygienists (ACGIH) adopted Threshold Limit Values (TLVs) for benzene at 0.5 ppm TWA and 2.5 ppm STEL.

Under New Jersey's Right-to-Know law, respiratory protection for benzene is discussed. As stated, improper use of respirators is dangerous. Respirators should only be used when there is a written respiratory program in place as described in the OSHA Respiratory Protection Standard (29 CFR 1910.134). The employer is to develop and implement a written respiratory protection program with required worksite-specific procedures and elements for required respirator use. This program must be administered by a suitably trained program administrator. For benzene:

- If there is a potential for exposure to 0.1 ppm, a NIOSH-approved half face respirator must be worn with an organic vapor cartridge (APF 10).
- If there is a potential for exposure to 0.5 ppm, a NIOSH-approved full face respirator must be worn with an organic vapor cartridge (APF 50).
- Where the potential exists for exposure to over 5 ppm, a NIOSH-approved air-supplied respirator with a full facepiece operated under pressure-demand or other positive-pressure mode must be used (APF 1,000).

Exposure Monitoring

Airborne exposure monitoring for benzene must be conducted in order to properly assess personal exposures and effectiveness of engineering controls. Initial exposure monitoring should be conducted by an industrial hygienist or person specifically trained and experienced in sampling techniques. Contact an AIHA Accredited Laboratory for advice on sampling methods.

Each employer with a place of employment where occupational exposures to benzene occur shall monitor each of these workplaces and work operations to determine accurately the airborne concentrations of benzene to which employees may be exposed. Representative 8-hour TWA employee exposures need to be determined on the basis of one sample or samples representing the full shift exposure for each job classification in each work area.

Unless air samples are taken frequently, the employer does not know the concentration and would not know how much of a protection factor is needed.

In providing consultation on work safety during oil clean-up operations following the Deepwater Horizon accident, OSHA has worked with a number of other government agencies to protect Gulf cleanup workers. OSHA partnered with the NIOSH to issue "Interim Guidance for Protecting Deepwater Horizon Response Workers and Volunteers" and recommend measures that should be taken to protect workers from a variety of different health hazards that these workers face. OSHA conceded that it recognizes that most of its PELs are outdated and inadequate measures of worker safety. In characterizing worker exposure, OSHA instead relies on more up-to-date recommended protective limits set by organizations such as NIOSH, the ACGIH, and the American Industrial Hygiene Association (AIHA), and not on the older, less protective PELS. Results of air monitoring are compared to the lowest known Occupational Exposure Limit for the listed contaminant for purposes of risk assessment and protective equipment recommendations.

Biomarkers of Exposure

Several tests can determine exposure to benzene. Benzene itself can be measured in breath, blood or urine, but such testing is usually limited to the first 24 hours post-exposure due to the relatively rapid removal of the chemical by exhalation or biotransformation. Most persons in developed countries have measureable baseline levels of benzene and other aromatic petroleum hydrocarbons in their blood. In the body, benzene is enzymatically converted to a series of oxidation products including muconic acid, phenylmercapturic acid, phenol, catechol, hydroquinone and 1,2,4-trihydroxybenzene. Most of these metabolites have some value as biomarkers of human exposure, since they accumulate in the urine in proportion to the extent and duration of exposure, and they may still be present for some days after exposure has ceased. The current ACGIH biological exposure limits for occupational exposure are 500 μg/g creatinine for muconic acid and 25 μg/g creatinine for phenylmercapturic acid in an end-of-shift urine specimen.

Methods of Excretion

Most inhaled benzene is not metabolized. Inhaled benzene is primarily expelled unchanged through exhalation. In a human study 16.4 to 41.6% of retained benzene was eliminated through the lungs within five to seven hours after a two- to three-hour exposure to 47 to 110 ppm and only 0.07 to 0.2% of the remaining benzene was excreted unchanged in the urine. After exposure to 63 to 405 mg/m3 of benzene for 1 to 5 hours, 51 to 87% was excreted in the urine as phenol over a period of 23 to 50 hours. In another human study, 30% of absorbed dermally applied benzene, which is primarily metabolized in the liver, was excreted as phenol in the urine.

Molecular Toxicology

The paradigm of toxicological assessment of benzene is slowly shifting towards the domain of molecular toxicology as it allows understanding of fundamental biological mechanisms in a better way. Glutathione seems to play an important role by protecting against benzene-induced DNA breaks and it is being identified as a new biomarker for exposure and effect. Benzene causes chromosomal aberrations in the peripheral blood leukocytes and bone marrow explaining the higher incidence of leukemia and multiple myeloma caused by chronic exposure. These aberrations can be monitored using fluorescent in situ hybridization (FISH) with DNA probes to assess the effects of benzene along with the hematological tests as markers of hematotoxicity. Benzene metabolism involves enzymes coded for by polymorphic genes. Studies have shown that genotype at these loci may influence susceptibility to the toxic effects of benzene exposure. Individuals carrying variant of NAD(P)H:quinone oxidoreductase 1 (NQO1), microsomal epoxide hydrolase (EPHX) and deletion of the glutathione S-transferase T1 (GSTT1) showed a greater frequency of DNA single-stranded breaks.

Biological Oxidation and Carcinogenic Activity

One way of understanding the carcinogenic effects of benzene is by examining the products of biological oxidation. Pure benzene, for example, oxidizes in the body to produce an epoxide, benzene oxide, which is not excreted readily and can interact with DNA to produce harmful mutations.

Summary

According to the Agency for Toxic Substances and Disease Registry (ATSDR) (2007), benzene is both an anthropogenically produced and naturally occurring chemical from processes that include: volcanic eruptions, wild fires, synthesis of chemicals such as phenol, production of synthetic fibers, and fabrication of rubbers, lubricants, pesticides, medications, and dyes. The major sources of benzene exposure are tobacco smoke, automobile service stations, exhaust from motor vehicles, and industrial emissions; however, ingestion and dermal absorption of benzene can also occur through contact with contaminated water. Benzene is hepatically metabolized and excreted in the urine.

Measurement of air and water levels of benzene is accomplished through collection via activated charcoal tubes, which are then analyzed with a gas chromatograph. The measurement of benzene in humans can be accomplished via urine, blood, and breath tests; however, all of these have their limitations because benzene is rapidly metabolized in the human body into by-products called metabolites.

OSHA regulates levels of benzene in the workplace. The maximum allowable amount of benzene in workroom air during an 8-hour workday, 40-hour workweek is 1 ppm. Because benzene can cause cancer, NIOSH recommends that all workers wear special breathing equipment when they are likely to be exposed to benzene at levels exceeding the recommended (8-hour) exposure limit of 0.1 ppm.Alkoxy group

In chemistry, the alkoxy group is an alkyl (carbon and hydrogen chain) group singular bonded to oxygen thus: R—O. The range of alkoxy groups is great, the simplest being methoxy (CH_3O—). An ethoxy group (CH_3CH_2O—) is found in the organic compound phenetol, $C_6H_5OCH_2CH_3$ which is also known as ethoxy benzene. Related to alkoxy groups are aryloxy groups, which have an aryl group singular bonded to oxygen such as the phenoxy group (C_6H_5O—).

An alkoxy or aryloxy group bonded to an alkyl or aryl (R^1—O—R^2) is an ether. If bonded to H it is an alcohol. An alkoxide (RO^-) is the ionic or salt form; it is a derivative of an alcohol where the proton has been replaced by a metal, typically sodium.

Toluene

Toluene, formerly known as toluol, is a clear, water-insoluble liquid with the typical smell of paint thinners. It is a mono-substituted benzene derivative, i.e., one in which a single hydrogen atom from the benzene molecule has been replaced by a univalent group, in this case CH_3.

It is an aromatic hydrocarbon that is widely used as an industrial feedstock and as a solvent. Like other solvents, toluene is sometimes also used as an inhalant drug for its intoxicating properties; however, inhaling toluene has potential to cause severe neurological harm. Toluene is an important organic solvent, but is also capable of dissolving a number of notable inorganic chemicals such as sulfur.

History

The name *toluene* was derived from the older name *toluol*, which refers to tolu balsam, an aromatic extract from the tropical Colombian tree *Myroxylon balsamum*, from which it was first isolated. It was originally named by Jons Jakob Berzelius.

Chemical Properties

Toluene reacts as a normal aromatic hydrocarbon towards electrophilic aromatic substitution. The methyl group makes it around 25 times more reactive than benzene in such reactions. It undergoes smooth sulfonation to give *p*-toluenesulfonic acid, and chlorination by Cl_2 in the presence of $FeCl_3$ to give ortho and para isomers of chlorotoluene. It undergoes nitration to give ortho and para nitrotoluene isomers, but if heated it can give dinitrotoluene and ultimately the explosive trinitrotoluene (TNT).

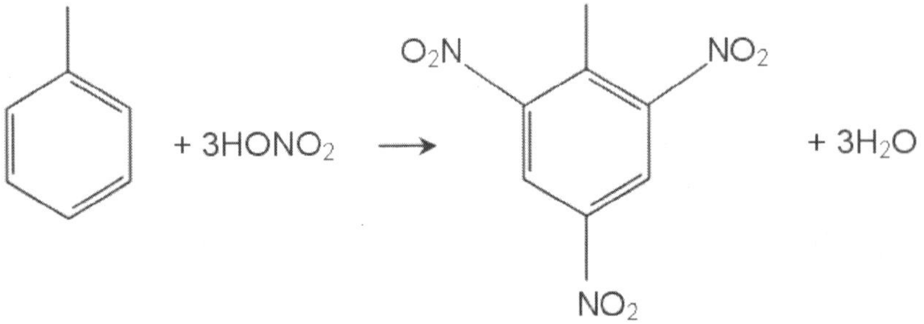

With other reagents the methyl side chain in toluene may react, undergoing oxidation. Reaction with potassium permanganate and diluted acid (e.g., sulfuric acid) or potassium permanganate with concentrated sulfuric acid, leads to benzoic acid, whereas reaction with chromyl chloride leads to benzaldehyde (Étard reaction). Halogenation can be performed under free radical conditions. For example, *N*-bromosuccinimide (NBS) heated with toluene in the presence of AIBN leads to benzyl bromide. Toluene can also be treated with elemental bromine in the presence of UV light (direct sunlight) to yield benzyl bromide. Toluene may also be brominated by treating it with HBr and H_2O_2 in the presence of light .

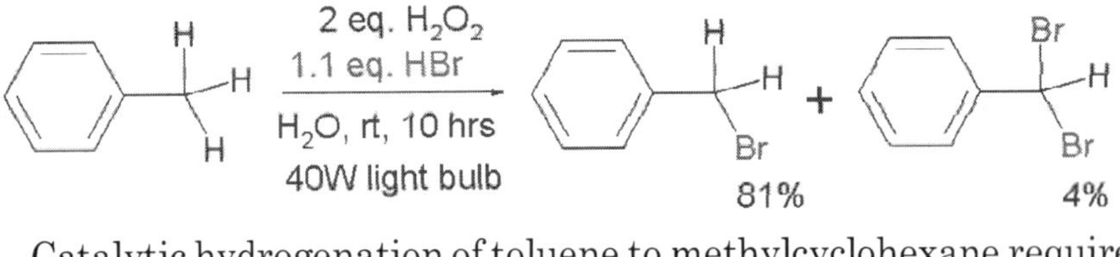

Catalytic hydrogenation of toluene to methylcyclohexane requires a high pressure of hydrogen to go to completion, because of the stability of the aromatic system. pK_a is approximately 45.

Production

Toluene occurs naturally at low levels in crude oil and is usually produced in the processes of making gasoline via a catalytic reformer, in an ethylene cracker or making coke from coal. Final separation (either via distillation or solvent extraction) takes place in a BTX plant.

Preparation

1. *From benzene (Friedel–Crafts reaction):* Benzene reacts with methyl chloride in presence of anhydrous aluminium chloride to form toluene. The formation follows an electrophilic substitution reaction mechanism:

$$CH_3Cl + AlCl_3 \rightarrow CH_3^+ + AlCl_4^-$$

$$C_6H_5H + CH_3^+ + AlCl_4^- \rightarrow C_6H_5CH_3 + HCl + AlCl_3$$

The following catalysts can be used in place of $AlCl_3$:

$$AlCl_3 > SbCl_3 > SnCl_4 > BF_3 > ZnCl_2 > HgCl_2$$

Note that the reaction is not very useful as the mono-alkyl derivative formed readily undergoes further alkylation at a still-greater speed to produce polysubstituted products.

2. *From bromobenzene (Wurtz-Fittig reaction):* The Wurtz-Fittig reaction is the reaction of an aryl halide and alkyl halide in presence of sodium metal to give substituted aromatic compounds.

When bromobenzene and methyl bromide react with sodium metal in dry ether solution, toluene is obtained.

$$C_6H_5Br + CH_3Br + 2Na \rightarrow C_6H_5CH_3 + 2NaBr$$

3. *From toluic acid (decarboxylation):* When sodium salt of toluic acid (o-, m-, p-) is heated with soda lime, toluene is obtained.

$C_6H_4CH_3COONa$ (sodium toluate) + NaOH $\rightarrow$ $C_6H_5CH_3$ (toluene) + Na_2CO_3

4. *From Cresol:* When cresol (o-, m-, p-) is distilled with zinc dust, toluene is obtained.

$C_6H_4CH_3OH$ (cresol) + Zn $\rightarrow$ $C_6H_5CH_3$ (toluene) + ZnO

5. *From Toluenesulfonic Acid:* When toluenesulfonic acid is treated with superheated steam or boiled with HCl, toluene is obtained.

$CH_3C_6H_4SO_3H$ (toluenesulfonic acid) + HOH (steam) $\rightarrow$ $C_6H_5CH_3$ (toluene) + H_2SO_4 (sulfuric acid)

6. *From Toluidine:* Toluidine is first diazotized with sodium nitrite ($NaNO_2$) and HCl at low temperature. The diazonium compound thus obtained is heated with alkaline stannous chloride ($SnCl_2$). This reaction gives toluene.

7. *From Grignard Reagent:* When phenyl magnesium bromide $(C_6H_5)MgBr$ is reacted with methyl bromide, toluene is obtained.
8. *From Aromatic Ketones:* Using a Friedel–Crafts reaction, when an acid chloride is reacted with an aromatic hydrocarbon in the presence of anhydrous aluminium chloride ($AlCl_3$), a mixed aliphatic/aromatic ketone is obtained. The ketone is then reduced with amalgamated zinc and concentrated HCl.

$$C_6H_6 + ClCOCH_3 \rightarrow C_6H_5COCH_3 \text{ (acetophenone)} + HCl$$

$$C_6H_5COCH_3 + 4H \rightarrow C_6H_5CH_2CH_3 \text{ (ethyl benzene)}$$

This method is used to produce alkyl benzenes other than toluene.

Uses

Toluene is a common solvent, able to dissolve paints, paint thinners, silicone sealants, many chemical reactants, rubber, printing ink, adhesives (glues), lacquers, leather tanners, and disinfectants. It can also be used as a fullerene indicator, and is a raw material for toluene diisocyanate (used in the manufacture of polyurethane foam) and TNT. In addition, it is used as a solvent to create a solution of carbon nanotubes. It is also used as a cement for fine polystyrene kits (by dissolving and then fusing surfaces) as it can be applied very precisely by brush and contains none of the bulk of an adhesive.

Industrial uses of toluene include dealkylation to benzene, and the disproportionation to a mixture of benzene and xylene in the BTX process. When oxidized it yields benzaldehyde and benzoic acid, two important intermediates in chemistry. It is also used as a carbon source for making Multi-Wall Carbon Nanotubes. Toluene can be used to break open red blood cells in order to extract hemoglobin in biochemistry experiments.

Toluene can be used as an octane booster in gasoline fuels used in internal combustion engines. Toluene at 86% by volume fueled all the turbo Formula 1 teams in the 1980s, first pioneered by the Honda team. The remaining 14% was a "filler" of n-heptane, to reduce the octane to meet Formula 1 fuel restrictions. Toluene at 100% can be used as a fuel for both two-stroke and four-stroke engines; however, due to the density of the fuel and other factors, the fuel does not vaporize easily unless preheated to 70 degrees Celsius (Honda accomplished this in their Formula 1 cars by routing the fuel lines through the muffler system to heat the fuel). Toluene also poses similar problems as alcohol fuels, as it eats through standard rubber fuel lines and has no lubricating properties, as standard gasoline does, which can break down fuel pumps and cause upper cylinder bore wear. In Australia, toluene has been found to have been illegally combined with petrol in fuel outlets for sale as standard vehicular fuel. Toluene attracts no fuel excise, while other fuels are taxed at over 40%, so fuel suppliers are able to profit from substituting the cheaper toluene for petrol.

This substitution is likely to affect engine performance and result in additional wear and tear. The extent of toluene substitution has not been determined. Toluene is another in a group of fuels that have recently been used as components for jet fuel surrogate blends. Toluene is used as a jet fuel surrogate for its content of aromatic compounds. Toluene has also been used as a coolant for its good heat transfer capabilities in sodium cold traps used in nuclear reactor system loops.

Toluene had also been used in the process of removing the cocaine from coca leaves in the production of Coca-Cola syrup.

Biology

Similar to many other solvents such as 1,1,1-trichloroethane and some alkylbenzenes, toluene has been shown to act as a non-competitive NMDA receptor antagonist and $GABA_A$ receptor positive allosteric modulator. It is abused as an inhalant likely on account of the euphoric and dissociative effects these actions produce. Additionally, toluene has been shown to display antidepressant-like effects in rodents in the forced swim test (FST) and the tail suspension test (TST).

Toxicology and Metabolism

Toluene should not be inhaled due to its health effects. Low to moderate levels can cause tiredness, confusion, weakness, drunken-type actions, memory loss, nausea, loss of appetite, and hearing and colour vision loss. These symptoms usually disappear when exposure is stopped. Inhaling high levels of toluene in a short time may cause light-headedness, nausea, or sleepiness. It can also cause unconsciousness, and even death. Toluene is, however, much less toxic than benzene, and has, as a consequence, largely replaced it as an aromatic solvent in chemical preparation.

Isotoluene

The isotoluenes in organic chemistry are the non-aromatic toluene isomers with an exocyclic double bond. They are of some academic interest in relation to aromaticity and isomerisation mechanisms.

The three basic isotoluenes are *ortho*-isotoluene or *5-methylene-1,3-cyclohexadiene* (here labelled 1); *para*-isotoluene (2); and *meta-*

isotoluene (3). Another structural isomer is the bicyclic compound *5-methylenebicyclo[2.2.0] hexene* (4).

The *o*- and *p*-isotoluenes isomerise to toluene, a reaction driven by aromatic stabilisation. It is estimated that these compounds are 23 kcal/mol less stable.

The isomerisation of *p*-isotoluene to toluene takes place at 100 °C in benzene with bimolecular reaction kinetics by an intermolecular free radical reaction. The intramolecular isomerisation, a 1,3-sigmatropic reaction , is unfavourable because a antarafacial mode is enforced. Other dimer radical reaction products are formed as well.

The *ortho*-isomer is found to isomerise at 60°C in benzene, also in a second order reaction. The proposed reaction mechanism is a concerted intermolecular ene reaction. The reaction product is either toluene or a mixture of dimerized ene reaction products, depending on the exact reaction conditions.

Isomer

In chemistry, isomers (from Greek isomeres; isos = "equal", meros = "part") are compounds with the same molecular formula but different structural formulas. Isomers do not necessarily share similar properties, unless they also have the same functional groups. There are many different classes of isomers, like stereoisomers, enantiomers, geometrical isomers, etc.. There are two main forms of isomerism: structural isomerism and stereoisomerism (spatial isomerism).

Forms

Structural Isomers: In structural isomers, sometimes referred to as *constitutional isomers,* the atoms and functional groups are joined together in different ways. Structural isomers have different IUPAC names and may or may not belong to the same functional group. This group includes chain isomerism whereby hydrocarbon chains have variable amounts of branching; position isomerism which deals with the position of a functional group on a chain; and functional group isomerism in which one functional group is split up into different ones.

For example, two position isomers would be 2-fluoropropane and 1-fluoropropane, illustrated on the right.

In skeletal isomers the main carbon chain is different between the two isomers. This type of isomerism is most identifiable in secondary and tertiary alcohol isomers.

Tautomers are structural isomers of the same chemical substance that spontaneously interconvert with each other, even when pure. They have different chemical properties, and consequently, distinct reactions characteristic to each form are observed. If the interconversion reaction is fast enough, tautomers cannot be isolated from each other. An example is when they differ by the position of a proton, such as in keto/enol tautomerism, where the proton is alternately on the carbon or oxygen.

Stereoisomers

In stereoisomers the bond structure is the same, but the geometrical positioning of atoms and functional groups in space differs.

Diastereomerism is again subdivided into "cis-trans isomers", which have restricted rotation within the molecule (typically isomers containing a double bond) and "conformational isomers" (conformers), which can rotate about one or more single bonds within the molecule.

An obsolete term for "cis-trans isomerism" is "geometric isomers".

For compounds with more than two substituents E-Z notation is used instead of cis and trans. If possible, *E* and *Z* (written in italic type) is also preferred in compounds with two substituents.

In octahedral coordination compounds *fac*- (with facial ligands) and *mer*- (with meridional ligands) isomers occur.

Note that although conformers can be referred to as stereoisomers, they are not stable isomers, since bonds in conformers can easily rotate thus converting one conformer to another which can be either diastereomeric or enantiomeric to the original one.

While structural isomers typically have different chemical properties, stereoisomers behave identically in most chemical reactions, except in their reaction with other stereoisomers. Enzymes however can distinguish between different enantiomers of a compound, and organisms often prefer one isomer over the other. Some stereoisomers also differ in the way they rotate polarized light.

Isomerisation

Isomerisation is the process by which one molecule is transformed into another molecule which has exactly the same atoms, but the atoms are rearranged. In some molecules and under some conditions, isomerisation occurs spontaneously. Many isomers are equal or roughly equal in bond energy, and so exist in roughly equal amounts, provided that they can interconvert relatively freely, that is the energy barrier between the two isomers is not too high. When the isomerisation occurs intramolecularly it is considered a rearrangement reaction.

An example of an organometallic isomerisation is the production of decaphenylferrocene, $[(\eta^5\text{-}C_5Ph_5)_2Fe]$ from its linkage isomer.

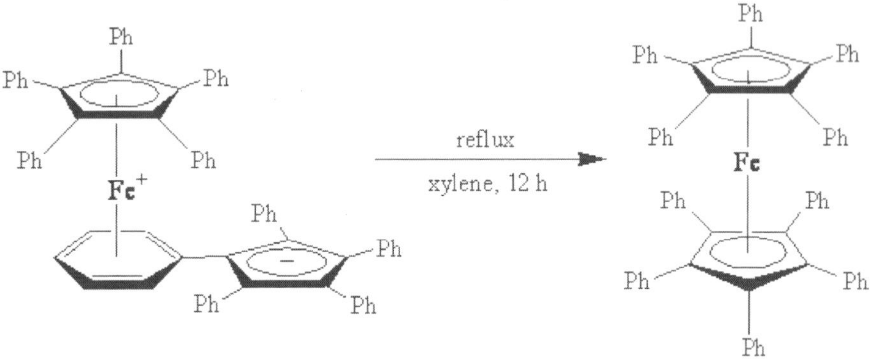

reflux
xylene, 12 h

Instances of Isomerization

- Isomerizations in hydrocarbon cracking. This is usually employed in organic chemistry, where fuels, such as pentane, a straight-chain isomer, are heated in the presence of a platinum catalyst. The resulting mixture of straight- and branched-chain isomers then have to be separated. An industrial process is also the isomerisation of n-butane into isobutane.

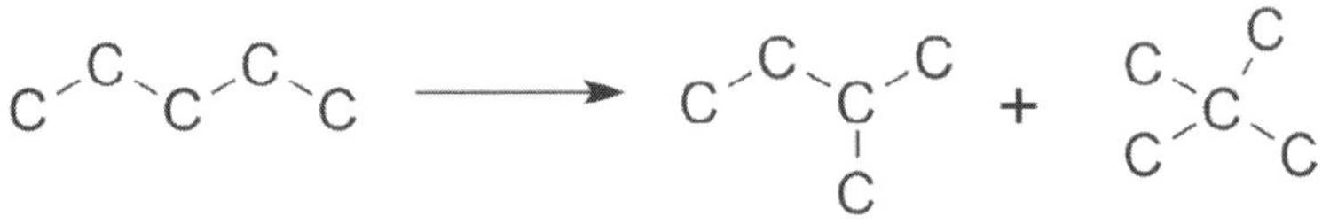

pentane 2-methylbutane 2,2-dimethylpropane

- Trans-cis isomerism. In certain compounds an interconversion of cis and trans isomers can be observed, for instance, with maleic acid and with azobenzene often by photoisomerization. Another example is the photochemical conversion of the trans isomer to the cis isomer of resveratrol :

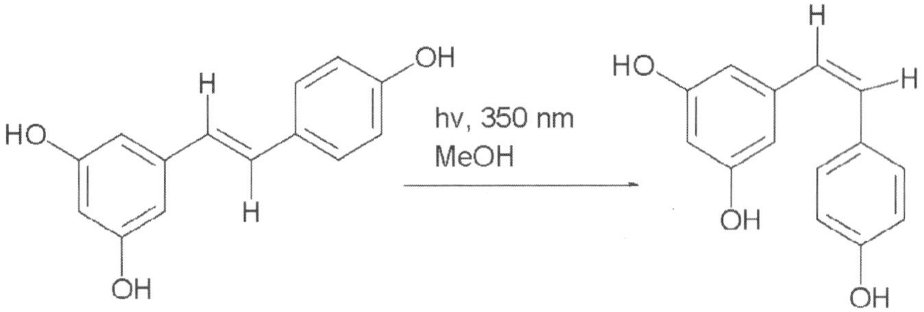

hv, 350 nm
MeOH

- Aldose-ketose isomerism in biochemistry.

- Isomerisations between conformational isomers. These take place without an actual rearrangement for instance inconversion of two cyclohexane conformations
- Fluxional molecules display rapid interconversion of isomers e.g. Bullvalene.
- valence isomerisation: the isomerisation of molecules which involve structural changes resulting only from a relocation of single and double bonds. If a dynamic equilibrium is established between the two isomers it is also referred to as valence tautomerism

The energy difference between two isomers is called isomerisation energy. Isomerisations with low energy difference both experimental and computational (in parentheses) are endothermic trans-cis isomerisation of 2-butene with 2.6 (1.2) kcal/mol, cracking of isopentane to n-pentane with 3.6 (4.0) kcal/mol or conversion of trans-2-butene to 1-butene with 2.6 (2.4) kcal/mol.

Examples

Propanol: A simple example of isomerism is given by propanol: it has the formula C_3H_8O (or C_3H_7OH) and occurs as two isomers: propan-1-ol (n-propyl alcohol; I) and propan-2-ol (isopropyl alcohol; II).

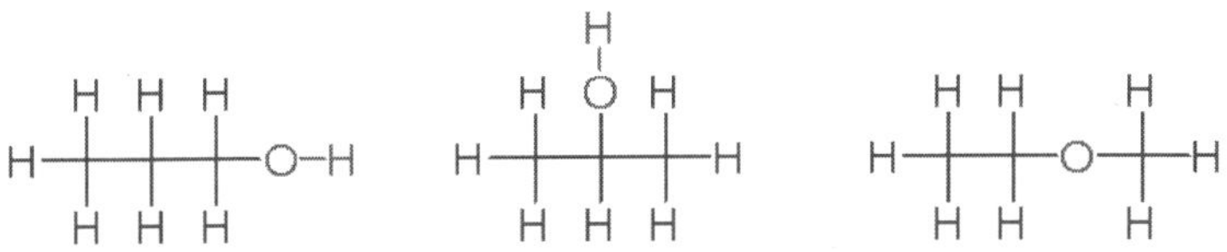

I II III

Note that the position of the oxygen atom differs between the two: it is attached to an end carbon in the first isomer, and to the center carbon in the second.

There is, however, another isomer of C_3H_8O which has significantly different properties: methoxyethane (methyl-ethyl-ether; III). Unlike the isomers of propanol, methoxyethane has an oxygen connected to two carbons rather than to one carbon and one hydrogen. This makes it an ether, not an alcohol, as it lacks a hydroxyl group, and has chemical properties more similar to other ethers than to either of the above alcohol isomers.

Examples of isomers having different medical properties can be easily found. For example, in the placement of methyl groups. In substituted xanthines, Theobromine, found in chocolate, is a vasodilator with some effects in common with caffeine, but if one of the two methyl groups is moved to a different position on the two-ring core, the isomer is theophylline, which has a variety of effects, including bronchodilation and anti-inflammatory action.

Another example of this occurs in the phenethylamine-based stimulant drugs. Phentermine is a non-chiral compound with a weaker effect than amphetamine. It is used as an appetite reducing medication and has mild or no stimulant properties. However, a different atomic arrangement gives dextromethamphetamine which is a stronger stimulant than amphetamine.

Allene and propyne are examples of isomers containing different bond types. Allene contains two double bonds, whereas propyne contains one triple bond.

Synthesis of Gumaric Acid

Industrial synthesis of fumaric acid proceeds via the cis-trans isomerization of maleic acid:

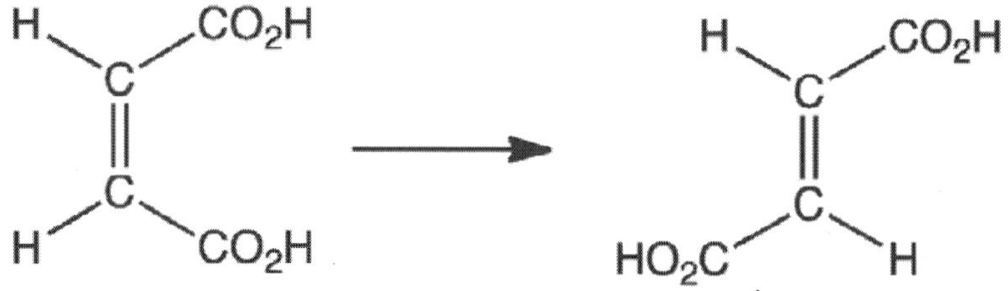

$$\begin{matrix} H & & CO_2H \\ & C & \\ & \| & \\ & C & \\ H & & CO_2H \end{matrix} \longrightarrow \begin{matrix} H & & CO_2H \\ & C & \\ & \| & \\ & C & \\ HO_2C & & H \end{matrix}$$

In medicinal chemistry and biochemistry, enantiomers are a special concern because they may possess quite different biological activity. The infamous case of thalidomide arose from the effects of the unwanted enantiomer. Many preparative procedure afford a mixture of equal amounts of both enantiomeric forms. In some cases, the enantiomers are separated by chromatography using chiral stationary phases. In other cases, enantioselective syntheses have been developed.

History

Isomerism was first noticed in 1827, when Friedrich Woehler prepared cyanic acid and noted that although its elemental composition was identical to fulminic acid (prepared by Justus von Liebig the previous year), its properties were quite different. This finding challenged the prevailing chemical understanding of the time, which held that chemical compounds could be different only when they had different elemental compositions. After additional discoveries of the same sort were made, such as Woehler's 1828 discovery that urea had the same atomic composition as the chemically distinct ammonium cyanate, Jons Jakob Berzelius introduced the term *isomerism* to describe the phenomenon. The individual molecules of each were the left and right optical stereoisomers, solutions of which rotate the plane of polarized light to the same degree but in opposite directions.

Other Types of Isomerism

Other types of isomerism exist outside this scope. Topological isomers called topoisomers are generally large molecules that wind about and form different shaped knots or loops. Molecules with topoisomers include catenanes and DNA. Topoisomerase enzymes can knot DNA and thus change its topology. There are also isotopomers or isotopic isomers that have the same numbers of each type of isotopic substitution but in chemically different positions. In nuclear physics, nuclear isomers are excited states of atomic nuclei. Spin isomers have differing distributions of spin among their constituent atoms.

Induced Gamma Emission

In physics, induced gamma emission (IGE) refers to the process of fluorescent emission of gamma rays from excited nuclei, usually involving a specific nuclear isomer. It is analogous to conventional fluorescence, which is defined as the emission of a photon (unit of light) by an excited electron in an atom or molecule. In the case of IGE, nuclear isomers can store significant amounts of excitation energy for times long enough for them to serve as nuclear fluorescent materials. There are over 800 known nuclear isomers but almost all are too intrinsically radioactive to be considered for applications. As of 2006 there were five proposed nuclear isomers that appeared to be physically capable of IGE fluorescence in safe arrangements: tantalum-180m, osmium-187m, platinum-186m, hafnium-178m2 and zinc-66m.

History

Induced gamma emission is an example of interdisciplinary research bordering on both nuclear physics and quantum electronics. Viewed as a nuclear reaction it would belong to a class in which only photons were involved in creating and destroying states of nuclear excitation. It is a class usually overlooked in traditional discussions. In 1939 Pontecorvo and Lazard reported the first example of this type of reaction. Indium was the target and in modern terminology describing nuclear reactions it would be written $^{115}In(\gamma,\gamma')^{115m}In$. The product nuclide carries an "m" to denote that it has a long enough half life (4.5 hr in this case) to qualify as being a nuclear isomer. That is what made the experiment possible in 1939 because the researchers had hours to remove the products from the irradiating environment and then to study them in a more appropriate location.

With projectile photons, momentum and energy can be conserved only if the incident photon, X-ray or gamma, has precisely the energy corresponding to the difference in energy between the initial state of the target nucleus and some excited state that is not too different in terms of quantum properties such as spin. There is no threshold behaviour and the incident projectile disappears and its energy is transferred into internal excitation of the target nucleus. It is a resonant process that is uncommon in nuclear reactions but normal in the excitation of fluorescence at the atomic level. Only as recently as 1988 was the resonant nature of this type of reaction finally proven. Such resonant reactions are more readily described by the formalities of atomic fluorescence and further development was facilitated by an interdisciplinary approach of IGE.

There is little conceptual difference in an IGE experiment when the target is a nuclear isomer. Such a reaction as $^{m}X(\gamma,\gamma')X$ where ^{m}X is one of the five candidates listed above, is only different because there are lower energy states for the product nuclide to enter after the reaction than there were at the start. Practical difficulties arise from the need to ensure safety from the spontaneous radioactive decay of nuclear isomers in quantities sufficient for experimentation. Lifetimes must be long enough that doses from the spontaneous decay from the targets always remain within safe limits. In 1988 Collins and coworkers reported the first excitation of IGE from a nuclear isomer. They excited fluorescence from the nuclear isomer tantalum-180m with x-rays produced by an external beam radiotherapy "linac". Results were surprising and considered to be controversial until the resonant states excited in the target were identified. Fully independent confirmation was reported by the Stuttgart Nuclear Group in 1999.

Distinctive Features

- If an incident photon is absorbed by an initial state of a target nucleus, that nucleus will be raised to a more energetic state of excitation. If that state can radiate its energy only during a transition back to the initial state, the result is a *scattering process*.
- If an incident photon is absorbed by an initial state of a target nucleus, that nucleus will be raised to a more energetic state of excitation. If there is a nonzero probability that sometimes that state will start a cascade of transitions as shown in the schematic, that state has been called a "gateway state" or "trigger level" or "intermediate state". One or more fluorescent photons are emitted, often with different delays after the initial absorption and the process is an example of IGE.

- If the initial state of the target nucleus is its ground (lowest energy) state, then the fluorescent photons will have less energy than that of the incident photon . Since the scattering channel is usually the strongest, it can "blind" the instruments being used to detect the fluorescence and early experiments preferred to study IGE by pulsing the source of incident photons while detectors were gated off and then concentrating upon any delayed photons of fluorescence when the instruments could be safely turned back on.
- If the initial state of the target nucleus is a nuclear isomer (starting with more energy than the ground) it can also support IGE. However in that case the schematic diagram is not simply the example seen for In but read from right to left with the arrows turned the other way. Such a "reversal" would require simultaneous (to within <0.25 ns) absorption of two incident photons of different energies to get from the 4 hr isomer back up to the "gateway state".

  Usually the study of IGE from a ground state to an isomer of the same nucleus teaches little about how the same isomer would perform if used as the initial state for IGE. In order to support IGE an energy for an incident photon would have to be found that would "match" the energy needed to reach some other gateway state not shown in the schematic that could launch its own cascade down to the ground state.
- If the target is a nuclear isomer storing a considerable amount of energy then IGE might produce a cascade that contains a transition that emits a photon with more energy than that of the incident photon. This would be the nuclear analog of upconversion in laser physics.
- If the target is a nuclear isomer storing a considerable amount of energy then IGE might produce a cascade through a pair of excited states whose lifetimes are "inverted" so that in a collection of such nuclei, population would build up in the longer lived upper level while emptying rapidly from the shorter lived lower member of the pair.

  The resulting inversion of population might support some form of coherent emission analogous to amplified spontaneous emission (ASE) in laser physics. If the physical dimensions of the collection of target isomer nuclei were long and thin, then a sort of "gamma ray laser" might result.

Potential Applications

- Since the IGE from ground state nuclei requires the absorption of very specific photon energies to produce delayed fluorescent photons that are easily counted, there is the possibility to construct energy-specific dosimeters by combining several different nuclides. This was demonstrated for the calibration of the radiation spectrum from the DNA-PITHON pulsed nuclear simulator. Such a dosimeter could be useful in radiation therapy where X-ray beams may contain many energies. Since photons of different energies deposit their effects at different depths in the tissue being treated, it could help calibrate how much of the total dose would be deposited in the actual target volume.
- In February 2003, the non-peer reviewed *New Scientist* wrote about the possibility of an IGE-powered airplane. The idea was to utilize ^{178m2}Hf (presumably due to its high energy to weight ratio) which would be triggered to release gamma rays that would heat air in a chamber for jet propulsion. This power source is apparently called a "quantum nucleonic reactor", although it is not clear if this name exists only in reference to the *New Scientist* article.
- It is partly this theoretical density that has made the entire IGE field so controversial. It has been suggested that the materials might be constructed to allow all of the stored energy to be released very quickly in a "burst". The density of gammas produced in this reaction would be high enough that it might allow them to be used to compress the fusion fuel of a fusion bomb. If this turns out to be the case, it might allow a fusion bomb to be constructed with no fissile material inside (i.e. a pure fusion weapon), and it is the control of the fissile material and the means for making it that underlies most attempts to stop nuclear proliferation. In fact, the possible energy release of the gammas alone would make IGE a potential high power "explosive" on its own, or a potential radiological weapon. Basic research remains in early stages but that has not deterred the worrying about these possibilities.

Societal Concerns

Due to the possibility, no matter how remote, of IGE being used as a shortcut to, or analog of, a nuclear bomb, IGE has become a "hot topic" in the arms control field, where IGE is one of a number of theoretical "shortcuts" that are often discussed together. For instance, the apparently mythical red mercury is another proposed mechanism to build a "mini-nuke", and it is not uncommon to see references to red mercury as being either a ballotechnic or IGE material. As a result, it is not uncommon to see confusion about ballotechnic materials being the same thing as IGE's.

Hafnium Controversy

The hafnium controversy is a debate over the possibility of 'triggering' rapid energy releases, via gamma ray emission, from a nuclear isomer of Hafnium, ^{178m2}Hf. The energy release is potentially 4 orders of magnitude (10,000) more energetic than a chemical reaction, but 4 orders of magnitude less than a nuclear reaction. In 1998, a group led by Carl Collins of the University of Texas at Dallas reported having successfully initiated such a trigger. Signal-to-noise ratios were small in those first experiments, and to date no other group has been able to duplicate these results.

Background

^{178m2}Hf is a particularly attractive candidate for *induced gamma emission* (hereafter "IGE") experiments, because of its high density of stored energy, 2.5 MeV per nucleus, and long 31-year half life for storing that energy. If radiation from some agent could "trigger" a release of that stored energy, the resulting cascade of gamma photons would have the best chance of finding a pair of excited states with the inverted lifetimes needed for stimulated emission.

While induced emission adds only power to a radiation field, stimulated emission adds coherence and the possibility to manipulate gamma ray coherence, even to a small degree would be interesting. The lifetime of the hafnium isomer is long enough for tractable amounts of material to be collected into experimental targets. Such samples would hold no hazards for personnel working with the material; 1 microgram of ^{178m2}Hf has an activity of only 40 microcuries (1.5 MBq).

A proposal to test the efficacy for "triggering" ^{178m2}Hf was approved by a NATO-Advanced Research Workshop(NATO-ARW) held in Predeal in 1995. Although the proposal was to use incident protons to bombard the target, α-particles were available when the first experiment was scheduled. It was done by a French, Russian, Romanian and American team. Results were said to be extraordinary, but the results were not published. Nevertheless, ^{178m2}Hf was implied to be of special importance to potential applications of IGE. A controversy quickly erupted.

Importance

- ^{178m2}Hf has the highest excitation energy of any comparably long-lived isomer. One gram of pure Hf-178-m2 would contain approximately 1330 megajoules of energy, the equivalent of exploding about 300 kilograms (660 pounds) of TNT. The half-life of Hf-178-m2 is 31 years or 1 Gs (gigasecond, 1,000,000,000 seconds) so that a gram's natural radioactivity is 1.6 TBq (terabecquerels) or roughly 40 Ci (curies). The activity is in a cascade of penetrating gamma rays, the most energetic of which is 0.574 MeV. Substantial shielding is needed before it is safe for people to be around.
- All of the energy released would be in the form of photons of X-rays and gamma rays.
- Discussions also indicate that the energy might be released very quickly, so that Hf-178-m2 could produce extremely high powers (on the order of exawatts).
- The characteristic scales of times for processes involved in applications would be favourable for consuming all of the initial radioactivity. The process for triggering a sample by IGE would use photons to trigger and produce photons as a product. The propagation of photons occurs at the speed of light while mechanical disassembly of the target would proceed with a velocity comparable to that of sound. Untriggered ^{178m2}Hf material might not be able to get away from a triggered event if the photons didn't interact first with the electrons.
- Both the proposal to the NATO-ARW and the fragmentary results from the subsequent experiment indicated that the energy of the photon needed to initiate IGE from ^{178m2}Hf would be less than 300 keV. Many economical sources of such low energy X-rays were available for delivering quite large fluxes to target samples of modest dimensions.
- Samples of ^{178m2}Hf were and remain available at low concentrations <0.1%.

Chronology of Notable Events

- Around 1997 the JASONS advisory group took testimony about the triggering of nuclear isomers. The JASON Defence Advisory Group published a relevant public report saying that they concluded that such a thing would be impossible and should not be attempted. Despite intervening publications in peer-reviewed journals of articles written by an international team reporting IGE from ^{178m2}Hf, around 2003 IDA took testimony, again from relevant scientists on matters of the credibility of reported results. The lead US member of the team, Prof. Carl Collins, that was publishing the successes did not testify.

- Around 2003, DARPA initiated exploratory research termed stimulated isomer energy release (SIER) and public interest was aroused, at both popular levels and at professional levels.
- The first focus of SIER was whether significant amounts of ^{178m2}Hf could be produced at acceptable costs for possible applications. A closed panel called HIPP was charged with the task and the conclusion was yes, it could. However, a scientist on that confidential DARPA HIPP review panel "leaked" prejudicial but preliminary concerns to the press. This unsubstianted assertion set into motion the subsequent cascade of inaccurate reports about the so-called "outrageous costs" of isomer triggering.
- Having satisfied the charge to the HIPP panel to explore the problem of production at acceptable cost, the SIER program turned to the matter of definitive confirmation of the reports of IGE from ^{178m2}Hf. A task of TRiggering Isomer Proof (TRIP) was mandated by DARPA and assigned to a completely independent team from those reporting success previously. The "gold standard" of Hafnium-isomer triggering was set as the Rusu dissertation. The TRIP experiment required independent confirmation of the Rusu Dissertation. It was successful, but could not be published.
- By 2006, the Collins team had published multiple papers supporting their initial observations of IGE from ^{178m2}Hf. Reprints (available at the link) of articles that were published after 2001 describe work conducted with tunable monochromatic X-ray beams from the synchrotron light sources SPring-8 in Hyogo and SLS in Villigen.
- By 2006 there were 2 articles that claimed to disprove possibilities for IGE from ^{178m2}Hf and three theoretical articles written by the same individual saying why it should not be possible to occur by the particular steps the author envisioned. The first two described synchrotron experiments in which the X-rays were not monochromatic.
- In 2007 Pereira et al. calculated a production cost of $1/J.
- February 29, 2008 DARPA distributed some of the 150 copies of the final report of the TRIP experiment that had independently

confirmed the "gold standard" of Hafnium-isomer triggering. Sustained by peer review, the 94 page report is distributed for official use only (FOUO) by the DARPA Technical Information Office, 3701 N. Fairfax Dr., Arlington, VA 22203 USA.

- October 9, 2008 LLNL released the 110 page evaluation of the DARPA TRIP experiment. Quoting from page 33, "The only experiment that shows statistical significance is the coincidence experiment described in the thesis by Rusu [131]." However, the report summary states, page 65: "Our conclusion is that the utilization of nuclear isomers for energy storage is impractical from the points of view of nuclear structure, nuclear reactions, and of prospects for controlled energy release. We note that the cost of producing the nuclear isomer is likely to be extraordinarily high, and that the technologies that would be required to perform the task are beyond anything done before and are difficult to cost at this time."
- In 2009 S.A. Karamian et al. published the results of a four nation team's experimental measurements at Dubna for the production of quantities of ^{178m2}Hf by spallation at energies as low as 80 MeV. Besides significantly lowering the projected cost of production, this experimental result proved the accessibility to sources of ^{178m2}Hf to be within the capabilities of the several idle cyclotron devices scattered around the world.

Isomeric Shift

The isomeric shift (also called isomer shift) is the shift on atomic spectral lines and gamma spectral lines, which occurs as a consequence of replacement of one nuclear isomer by another. It is usually called isomeric shift on atomic spectral lines and Mossbauer isomeric shift respectively. If the spectra have also hyperfine structure the shift refers to the center of gravity of the spectra. The isomeric shift provides important information about the nuclear structure and the physical, chemical or biological environment of atoms. More recently the effect has also been proposed as a tool in the search for the time variation of fundamental constants of nature.

The Isomeric Shift on Atomic Spectral Lines

The isomeric shift on atomic spectral lines is the energy or frequency shift in atomic spectra, which occurs when one replaces one nuclear isomer by another. The effect was predicted by Richard M. Weiner in 1956 whose calculations showed that it should be measurable by atomic (optical) spectroscopy (cf. also). It was observed experimentally for the first time in 1958. The theory of the atomic isomeric shift developed in is also used in the interpretation of the Mossbauer isomeric shift.

Terminology

The notion of isomer appears also in other fields such as chemistry and meteorology. Therefore in the first papers devoted to this effect the name *nuclear isomeric shift on spectral lines* was used. Before the discovery of the Mossbauer effect, the isomeric shift referred exclusively to atomic spectra; this explains the absence of the word *atomic* in the initial definition of the effect. Subsequently the isomeric shift was also observed in gamma spectroscopy through the Mossbauer effect and was called Mossbauer isomeric shift. For further details on the history of the isomeric shift and the terminology used cf. Refs.,

Isotopic versus Isomeric Shift on Atomic Spectral Lines

Atomic spectral lines are due to transitions of electrons between different atomic energy levels E, followed by emission of photons. Atomic levels are a manifestation of the electromagnetic interaction between electrons and nuclei. The energy levels of two atoms the nuclei of which are different isotopes of the same element are shifted one with respect to the other, despite the fact that the electric charges Z of the two isotopes are identical. This is so because isotopes differ by the number of neutrons and therefore the masses and volumes of two isotopes are different; these differences give rise to the isotopic shift on atomic spectral lines.

In the case of two nuclear isomers the number of protons and the number of neutrons are identical, but the quantum states and in particular the energy levels of the two nuclear isomers differ. This difference induces a difference in the electric charge distributions of two isomers and thus a difference $\delta\varphi$ in the corresponding electrostatic nuclear potentials φ, which ultimately leads to a difference ΔE in the atomic energy levels. The isomeric shift on atomic spectral lines is then given by

$$\Delta E = -e \int \delta\varphi \, |\psi|^2 d\tau$$

where ψ is the wave function of the electron involved in the transition, e its electric charge and the integration is performed over the electron coordinates. The isotopic and the isomeric shift are similar in the sense that both are effects in which the finite size of the nucleus manifests itself and both are due to a difference in the electromagnetic interaction energy between the electrons and the nucleus of the atom. The isotopic shift had been known decades before the isomeric shift and it provided useful but limited information about atomic nuclei. Unlike the isomeric shift, the isotopic shift was at first discovered in experiment and then interpreted theoretically (cf. also). While in the case of the isotopic shift the determination of the interaction energy between electrons and nuclei is a relatively simple electromagnetic problem, for isomers the problem is more involved, since it is the strong interaction, which accounts for the isomeric excitation of the nucleus and thus for the difference of charge distributions of the two isomeric states.

This circumstance explains in part why the nuclear isomeric shift was not discovered earlier: the appropriate nuclear theory and in particular the nuclear shell model were developed only in the late 1940s and early 1950s. As to the experimental observation of this shift, it also had to await the development of a new technique, that permitted spectroscopy with isomers, which are metastable nuclei. This too happened only in the 1950s. While the isomeric shift is sensitive to the internal structure of the nucleus, the isotopic shift is (in a good approximation) not. Therefore the nuclear physics information, which can be obtained from the investigation of the isomeric shift, is superior to that which can be obtained from isotopic shift studies. The measurements through the isomeric shift of e.g. the difference of nuclear radii of the excited and ground state constitute one of the most sensitive tests of nuclear models. Moreover, combined with the Mossbauer effect, the isomeric shift constitutes at present a unique tool in many other fields, besides physics.

The Isomeric Shift and the Nuclear Shell Model

According to the nuclear shell model there exists a class of isomers for which, in a first approximation, it is sufficient to consider one single nucleon, called the "optical" nucleon, to get an estimate of the difference between the charge distributions of the two isomer states, the rest of the nucleons being *filtered out*. This applies in particular for isomers in odd proton-even neutron nuclei, near closed shells. In115, for which the effect was calculated in, is such an example. The result of the calculation was that the isomeric shift on atomic spectral lines, although rather small, turned out to be two orders of magnitude bigger than a typical natural line width, which constitutes the limit of optical measurability.

The shift measured three years later in Hg197 was quite close to that calculated for In115, although in Hg197, unlike in In115, the optical nucleon is a neutron and not a proton and the electron-free neutron interaction is much smaller than the electron-free proton interaction.

This is a consequence of the fact that the optical nucleons are not free but bound particles. Thus the results of could be explained within the theory of by associating with the odd optical neutron an effective electric charge of Z/A.

The Mossbauer Isomeric Shift

The Mossbauer isomeric shift is the shift seen in gamma ray spectroscopy when one compares two different nuclear isomeric states in two different physical, chemical or biological environments, and is due to the combined effect of the recoil-free Mossbauer transition between the two nuclear isomeric states and the transition between two atomic states in those two environments.

The isomeric shift on atomic spectral lines depends on the electron wave function ø and on the difference $\delta\varphi$ of electrostatic potentials φ of the two isomeric states.

For a given nuclear isomer in two different physical or chemical environments (different physical phases or different chemical combinations) the electron wave functions are also different. Therefore on top of the isomeric shift on atomic spectral lines, which is due to the difference of the two nuclear isomer states, there will be a shift between the two environments (because of the experimental arrangement, these are called source (s) and absorber (a)). This combined shift is the Mossbauer isomeric shift and it is described mathematically by the same formalism as the nuclear isomeric shift on atomic spectral lines, except that instead of one electron wave function, that in the source ψ_{source}, one deals with the difference between the electron wave function in the source ψ_{source} and the electron wave function in the absorber $\psi_{absorber}$:

$$\Delta E_{M\ddot{o}ssbauer} = \Delta E_{source} - \Delta E_{absorber} = -e\int \delta\varphi \left[|\psi_{source}|^2 - |\psi_{absorber}|^2 \right] d\tau$$

The first measurement of the isomeric shift in gamma spectroscopy with the help of the Mossbauer effect was reported in 1960, two years after its first experimental observation in atomic spectroscopy. By measuring this shift one obtains important and extremely precise information, both about the nuclear isomer states and about the physical, chemical or biological environment of the atoms, represented by the electronic wave functions.

Under its Mossbauer variant, the isomeric shift has found important applications in domains as different as Atomic Physics, Solid State Physics, Nuclear Physics, Chemistry, Biology, Metallurgy, Mineralogy, Geology, and Lunar research. For further literature cf. also Ref.

The nuclear isomeric shift has also been observed in muonic atoms, that is atoms in which a muon is captured by the excited nucleus and makes a transition from an atomic excited state to the atomic ground state in a time which is short compared to the lifetime of the excited isomeric nuclear state.

Chirality

The feature that is most often the cause of chirality in molecules is the presence of an asymmetric carbon atom.The term chiral in general is used to describe an object that is non-superimposable on its mirror image. Achiral (not chiral) objects are objects that are identical to their mirror image. Human hands are perhaps the most universally recognized example of chirality:

The left hand is a non-superimposable mirror image of the right hand; no matter how the two hands are oriented, it is impossible for all the major features of both hands to coincide. This difference in symmetry becomes obvious if someone attempts to shake the right hand of a person using his left hand, or if a left-handed glove is placed on a right hand. The term *chirality* is derived from the Greek word for hand, (cheir). It is a mathematical approach to the concept of "handedness".

In chemistry, chirality usually refers to molecules. Two mirror images of a chiral molecule are called enantiomers or optical isomers. Pairs of enantiomers are often designated as "right-" and "left-handed".

Molecular chirality is of interest because of its application to stereochemistry in inorganic chemistry, organic chemistry, physical chemistry, biochemistry, and supramolecular chemistry.

History

The term *optical activity* is derived from the interaction of chiral materials with polarized light. A solution of the (–)-form of an optical isomer rotates the plane of polarization of a beam of polarized light in a counterclockwise direction (levorotatory), vice-versa for the (+) (dextrorotatory) optical isomer. The property was first observed by Jean-Baptiste Biot in 1815, and gained considerable importance in the sugar industry, analytical chemistry, and pharmaceuticals. Louis Pasteur deduced in 1848 that this phenomenon has a molecular basis. Artificial composite materials displaying the analog of optical activity but in the microwave region were introduced by J.C. Bose in 1898, and gained considerable attention from the mid-1980s. The term *chirality* itself was coined by Lord Kelvin in 1873.

The word "racemic" (mix of both chiralities) is derived from the Latin word "racemus" for "bunch of grapes"; the term having its origins in the work of Louis Pasteur who isolated racemic tartaric acid from wine.

Symmetry

The symmetry of a molecule (or any other object) determines whether it is chiral. A molecule is *achiral* (not chiral) when an improper rotation, that is a combination of a rotation and a reflection in a plane, perpendicular to the axis of rotation, results in the same molecule - see chirality (mathematics). An equivalent definition is that a chiral molecule lacks a plane of symmetry. For tetrahedral molecules, the molecule is chiral if all four substituents are different.

A chiral molecule is not necessarily asymmetric (devoid of any symmetry element), as it can have, for example, rotational symmetry.

Naming conventions

By configuration: R- and S-: For chemists, the *R* / *S* system is the most important nomenclature system for denoting enantiomers, which does not involve a reference molecule such as glyceraldehyde. It labels each chiral center *R* or *S* according to a system by which its substituents are each assigned a *priority*, according to the Cahn–Ingold–Prelog priority rules (CIP), based on atomic number. If the center is oriented so that the lowest-priority of the four is pointed away from a viewer, the viewer will then see two possibilities: If the priority of the remaining three substituents decreases in clockwise direction, it is labelled *R* (for *Rectus*, Latin for right), if it decreases in counterclockwise direction, it is *S* (for *Sinister*, Latin for left).

This system labels each chiral center in a molecule (and also has an extension to chiral molecules not involving chiral centers). Thus, it has greater generality than the D/L system, and can label, for example, an (*R*,*R*) isomer versus an (*R*,*S*) — diastereomers.

The *R* / *S* system has no fixed relation to the (+)/(–) system. An *R* isomer can be either dextrorotatory or levorotatory, depending on its exact substituents.

The *R* / *S* system also has no fixed relation to the D/L system. For example, the side-chain one of serine contains a hydroxyl group, -OH. If a thiol group, -SH, were swapped in for it, the D/L labelling would, by its definition, not be affected by the substitution. But this substitution would invert the molecule's *R* / *S* labelling, because the CIP priority of CH_2OH is lower than that for CO_2H but the CIP priority of CH_2SH is higher than that for CO_2H.

For this reason, the D/L system remains in common use in certain areas of biochemistry, such as amino acid and carbohydrate chemistry, because it is convenient to have the same chiral label for all of the commonly occurring structures of a given type of structure in higher organisms. In the D/L system, they are nearly all consistent - naturally occurring amino acids are nearly all L, while naturally occurring carbohydrates are nearly all D. In the *R* / *S* system, they are mostly *S*, but there are some common exceptions.

By Optical Activity: (+)- and (-)-

An enantiomer can be named by the direction in which it rotates the plane of polarized light. If it rotates the light clockwise (as seen by a viewer towards whom the light is travelling), that enantiomer is labelled (+). Its mirror-image is labelled (-). The (+) and (-) isomers have also been termed *d*- and *l*-, respectively (for *dextrorotatory* and *levorotatory*). Naming with *d*- and *l*- is easy to confuse with D- and L-labelling and is therefore strongly discouraged by IUPAC.

By configuration: D- and L-

An optical isomer can be named by the spatial configuration of its atoms. The D/L system does this by relating the molecule to glyceraldehyde. Glyceraldehyde is chiral itself, and its two isomers are labelled D and L (typically typeset in SMALL CAPS in published work). Certain chemical manipulations can be performed on glyceraldehyde without affecting its configuration, and its historical use for this purpose (possibly combined with its convenience as one of the smallest commonly used chiral molecules) has resulted in its use for nomenclature. In this system, compounds are named by analogy to glyceraldehyde, which, in general, produces unambiguous designations, but is easiest to see in the small biomolecules similar to glyceraldehyde. One example is the amino acid alanine, which has two optical isomers, and they are labelled according to which isomer of glyceraldehyde they come from. On the other hand, glycine, the amino acid derived from glyceraldehyde, has no optical activity, as it is not chiral (achiral). Alanine, however, is chiral.

The D/L labelling is unrelated to (+)/(-); it does not indicate which enantiomer is dextrorotatory and which is levorotatory. Rather, it says that the compound's stereochemistry is related to that of the dextrorotatory or levorotatory enantiomer of glyceraldehyde—the dextrorotatory isomer of glyceraldehyde is, in fact, the D- isomer. Nine of the nineteen L-amino acids commonly found in proteins are dextrorotatory (at a wavelength of 589 nm), and D-fructose is also referred to as levulose because it is levorotatory.

A rule of thumb for determining the D/L isomeric form of an amino acid is the "CORN" rule. The groups:

COOH, R, NH2 and H (where R is a variant carbon chain) are arranged around the chiral center carbon atom. Starting with the hydrogen atom away from the viewer, if these groups are arranged clockwise around the carbon atom, then it is the D-form. If counter-clockwise, it is the L-form.

Nomenclature

- Any non-racemic chiral substance is called scalemic.
- A chiral substance is enantiopure or homochiral when only one of two possible enantiomers is present.
- A chiral substance is enantioenriched or heterochiral when an excess of one enantiomer is present but not to the exclusion of the other.
- Enantiomeric excess or ee is a measure for how much of one enantiomer is present compared to the other. For example, in a sample with 40% ee in R, the remaining 60% is racemic with 30% of R and 30% of S, so that the total amount of R is 70%.

Stereogenic Centres

In general, chiral molecules have point chirality at a single *stereogenic* atom, which has four different substituents. The two enantiomers of such compounds are said to have different absolute configurations at this center. This center is thus stereogenic (i.e., a grouping within a molecular entity that may be considered a focus of stereoisomerism).

Normally when an atom has four different substituents, it is chiral. However in rare cases, two of the ligands differ from each other by being mirror images of each other. When this happens, the mirror image of the molecule is identical to the original, and the molecule is achiral. This is called pseudochirality.

A molecule can have multiple chiral centres without being chiral overall if there is a symmetry between the two (or more) chiral centres themselves. Such a molecule is called a meso compound.

It is also possible for a molecule to be chiral without having actual point chirality. Common examples include 1,1'-bi-2-naphthol (BINOL) and 1,3-dichloro-allene, which have axial chirality, (*E*)-cyclooctene, which has planar chirality, and certain calixarenes and fullerenes, which have inherent chirality.

It is important to keep in mind that molecules have considerable flexibility and thus, depending on the medium, may adopt a variety of different conformations. These various conformations are themselves almost always chiral. When assessing chirality, a time-averaged structure is considered and for routine compounds, one should refer to the most symmetric possible conformation.

When the optical rotation for an enantiomer is too low for practical measurement, it is said to exhibit cryptochirality. Even isotopic differences must be considered when examining chirality. Replacing one of the two ^{1}H atoms at the CH_2 position of benzyl alcohol with a deuterium (^{2}H) makes that carbon a stereocentre. The resulting benzyl-α-*d* alcohol exists as two distinct enantiomers, which can be assigned by the usual stereochemical naming conventions. The *S* enantiomer has $[\alpha]_D = +0.715°$.

Properties of enantiomers

Normally, the two enantiomers of a molecule behave identically to each other. For example, they will migrate with identical R_f in thin layer chromatography and have identical retention time in HPLC. Their NMR and IR spectra are identical. However, enantiomers behave differently in the presence of other chiral molecules or objects. For example, enantiomers do not migrate identically on chiral chromatographic media, such as quartz or standard media that have been chirally modified. The NMR spectra of enantiomers are affected differently by single-enantiomer chiral additives such as Eufod.

Chiral compounds rotate plane polarized light. Each enantiomer will rotate the light in a different sense, clockwise or counterclockwise. Molecules that do this are said to be optically active.

Characteristically, different enantiomers of chiral compounds often taste and smell differently and have different effects as drugs – see below. These effects reflect the chirality inherent in biological systems.

One chiral 'object' that interacts differently with the two enantiomers of a chiral compound is circularly polarised light: An enantiomer will absorb left- and right-circularly polarised light to differing degrees. This is the basis of circular dichroism (CD) spectroscopy. Usually the difference in absorptivity is relatively small (parts per thousand). CD spectroscopy is a powerful analytical technique for investigating the secondary structure of proteins and for determining the absolute configurations of chiral compounds, in particular, transition metal complexes. CD spectroscopy is replacing polarimetry as a method for characterising chiral compounds, although the latter is still popular with sugar chemists.

In Biology

Many biologically active molecules are chiral, including the naturally occurring amino acids (the building blocks of proteins) and sugars. In biological systems, most of these compounds are of the same chirality: most amino acids are L and sugars are D. Typical naturally occurring proteins, made of L amino acids, are known as *left-handed proteins*, whereas D amino acids produce *right-handed proteins.*

The origin of this homochirality in biology is the subject of much debate. Most scientists believe that Earth life's "choice" of chirality was purely random, and that if carbon-based life forms exist elsewhere in the universe, their chemistry could theoretically have opposite chirality. However, there is some suggestion that early amino acids could have formed in comet dust.

In this case, circularly polarised radiation (which makes up 17% of stellar radiation) could have caused the selective destruction of one chirality of amino acids, leading to a selection bias which ultimately resulted in all life on Earth being homochiral.

Enzymes, which are chiral, often distinguish between the two enantiomers of a chiral substrate. Imagine an enzyme as having a glove-like cavity that binds a substrate. If this glove is right-handed, then one enantiomer will fit inside and be bound, whereas the other enantiomer will have a poor fit and is unlikely to bind. D-form amino acids tend to taste sweet, whereas L-forms are usually tasteless. Spearmint leaves and caraway seeds, respectively, contain L-carvone and D-carvone - enantiomers of carvone. These smell different to most people because our olfactory receptors also contain chiral molecules that behave differently in the presence of different enantiomers.

Chirality is important in context of ordered phases as well, for example the addition of a small amount of an optically active molecule to a nematic phase (a phase that has long range orientational order of molecules) transforms that phase to a chiral nematic phase (or cholesteric phase). Chirality in context of such phases in polymeric fluids has also been studied in this context.

The relative abundances of each of the different D-isomers of several amino acids have recently been quantified by collecting experimentally reported data from the proteome across all organisms in the Swiss-Prot database.

The D-isomers observed experimentally were found to occur very rarely as shown in the following table in the database of protein sequences containing over 187 million amino acids.

D-amino acid	*# of Times Experimentally Observed*
D-alanine	664
D-serine	114
D-methionine	19
D-phenylalanine	15
D-valine	8
D-tryptophan	7
D-leucine	6
D-asparagine	2
D-threonine	2

Inorganic Chemistry

Many coordination compounds are chiral. At one time, the chirality was associated with organic chemistry, but this misconception was overthrown by the resolution of a purely inorganic compound by Alfred Werner. Specifically, the cobalt hydroxo complex called hexol is significant as being the first compound devoid of carbon to display optical activity.

A famous example is tris(bipyridine)ruthenium(II) complex in which the three bipyridine ligands adopt a chiral propeller-like arrangement. It is now appreciated that chirality is pervasive in inorganic chemistry, an example from the mineral kingdom being quartz.

Chirality of Compounds with a Stereogenic "Lone Pair"

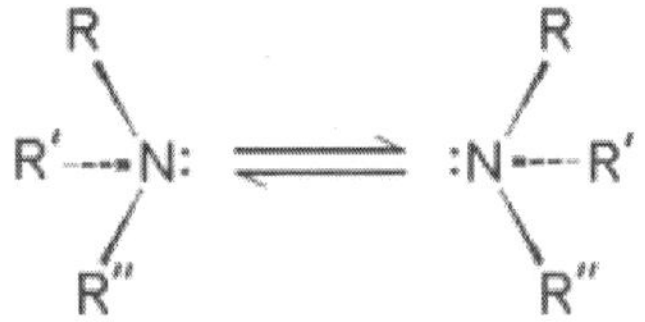

When a nonbonding pair of electrons, a lone pair, occupies space, chirality can result. The effect is pervasive in certain amines, phosphines, sulfonium ions, sulfoxides, and even carbanions. The main requirement is that aside from the lone pair, the other three substituents differ mutually. Chiral phosphine ligands are useful in asymmetric synthesis.

Chiral amines are special in the sense that the enantiomers can rarely be separated. The energy barrier for the inversion of the stereocenter is generally about 30 kJ/mol, which means that the two stereoisomers rapidly interconvert at room temperature. As a result, such chiral amines cannot be resolved into individual enantiomers unless some of the substituents are constrained in cyclic structures as in Troger's base.

Chemical Chirality in Popular Fiction

Although little was known about chemical chirality in the time of Lewis Carroll, his work Through the Looking-glass contains a prescient reference to the differing biological activities of enantiomeric drugs: "Perhaps Looking-glass milk isn't good to drink," Alice said to her cat.

In the Dorothy L. Sayers novel The Documents in the Case a murder is committed that is designed to appear to be accidental death caused by eating poisonous mushrooms containing muscarine. The case is proved to be murder because the muscarine found in the deceased's stomach is racemic and therefore synthetic.

In James Blish's Star Trek novella Spock Must Die! the tachyon 'mirrored' Mr Spock is later discovered to have stolen chemical reagents from the medical bay and to have been using them to convert certain amino acids to opposite-chirality isomers, since the mirrored Mr Spock's metabolism is reversed, and, hence, must process the opposite polarity of these isomers. In Larry Niven's Destiny's Road, the title planet's indigenous life is based upon right-handed proteins. When human colonists arrive from Earth via a generation ship, extreme measures are taken to permit the colony's survival.

A peninsula is sterilized with a lander's fusion drive, creating the titular "road" out of fused bedrock. The area is then reseeded with Earth life to provide the colonists with food. Though the soil lacks potassium due to other factors, necessitating supplements that produce a hydraulic empire common to Niven's fiction, the colony otherwise prospers. Native viruses and bacteria cannot infect colonists, resulting in longer lifespans. Sealife quickly recovers, and is consumed by the colonists as a "diet" food, as their digestive systems cannot metabolize it into fat.

Marti Steussy's "Dreams of Dawn" (1988) has a similar premise, where the locals evolved based on right-handed amino acids.

In the Trauma Center series of games, doctors test for a "chiral reaction" in order to determine whether or not a patient is infected with "Gangliated Utrophin Immuno Latency Toxin," a fictional, parasitic pathogen more commonly referred to as *G.U.I.L.T.* A positive reaction means the patient is infected, while a negative reaction means the patient has either been cured or is not infected.

In the video game Mass Effect, the turian and quarian alien races have biology based upon right-handed amino acids. Because of this foods from other races which have life forms based upon left-handed amino acids have no nutritional value and may cause fatal allergic reactions.

The denouement of Poul Anderson's After Doomsday relies partly on chirality.

The plot of Roger Zelazny's Doorways in the Sand centres around a device called the Rhennius Machine, any object passed through which will emerge its complete chiral opposite, down to the molecular level.

Chapter 7

Alkylation

Alkylation is the transfer of an alkyl group from one molecule to another. The alkyl group may be transferred as an alkyl carbocation, a free radical, a carbanion or a carbene (or their equivalents). Alkylating agents are widely used in chemistry because the alkyl group is probably the most common group encountered in organic molecules.

Many biological target molecules or their synthetic precursors are composed of an alkyl chain with specific functional groups in a specific order. Selective alkylation, or adding parts to the chain with the desired functional groups, is used, especially if there is no commonly available biological precursor. Alkylation with only one carbon is termed methylation.

In oil refining contexts, alkylation refers to a particular alkylation of isobutane with olefins. It is a major aspect of the upgrading of petroleum.

In medicine, alkylation of DNA is used in chemotherapy to damage the DNA of cancer cells. Alkylation is accomplished with the class of drugs called alkylating antineoplastic agents.

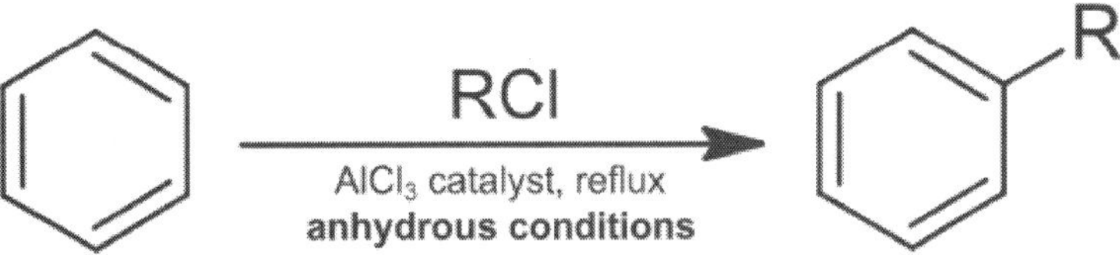

Figure: *Benzene Friedel-Crafts alkylation.*

Alkylating Agents

Alkylating agents are classified according to their nucleophilic or electrophilic character.

Nucleophilic Alkylating Agents

Nucleophilic alkylating agents deliver the equivalent of an alkyl anion (carbanion). Examples include the use of organometallic compounds such as Grignard (organomagnesium), organolithium, organocopper, and organosodium reagents. These compounds typically can add to an electron-deficient carbon atom such as at a carbonyl group. Nucleophilic alkylating agents can also displace halide substituents on a carbon atom. In the presence of catalysts, they also alkylate alkyl and aryl halides, as exemplified by Suzuki couplings.

Electrophilic Alkylating Agents

Electrophilic alkylating agents deliver the equivalent of an alkyl cation. Examples include the use of alkyl halides with a Lewis acid catalyst to alkylate aromatic substrates in Friedel-Crafts reactions. Alkyl halides can also react directly with amines to form C-N bonds; the same holds true for other nucleophiles such as alcohols, carboxylic acids, thiols, etc.

Electrophilic, soluble alkylating agents are often very toxic, due to their ability to alkylate DNA. They should be handled with proper PPE. This mechanism of toxicity is also responsible for the ability of some alkylating agents to perform as anti-cancer drugs in the form of alkylating antineoplastic agents, and also as chemical weapons such as mustard gas. Alkylated DNA either does not coil or uncoil properly, or cannot be processed by information-decoding enzymes. This results in cytotoxicity with the effects of inhibition the growth of the cell, initiation of programmed cell death or apoptosis. However, mutations are also triggered, including carcinogenic mutations, explaining the higher incidence of cancer after exposure.

Alcohols and phenols can be alkylated to give alkyl ethers:

$$R-OH + R'-X \rightarrow R-O-R' + H-X$$

The produced acid HX is removed with a base, or, alternatively, the alcohol is deprotonated first to give an alkoxide or phenoxide. For example, dimethyl sulfate alkylates the sodium salt of phenol to give anisole, the methyl ether of phenol. The dimethyl sulfate is dealkylated to sodium methylsulfate.

$$Ph-O^{-}Na^{+} + Me_2SO_4 \rightarrow Ph-O-Me + Na^{+}MeSO_4^{-}$$

On the contrary, the alkylation of amines introduces the problem that the alkylation of an amine makes it *more* nucleophilic. Thus, when an electrophilic alkylating agent is introduced to a primary amine, it will preferentially alkylate all the way to a quaternary ammonium cation.

$R-NH_2 \rightarrow R-NH-R' \rightarrow R-N(R')_2 \rightarrow R-N(R')_3^+$ (alkylating agent omitted for clarity)

If the quaternary ammonium is not the desired product, more circuitious routes such as reductive amination are necessary.

Carbene Alkylating Agents

Carbenes are extremely reactive and are known to attack even unactivated C-H bonds. Carbenes can be generated by elimination of a diazo group. A metal can form a carbene equivalent called a transition metal carbene complex.

In Biology

Methylation is the most common type of alkylation, being associated with the transfer of a methyl group. Methylation is distinct from alkylation in that it is specifically the transfer of one carbon, whereas alkylation can refer to the transfer of long chain carbon groups. Methylation in nature is typically effected by vitamin B12-derived enzymes, where thc mcthyl group is carried by cobalt. In methanogenesis, coenzyme M is methylated by tetrahydromethanopterin.

Electrophilic compounds may alkylate different nucleophiles in the body. The toxicity, carcinogenity, and paradoxically, cancer cell-killing abilities of different DNA alkylating agents are an example.

Oil Refining

In a standard oil refinery process, isobutane is alkylated with low-molecular-weight alkenes (primarily a mixture of propene and butene) in the presence of a strong acid catalyst, either sulfuric acid or hydrofluoric acid. In an oil refinery it is referred to as a sulfuric acid alkylation unit (SAAU) or a hydrofluoric alkylation unit, (HFAU). Refinery workers may simply refer to it as the alky or alky unit. The catalyst protonates the alkenes (propene, butene) to produce reactive carbocations, which alkylate isobutane. The reaction is carried out at mild temperatures (0 and 30 °C) in a two-phase reaction. Because the reaction is exothermic, cooling is needed: SAAU plants require lower temperatures so the cooling medium needs to be chilled, for HFAU normal refinery cooling water will suffice. It is important to keep a high ratio of isobutane to alkene at the point of reaction to prevent side reactions which produces a lower octane product, so the plants have a high recycle of isobutane back to feed. The phases separate spontaneously, so the acid phase is vigorously mixed with the hydrocarbon phase to create sufficient contact surface.

The product is called alkylate and is composed of a mixture of high-octane, branched-chain paraffinic hydrocarbons (mostly isoheptane and isooctane). Alkylate is a premium gasoline blending stock because it has exceptional antiknock properties and is clean burning. Alkylate is also a key component of avgas. The octane number of the alkylate depends mainly upon the kind of alkenes used and upon operating conditions. For example, isooctane results from combining butylene with isobutane and has an octane rating of 100 by definition. There are other products in the alkylate, so the octane rating will vary accordingly.

Since crude oil generally contains only 10 to 40 percent of hydrocarbon constituents in the gasoline range, refineries use a fluid catalytic cracking process to convert high molecular weight hydrocarbons into smaller and more volatile compounds, which are then converted into liquid gasoline-size hydrocarbons. Alkylation processes transform low molecular-weight alkenes and iso-paraffin molecules into larger iso-paraffins with a high octane number.

Combining cracking, polymerization, and alkylation can result in a gasoline yield representing 70 percent of the starting crude oil. More advanced processes, such as cyclicization of paraffins and dehydrogenation of naphthenes forming aromatic hydrocarbons in a catalytic reformer, have also been developed to increase the octane rating of gasoline. Modern refinery operation can be shifted to produce almost any fuel type with specified performance criteria from a single crude feedstock.

In the entire range of refinery processes, alkylation is a very important process that enhances the yield of high-octane gasoline. However, not all refineries have an alkylation plant. The oil and gas journal annual survey of worldwide refining capacities for January 2007 lists many countries with no alkylation plants at their refineries.

Refineries examine whether it makes sense economically to install alkylation units. Alkylation units are complex, with substantial economy of scale. In addition to a suitable quantity of feedstock, the price spread between the value of alkylate product and alternate feedstock disposition value must be large enough to justify the installation. Alternative outlets for refinery alklylation feedstocks include sales as LPG, blending of C4 streams directly into gasoline and feedstocks for chemical plants. Local market conditions vary widely between plants. Variation in the RVP specification for gasoline between countries and between seasons dramatically impacts the amount of butane streams that can be blended directly into gasoline. The transportation of specific types of LPG streams can be expensive so local disparities in economic conditions are often not fully mitigated by cross market movements of alkylation feedstocks.

The availability of a suitable catalyst is also an important factor in deciding whether to build an alkylation plant. If sulfuric acid is used, significant volumes are needed. Access to a suitable plant is required for the supply of fresh acid and the disposition of spent acid. If a sulfuric acid plant must be constructed specifically to support an alkylation unit, such construction will have a significant impact on both the initial requirements for capital and ongoing costs of operation.

Alternatively it is possible to install a WSA Process unit to regenerate the spent acid. No drying of the gas takes place. This means that there will be no loss of acid, no acidic waste material and no heat is lost in process gas reheating. The selective condensation in the WSA condenser ensures that the regenerated fresh acid will be 98% w/w even with the humid process gas. It is possible to combine spent acid regeneration with disposal of hydrogen sulfide by using the hydrogen sulfide as a fuel.

The second main catalyst option is hydrofluoric acid. Rates of consumption for HF acid in alkylation plants are much lower than for sulfuric acid. HF acid plants can process a wider range of feedstock mix with propylenes and butylenes. HF plants also produce alkylate with better octane rating than sulfuric plants. However, due to the hazardous nature of the material, HF acid is produced at very few locations and transportation must be managed rigorously.

Hydrodealkylation

Hydrodealkylation is a chemical reaction that often involves reacting an aromatic hydrocarbon, such as toluene, in the presence of hydrogen gas to form a simpler aromatic hydrocarbon devoid of functional groups. An example is the conversion of 1,2,4-trimethylbenzene to xylene. This chemical process usually occurs at high temperature, at high pressure, or in the presence of a catalyst. These are predominantly transition metals, such as chromium or molybdenum.

Examples

- Toluene hydrodealkylation to benzene
- Transalkylation.

Acetyl

In organic chemistry, acetyl is a functional group, the acyl with chemical formula $COCH_3$. It is sometimes represented by the symbol Ac (not to be confused with the element actinium). The acetyl group contains a methyl group single-bonded to a carbonyl.

The carbonyl centre of an acyl radical has one nonbonded electron with which it forms a chemical bond to the remainder *R* of the molecule. In IUPAC nomenclature, acetyl is called ethanoyl, although this term is rarely heard. The acetyl moiety is a component of many organic compounds, including the neurotransmitter acetylcholine, acetyl-CoA, acetylcysteine, and the analgesics acetaminophen and acetylsalicylic acid (better known as aspirin).

Acetylation

In Nature

The introduction of an acetyl group into a molecule is called acetylation. In biological organisms, acetyl groups are commonly transferred from acetyl-CoA to coenzyme A (CoA). Acetyl-CoA is an intermediate both in the biological synthesis and in the breakdown of many organic molecules.

Histones and other proteins are often modified by acetylation. For example, on the DNA level, histone acetylation by acetyltransferases (HATs) causes an expansion of chromatin architecture, allowing for genetic transcription to occur. However, removal of the acetyl group by histone deacetylases (HDACs) condenses DNA structure, thereby preventing transcription. In addition to HDACs, Methyl group additions are able to bind DNA resulting in DNA methylation, and this is another common way to block DNA acetylation and inhibit gene transcription.

Synthetic Organic and Pharmaceutical Chemistry

Acetylation can be achieved using a variety of methods, the most common one being via the use of acetic anhydride or acetyl chloride, often in the presence of a tertiary or aromatic amine base. A typical acetylation is the conversion of glycine to acetylglycine:

$$H_2NCH_2CO_2H + (CH_3CO)_2O \rightarrow CH_3C(O)NHCH_2CO_2H + CH_3CO_2H$$

Pharmacology

Acetylated organic molecules exhibit increased ability to cross the blood-brain barrier. Acetylation helps a given drug reach the brain more quickly, making the drug's effects more intense and increasing the effectiveness of a given dose. The acetyl group in acetylsalicylic acid (aspirin) enhances its effectiveness relative to the natural anti-inflammatant salicylic acid. In similar manner, acetylation converts the natural painkiller morphine into the far more potent heroin (diacetylmorphine).

The supplement industry touts acetyl-L-carnitine as being more effective than other preparations of carnitine. Acetylation of resveratrol holds promise as one of the first anti-radiation medicines for human populations.

Acetoxy Group

Acetoxy group, abbreviated AcO or OAc, is a chemical functional group of the structure CH_3-C(=O)-O-. It differs from the acetyl group CH_3-C(=O)- by the presence of one additional oxygen atom. The name acetoxy is the short form of *acetyl-oxy*.

Functionality

An acetoxy group may be used as a protection for an alcohol functionality in a synthetic route although the protecting group itself is called an acetyl group.

Alcohol Protection

There are several options of introducing an acetoxy functionality in a molecule from an alcohol (in effect protecting the alcohol by acetylation):

- Acetyl Halide, such as Acetyl chloride in the presence of a base like triethylamine
- Activated ester form of acetic acid , such as a N-hydroxysuccinimide ester, although this is not advisable due to higher costs and difficulties.
- Acetic anhydride in the presence of base with a catalyst such as pyridine with a bit of DMAP added.

An alcohol is not a particularly strong nucleophile and, when present, more powerful nucleophiles like amines will react with the above mentioned reagents in preference to the alcohol.

Alcohol Deprotection

For deprotection (regeneration of the alcohol)

- Aqueous base (pH >9)
- Aqueous acid (pH <2), may have to be heated

- Anhydrous base such as sodium methoxide in methanol. Very useful when a methyl ester of a carboxylic acid is also present in the molecule, as it will not hydrolyze it like an aqueous base would. (Same also holds with an ethoxide in ethanol with ethyl esters)

Histone Acetylation and Deacetylation

In histone acetylation and deacetylation, the histones are acetylated and deacetylated on lysine residues in the N-terminal tail and on the surface of the nucleosome core as part of gene regulation. Typically, these reactions are catalyzed by enzymes with "histone acetyltransferase" (HAT) or "histone deacetylase" (HDAC) activity. The source of the acetyl group in histone acetylation is Acetyl-Coenzyme A, and in histone deacetylation the acetyl group is transferred to Coenzyme A.

Acetylated histones and nucleosomes represent a type of epigenetic tag within chromatin. Acetylation brings in a negative charge, acting to neutralize the positive charge on the histones and decreases the interaction of the N termini of histones with the negatively charged phosphate groups of DNA.

As a consequence, the condensed chromatin is transformed into a more relaxed structure which is associated with greater levels of gene transcription. This relaxation can be reversed by HDAC activity. Relaxed, transcriptionally active DNA is referred to as euchromatin. More condensed (tightly packed) DNA is referred to as heterochromatin. Condensation can be brought about by processes including deacetylation and methylation; the action of methylation is indirect and has no effect upon charge.

This charge neutralization model has been challenged by recent studies, according to which transcriptionally active genes are correlated with rapid turnover of histone acetylation. This requires that the HATs and HDACs must act continuously on the affected histone tail. Methylation at a specific lysine residue (K4) is involved in targeting histone tails for continuous acetylation and deacetylation.

Histone Methylation

Histone methylation is the modification of certain amino acids in a histone protein by the addition of one, two, or three methyl groups. In the cell nucleus, DNA is wound around histones. Methylation and demethylation of histones turns the genes in DNA "off" and "on", respectively, either by loosening their tails, thus, allowing transcription factors and other proteins to access the DNA or by encompassing their tails around the DNA, thus, restricting access to the DNA. This is true in most cases.

Function

This modification alters the properties of the nucleosome and affects its interactions with other proteins.

- Histone methylation is generally associated with transcriptional repression.
- However, methylation of some lysine and arginine residues of histones results in transcriptional activation. Examples include methylation of lysine 4 of histone 3 (H3K4), and arginine (R) residues on H3 and H4.

Methylation

In the chemical sciences, methylation denotes the addition of a methyl group to a substrate or the substitution of an atom or group by a methyl group. Methylation is a form of alkylation with, to be specific, a methyl group, rather than a larger carbon chain, replacing a hydrogen atom. These terms are commonly used in chemistry, biochemistry, soil science, and the biological sciences.

In biological systems, methylation is catalyzed by enzymes; such methylation can be involved in modification of heavy metals, regulation of gene expression, regulation of protein function, and RNA metabolism. Methylation of heavy metals can also occur outside of biological systems. Chemical methylation of tissue samples is also one method for reducing certain histological staining artifacts.

In Biology

Epigenetics: Methylation contributing to epigenetic inheritance can occur through either DNA methylation or protein methylation.

DNA methylation in vertebrates typically occurs at CpG sites (cytosine-phosphate-guanine sites, that is, where a cytosine is directly followed by a guanine in the DNA sequence). This methylation results in the conversion of the cytosine to 5-methylcytosine. The formation of Me-CpG is catalyzed by the enzyme DNA methyltransferase. Human DNA has about 80%-90% of CpG sites methylated, but there are certain areas, known as CpG islands, that are GC-rich (made up of about 65% CG residues), wherein none are methylated. These are associated with the promoters of 56% of mammalian genes, including all ubiquitously expressed genes. One to two percent of the human genome are CpG clusters, and there is an inverse relationship between CpG methylation and transcriptional activity.

Protein methylation typically takes place on arginine or lysine amino acid residues in the protein sequence. Arginine can be methylated once (monomethylated arginine) or twice, with either both methyl groups on one terminal nitrogen (asymmetric dimethylated arginine) or one on both nitrogens (symmetric dimethylated arginine) by peptidylarginine methyltransferases (PRMTs). Lysine can be methylated once, twice or three times by lysine methyltransferases. Protein methylation has been most-studied in the histones. The transfer of methyl groups from S-adenosyl methionine to histones is catalyzed by enzymes known as histone methyltransferases. Histones that are methylated on certain residues can act epigenetically to repress or activate gene expression. Protein methylation is one type of post-translational modification.

Embryonic Development

During the development of germ cells their genomes are demethylated, while chromosomes in the somatic cells retain the parental methylation patterns. After that, a *De novo* methylation of the germ cells occurs, modifying and adding epigenetic information to the genome based on the sex of the individual.

After fertilization of an oocyte and formations of a zygote, its combined genome is demethylated and remethylated again (with the exception of the imprinted genes). By blastula stage, the methylation of the embryonic cells is complete.

The process of demethylation/remethylation is referred to as "reprogramming". The importance of methylation was shown in knockout mutants without DNA methyltransferase, which all died at the morula stage.

Postnatal Development

Increasing evidence is revealing a role of methylation in the interaction of environmental factors with genetic expression. Differences in maternal care during the first 6 days of life in the rat induce differential methylation patterns in some promoter regions, thus influencing gene expression. Furthermore, even-more-dynamic processes such as interleukin signalling have been shown to be regulated by methylation.

Research in humans has shown that repeated high level activation of the body's stress system, especially in early childhood, can alter methylation processes and lead to changes in the chemistry of the individual's DNA. The chemical changes can disable genes and prevent the brain from properly regulating its response to stress. Researchers and clinicians have drawn a link between this neurochemical disregulation and the development of chronic health problems such as depression, obesity, diabetes, hypertension, and coronary artery disease.

Cancer

The pattern of methylation has recently become an important topic for research. Studies have found that in normal tissue, methylation of a gene is mainly localized to the coding region, which is CpG-poor. In contrast, the promoter region of the gene is unmethylated, despite a high density of CpG islands in the region.

Neoplasia is characterized by "methylation imbalance" where genome-wide hypomethylation is accompanied by localized hypermethylation and an increase in expression of DNA methyltransferase. The overall methylation state in a cell might also be a precipitating factor in carcinogenesis as evidence suggests that genome-wide hypomethylation can lead to chromosome instability and increased mutation rates. The methylation state of some genes can be used as a biomarker for tumorigenesis. For instance, hypermethylation of the pi-class glutathione S-transferase gene (GSTP1) appears to be a promising diagnostic indicator of prostate cancer.

In cancer, the dynamics of genetic and epigenetic gene silencing are very different. Somatic genetic mutation leads to a block in the production of functional protein from the mutant allele. If a selective advantage is conferred to the cell, the cells expand clonally to give rise to a tumour in which all cells lack the capacity to produce protein.

In contrast, epigenetically mediated gene silencing occurs gradually. It begins with a subtle decrease in transcription, fostering a decrease in protection of the CpG island from the spread of flanking heterochromatin and methylation into the island. This loss results in gradual increases of individual CpG sites, which vary between copies of the same gene in different cells.

Bacterial Host Defence

In addition, adenosine or cytosine methylation is part of the restriction modification system of many bacteria. Bacterial DNAs are methylated periodically throughout the genome. A methylase is the enzyme that recognizes a specific sequence and methylates one of the bases in or near that sequence.

Foreign DNAs (which are not methylated in this manner) that are introduced into the cell are degraded by sequence-specific restriction enzymes. Bacterial genomic DNA is not recognized by these restriction enzymes. The methylation of native DNA acts as a sort of primitive immune system, allowing the bacteria to protect themselves from infection by bacteriophage. These restriction enzymes are the basis of restriction fragment length polymorphism (RFLP) testing, used to detect DNA polymorphisms.

In Chemistry

The term methylation in organic chemistry refers to the alkylation process used to describe the delivery of a CH_3 group. This is commonly performed using *electrophilic* methyl sources - iodomethane, dimethyl sulfate, dimethyl carbonate, or less commonly with the more powerful (and more dangerous) methylating reagents of methyl triflate or methyl fluorosulfonate (magic methyl), which all react via S_N2 nucleophilic substitution. For example a carboxylate may be methylated on oxygen to give a methyl ester, an alkoxide salt RO^- may be likewise methylated to give an ether, $ROCH_3$, or a ketone enolate may be methylated on carbon to produce a new ketone.

R–C(=O)OH —(CH_3I; K_2CO_3, CH_3OH)→ R–C(=O)O–CH_3

C_6H_5OH —(CH_3I; Li_2CO_3, DMF)→ C_6H_5O–CH_3

On the other hand, the methylation may involve use of *nucleophilic* methyl compounds such as methyllithium (CH_3Li) or Grignard reagents (CH_3MgX). For example, CH_3Li will methylate acetone, adding across the carbonyl (C=O) to give the lithium alkoxide of *tert*-butanol:

Acetone + "CH_3^-" —($CH_3^-Li^+$)→ O^- Li^+ / CH_3

Purdie Methylation

Purdie methylation is a specific method for the methylation at oxygen of carbohydrates using iodomethane and silver oxide.

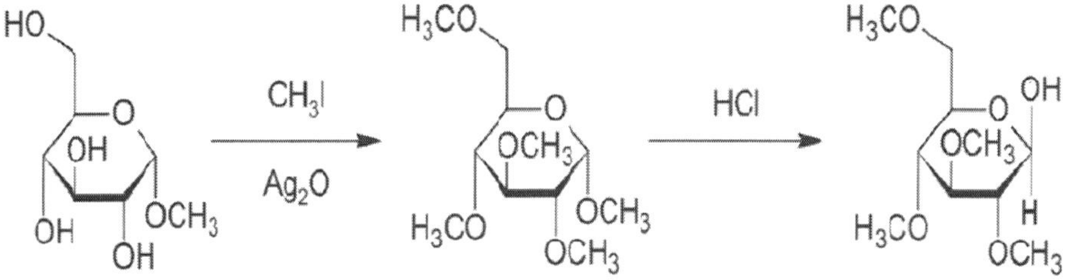

Bisulfite Sequencing

Bisulfite sequencing is the use of bisulfite treatment of DNA to determine its pattern of methylation. DNA methylation was the first discovered epigenetic mark, and remains the most studied. In animals it predominantly involves the addition of a methyl group to the carbon-5 position of cytosine residues of the dinucleotide CpG, and is implicated in repression of transcriptional activity.

Treatment of DNA with bisulfite converts cytosine residues to uracil, but leaves 5-methylcytosine residues unaffected. Thus, bisulfite treatment introduces specific changes in the DNA sequence that depend on the methylation status of individual cytosine residues, yielding single- nucleotide resolution information about the methylation status of a segment of DNA. Various analyses can be performed on the altered sequence to retrieve this information. The objective of this analysis is therefore reduced to differentiating between single nucleotide polymorphisms (cytosines and thymidine) resulting from bisulfite conversion.

Methods

Bisulfite sequencing applies routine sequencing methods on bisulfite-treated genomic DNA to determine methylation status at CpG dinucleotides. Other non-sequencing strategies are also employed to interrogate the methylation at specific loci or at a genome-wide level. All strategies assume that bisulfite-induced conversion of unmethylated cytosines to uracil is complete, and this serves as the basis of all subsequent techniques.

Ideally, the method used would determine the methylation status separately for each allele. Alternative methods to bisulfite sequencing include Combined Bisulfite Restriction Analysis and methylated DNA immunoprecipitation (MeDIP).

Methodologies to analyze bisulfite-treated DNA are continuously being developed. To summarize these rapidly evolving methologies, numerous review articles have been written.

The methodologies can be generally divided into strategies based on methylation-specific PCR (MSP), and strategies employing polymerase chain reaction (PCR) performed under non-methylation-specific conditions. Microarray-based methods use PCR based on non-methylation-specific conditions also.

Non-methylation-specific PCR based Methods

Direct Sequencing

The first reported method of methylation analysis using bisulfite-treated DNA utilized PCR and standard dideoxynucleotide DNA sequencing to directly determine the nucleotides resistant to bisulfite conversion. Primers are designed to be strand-specific as well as bisulfite-specific (i.e., primers containing non-CpG cytosines such that they are not complementary to non-bisulfite-treated DNA), flanking (but not involving) the methylation site of interest. Therefore, it will amplify both methylated and unmethylated sequences, in contrast to methylation-specific PCR. All sites of unmethylated cytosines are displayed as thymines in the resulting amplified sequence of the sense strand, and as adenines in the amplified antisense strand. This technique required cloning of the PCR product prior to sequencing for adequate sensitivity, and therefore was a very labour-intensive method unsuitable for higher throughput. Alternatively, nested PCR methods can be used to enhance the product for sequencing.

All subsequent DNA methylation analysis techniques using bisulfite-treated DNA is based on this report by Frommer et al.. Although most other modalities are not true sequencing-based techniques, the term "bisulfite sequencing" is often used to describe bisulfite-conversion DNA methylation analysis techniques in general.

Pyrosequencing

Pyrosequencing has also been used to analyze bisulfite-treated DNA without using methylation-specific PCR. Following PCR amplification of the region of interest, Pyrosequencing is used to determine the bisulfite-converted sequence of specific CpG sites in the region. The ratio of C-to-T at individual sites can be determined quantitatively based on the amount of C and T incorporation during the sequence extension. The main limitation of this method is the cost of the technology. However, Pyrosequencing does well allow for extension to high-throughput screening methods. A further improvement to this technique was recently described by Wong et al., which uses allele-specific primers that incorporate single-nucleotide polymorphisms into the sequence of the sequencing primer, thus allowing for separate analysis of maternal and paternal alleles. This technique is of particular usefulness for genomic imprinting analysis.

Methylation-sensitive Single-strand Conformation Analysis (MS-SSCA)

This method is based on the single-strand conformation polymorphism analysis (SSCA) method developed for single-nucleotide polymorphism (SNP) analysis. SSCA differentiates between single-stranded DNA fragments of identical size but distinct sequence based on differential migration in non-denaturating electrophoresis. In MS-SSCA, this is used to distinguish between bisulfite-treated, PCR-amplified regions containing the CpG sites of interest. Although SSCA lacks sensitivity when only a single nucleotide difference is present, bisulfite treatment frequently makes a number of C-to-T conversions in most regions of interest, and the resulting sensitivity approaches 100%. MS-SSCA also provides semi-quantitative analysis of the degree of DNA methylation based on the ratio of band intensities. However, this method is designed to assess all CpG sites as a whole in the region of interest rather than individual methylation sites.

High Resolution Melting Analysis (HRM)

A further method to differentiate converted from unconverted bisulfite-treated DNA is using high-resolution melting analysis (HRM), a real-time PCR-based technique initially designed to distinguish SNPs. The PCR amplicons are analyzed directly by temperature ramping and resulting liberation of an intercalating fluorescent dye during melting. The degree of methylation, as represented by the C-to-T content in the amplicon, determines the rapidity of melting and consequent release of the dye. This method allows direct quantitation in a single-tube assay, but assesses methylation in the amplified region as a whole rather than at specific CpG sites.

Methylation-sensitive Single-nucleotide Primer Extension (MS-SnuPE)

MS-SnuPE employs the primer extension method initially designed for analyzing single-nucleotide polymorphisms. DNA is bisulfite-converted, and bisulfite-specific primers are annealed to the sequence up to the base pair immediately before the CpG of interest. The primer is allowed to extend one base pair into the C (or T) using DNA polymerase terminating dideoxynucleotides, and the ratio of C to T is determined quantitatively.

A number of methods can be used to determine this C:T ratio. At the beginning, MS-SnuPE relied on radioactive ddNTPs as the reporter of the primer extension. Fluorescence-based methods or Pyrosequencing can also be used. However, matrix-assisted laser desorption ionization/ time-of-flight (MALDI-TOF) mass spectrometry analysis to differentiate between the two polymorphic primer extension products can be used, in essence, based on the GOOD assay designed for SNP genotyping. Ion pair reverse-phase high-performance liquid chromatography (IP-RP-HPLC) has also been used to distinguish primer extension products.

Base-specific Cleavage/MALDI-TOF

A recently described method by Ehrich et al. further takes advantage of bisulfite-conversions by adding a base-specific cleavage step to enhance the information gained from the nucleotide changes. By first using in vitro transcription of the region of interest into RNA (by adding an RNA polymerase promoter site to the PCR primer in the initial amplification), RNase A can be used to cleave the RNA transcript at base-specific sites.

As RNase A cleaves RNA specifically at cytosine and uracil ribonucleotides, base-specificity is achieved by adding incorporating cleavage-resistant dTTP when cytosine-specific (C-specific) cleavage is desired, and incorporating dCTP when uracil-specific (U-specific) cleavage is desired. The cleaved fragments can then be analyzed by MALDI-TOF.

Bisulfite treatment results in either introduction/removal of cleavage sites by C-to-U conversions or shift in fragment mass by G-to-A conversions in the amplified reverse strand. C-specific cleavage will cut specifically at all methylated CpG sites. By analyzing the sizes of the resulting fragments, it is possible to determine the specific pattern of DNA methylation of CpG sites within the region, rather than determining the extent of methylation of the region as a whole. This method demonstrated efficacy for high-throughput screening, allowing for interrogation of numerous CpG sites in multiple tissues in a cost-efficient manner.

Methylation-specific PCR (MSP)

This alternative method of methylation analysis also uses bisulfite-treated DNA but avoids the need to sequence the area of interest. Instead, primer pairs are designed themselves to be "methylated-specific" by including sequences complementing only unconverted 5-

methylcytosines, or, on the converse, "unmethylated-specific", complementing thymines converted from unmethylated cytosines. Methylation is determined by the ability of the specific primer to achieve amplification.

This method is particularly useful to interrogate CpG islands with possibly high methylation density, as increased numbers of CpG pairs in the primer increase the specificity of the assay. Placing the CpG pair at the 3'-end of the primer also improves the sensitivity. The initial report using MSP described sufficient sensitivity to detect methylation of 0.1% of alleles. In general, MSP and its related protocols are considered to be the most sensitive when interrogating the methylation status at a specific locus.

The MethyLight method is based on MSP, but provides a quantitative analysis using real-time PCR. Methylated-specific primers are used, and a methylated-specific fluorescence reporter probe is also used that anneals to the amplified region. In alternative fashion, the primers or probe can be designed without methylation specificity if discrimination is needed between the CpG pairs within the involved sequences. Quantitation is made in reference to a methylated reference DNA. A modification to this protocol to increase the specificity of the PCR for successfully bisulfite-converted DNA (ConLight-MSP) uses an additional probe to bisulfite-unconverted DNA to quantify this non-specific amplification.

Further methodology using MSP-amplified DNA analyzes the products using melting curve analysis (Mc-MSP). This method amplifies bisulfite-converted DNA with both methylated-specific and unmethylated-specific primers, and determines the quantitative ratio of the two products by comparing the differential peaks generated in a melting curve analysis. A high-resolution melting analysis method that uses both real-time quantification and melting analysis has been introduced, in particular, for sensitive detection of low-level methylation.

Microarray-based Methods

Microarray-based methods are a logical extension of the technologies available to analyze bisulfite-treated DNA to allow for genome-wide analysis of methylation. Oligonucleotide microarrays are designed using pairs of oligonucleotide hybridization probes targeting CpG sites of interest. One is complementary to the unaltered methylated sequence, and the other is complementary to the C-to-U-converted unmethylated sequence. The probes are also bisulfite-specific to prevent binding to DNA incompletely converted by bisulfite. The Illumina Methylation Assay is one such assay that applies the bisulfite sequencing technology on a microarray level to generate genome-wide methylation data.

Limitations

Incomplete Conversion: Bisulfite sequencing relies on the conversion of every single unmethylated cytosine residue to uracil. If conversion is incomplete, the subsequent analysis will incorrectly interpret the unconverted unmethylated cytosines as methylated cytosines, resulting in false positive results for methylation. Only cytosines in single-stranded DNA are susceptible to attack by bisulfite, therefore denaturation of the DNA undergoing analysis is critical. It is important to ensure that reaction parameters such as temperature and salt concentration are suitable to maintain the DNA in a single-stranded conformation and allow for complete conversion. Embedding the DNA in agarose gel has been reported to improve the rate of conversion by keeping strands of DNA physically separate.

Degradation of DNA during Bisulfite Treatment

A major challenge in bisulfite sequencing is the degradation of DNA that takes place concurrently with the conversion. The conditions necessary for complete conversion, such as long incubation times, elevated temperature, and high bisulfite concentration, can lead to the degradation of about 90% of the incubated DNA. Given that the starting amount of DNA is often limited, such extensive degradation can be problematic.

The degradation occurs as depurinations resulting in random strand breaks. Therefore the longer the desired PCR amplicon, the more limited the number of intact template molecules will likely be. This could lead to the failure of the PCR amplification, or the loss of quantitatively accurate information on methylation levels resulting from the limited sampling of template molecules. Thus, it is important to assess the amount of DNA degradation resulting from the reaction conditions employed, and consider how this will affect the desired amplicon. Techniques can also be used to minimize DNA degradation, such as cycling the incubation temperature.

Other Concerns

A potentially significant problem following bisulfite treatment is incomplete desulfonation of pyrimidine residues due to inadequate alkalization of the solution. This may inhibit some DNA polymerases, rendering subsequent PCR difficult. However, this situation can be avoided by monitoring the pH of the solution to ensure that desulphonation will be complete.

A final concern is that bisulfite treatment greatly reduces the level of complexity in the sample, which can be problematic if multiple PCR reactions are to be performed (2006). Primer design is more difficult, and inappropriate cross-hybridization is more frequent.

Applications: Genome-wide Methylation Analysis

The advances in bisulfite sequencing have led to the possibility of applying them at a genome-wide scale, where, previously, global measure of DNA methylation was feasible only using other techniques, such as Restriction landmark genomic scanning.

The mapping of the human epigenome is seen by many scientists as the logical follow-up to the completion of the Human Genome Project. This epigenomic information will be important in understanding how the function of the genetic sequence is implemented and regulated. Since the epigenome is less stable than the genome, it is thought to be important in gene-environment interactions.

Epigenomic mapping is inherently more complex than genome sequencing, however, since the epigenome is much more variable than the genome. While an individual has only one genome, one's epigenome varies with age, differs between tissues, is altered by environmental factors, and shows aberrations in diseases.

Such rich epigenomic mapping, however, representing different ages, tissue types, and disease states, would yield valuable information on the normal function of epigenetic marks as well as the mechanisms leading to aging and disease.

Direct benefits of epigenomic mapping include probable advances in cloning technology. It is believed that failures to produce cloned animals with normal viability and lifespan result from inappropriate patterns of epigenetic marks. Also, aberrant methylation patterns are well characterized in many cancers. Global hypomethylation results in decreased genomic stability, while local hypermethylation of tumour suppressor gene promoters often accounts for their loss of function. Specific patterns of methylation are indicative of specific cancer types, have prognostic value, and can help to guide the best course of treatment.

Large-scale epigenome mapping efforts are under way around the world and have been organized under the Human Epigenome Project.

This is based on a multi-tiered strategy, whereby bisulfite sequencing is used to obtain high-resolution methylation profiles for a limited number of reference epigenomes, while less thorough analysis is performed on a wider spectrum of samples. This approach is intended to maximize the insight gained from a given amount of resources, as high-resolution genome-wide mapping remains a costly undertaking.

DNA Methylation

DNA methylation is a biochemical process that is important for normal development in higher organisms. It involves the addition of a methyl group to the 5 position of the cytosine pyrimidine ring or the number 6 nitrogen of the adenine purine ring (cytosine and adenine are two of the four bases of DNA). This modification can be inherited through cell division.

DNA methylation is a crucial part of normal organismal development and cellular differentiation in higher organisms. DNA methylation stably alters the gene expression pattern in cells such that cells can "remember where they have been" or decrease gene expression; for example, cells programmed to be pancreatic islets during embryonic development remain pancreatic islets throughout the life of the organism without continuing signals telling them that they need to remain islets.

DNA methylation is typically removed during zygote formation and re-established through successive cell divisions during development. However, the latest research shows that hydroxylation of methyl group occurs rather than complete removal of methyl groups in zygote. Some methylation modifications that regulate gene expression are inheritable and are referred to as epigenetic regulation.

In addition, DNA methylation suppresses the expression of viral genes and other deleterious elements that have been incorporated into the genome of the host over time. DNA methylation also forms the basis of chromatin structure, which enables cells to form the myriad characteristics necessary for multicellular life from a single immutable sequence of DNA. DNA methylation also plays a crucial role in the development of nearly all types of cancer.

DNA methylation at the 5 position of cytosine has the specific effect of reducing gene expression and has been found in every vertebrate examined. In adult somatic tissues, DNA methylation typically occurs in a CpG dinucleotide context; non-CpG methylation is prevalent in embryonic stem cells.

In Mammals

DNA methylation is essential for normal development and is associated with a number of key processes including genomic imprinting, X-chromosome inactivation, suppression of repetitive elements, and carcinogenesis.

Between 60% and 90% of all CpGs are methylated in mammals. Methylated C residues spontaneously deaminate to form T residues over evolutionary time; hence CpG dinucleotides steadily mutate to TpG dinucleotides, which is evidenced by the under-representation of CpG dinucleotides in the human genome (they occur at only 21% of the expected frequency). (On the other hand, spontaneous deamination of unmethylated C residues gives rise to U residues, a mutation that is quickly recognized and repaired by the cell.)

Unmethylated CpGs are often grouped in clusters called *CpG islands*, which are present in the 5' regulatory regions of many genes. In many disease processes, such as cancer, gene promoter CpG islands acquire abnormal hypermethylation, which results in transcriptional silencing that can be inherited by daughter cells following cell division.

Alterations of DNA methylation have been recognized as an important component of cancer development. Hypomethylation, in general, arises earlier and is linked to chromosomal instability and loss of imprinting, whereas hypermethylation is associated with promoters and can arise secondary to gene (oncogene suppressor) silencing, but might be a target for epigenetic therapy.

DNA methylation may affect the transcription of genes in two ways. First, the methylation of DNA itself may physically impede the binding of transcriptional proteins to the gene, and second, and likely more important, methylated DNA may be bound by proteins known as methyl-CpG-binding domain proteins (MBDs).

MBD proteins then recruit additional proteins to the locus, such as histone deacetylases and other chromatin remodelling proteins that can modify histones, thereby forming compact, inactive chromatin, termed heterochromatin.

This link between DNA methylation and chromatin structure is very important. In particular, loss of methyl-CpG-binding protein 2 (MeCP2) has been implicated in Rett syndrome; and methyl-CpG-binding domain protein 2 (MBD2) mediates the transcriptional silencing of hypermethylated genes in cancer.

Research has suggested that long-term memory storage in humans may be regulated by DNA methylation.

In Cancer

DNA methylation is an important regulator of gene transcription and a large body of evidence has demonstrated that aberrant DNA methylation is associated with unscheduled gene silencing, and the genes with high levels of 5-methylcytosine in their promoter region are transcriptionally silent.

DNA methylation is essential during embryonic development, and in somatic cells, patterns of DNA methylation are generally transmitted to daughter cells with a high fidelity. Aberrant DNA methylation patterns have been associated with a large number of human malignancies and found in two distinct forms: hypermethylation and hypomethylation compared to normal tissue.

Hypermethylation is one of the major epigenetic modifications that repress transcription via promoter region of tumour suppressor genes. Hypermethylation typically occurs at CpG islands in the promoter region and is associated with gene inactivation. Global hypomethylation has also been implicated in the development and progression of cancer through different mechanisms.

DNA Methyltransferases

In mammalian cells, DNA methylation occurs mainly at the C5 position of CpG dinucleotides and is carried out by two general classes of enzymatic activities – maintenance methylation and *de novo* methylation.

Maintenance methylation activity is necessary to preserve DNA methylation after every cellular DNA replication cycle. Without the DNA methyltransferase (DNMT), the replication machinery itself would produce daughter strands that are unmethylated and, over time, would lead to passive demethylation. DNMT1 is the proposed maintenance methyltransferase that is responsible for copying DNA methylation patterns to the daughter strands during DNA replication. Mouse models with both copies of DNMT1 deleted are embryonic lethal at approximately day 9, due to the requirement of DNMT1 activity for development in mammalian cells.

It is thought that DNMT3a and DNMT3b are the *de novo* methyltransferases that set up DNA methylation patterns early in development. DNMT3L is a protein that is homologous to the other DNMT3s but has no catalytic activity. Instead, DNMT3L assists the *de novo* methyltransferases by increasing their ability to bind to DNA and stimulating their activity. Finally, DNMT2 (TRDMT1) has been identified as a DNA methyltransferase homologue, containing all 10 sequence motifs common to all DNA methyltransferases; however, DNMT2 (TRDMT1) does not methylate DNA but instead methylates cytosine-38 in the anticodon loop of aspartic acid transfer RNA.

Since many tumour suppressor genes are silenced by DNA methylation during carcinogenesis, there have been attempts to re-express these genes by inhibiting the DNMTs. 5-Aza-2'-deoxycytidine (decitabine) is a nucleoside analog that inhibits DNMTs by trapping them in a covalent complex on DNA by preventing the β-elimination step of catalysis, thus resulting in the enzymes' degradation. However, for decitabine to be active, it must be incorporated into the genome of the cell, which can cause mutations in the daughter cells if the cell does not die. In addition, decitabine is toxic to the bone marrow, which limits the size of its therapeutic window. These pitfalls have led to the development of antisense RNA therapies that target the DNMTs by degrading their mRNAs and preventing their translation. However, it is currently unclear whether targeting DNMT1 alone is sufficient to reactivate tumour suppressor genes silenced by DNA methylation.

In Plants

Significant progress has been made in understanding DNA methylation in the model plant *Arabidopsis thaliana*. DNA methylation in plants differs from that of mammals: while DNA methylation in mammals mainly occurs on the cytosine nucleotide in a CpG site, in plants the cytosine can be methylated at CpG, CpHpG, and CpHpH sites, where H represents any nucleotide but guanine.

The principal *Arabidopsis* DNA methyltransferase enzymes, which transfer and covalently attach methyl groups onto DNA, are DRM2, MET1, and CMT3. Both the DRM2 and MET1 proteins share significant homology to the mammalian methyltransferases DNMT3 and DNMT1, respectively, whereas the CMT3 protein is unique to the plant kingdom. There are currently two classes of DNA methyltransferases:

1) the *de novo* class, or enzymes that create new methylation marks on the DNA; and
2) a maintenance class that recognizes the methylation marks on the parental strand of DNA and transfers new methylation to the daughters strands after DNA replication.

DRM2 is the only enzyme that has been implicated as a *de novo* DNA methyltransferase. DRM2 has also been shown, along with MET1 and CMT3 to be involved in maintaining methylation marks through DNA replication. Other DNA methyltransferases are expressed in plants but have no known function.

It is not clear how the cell determines the locations of *de novo* DNA methylation, but evidence suggests that, for many (though not all) locations, RNA-directed DNA methylation (RdDM) is involved. In RdDM, specific RNA transcripts are produced from a genomic DNA template, and this RNA forms secondary structures called double-stranded RNA molecules.

The double-stranded RNAs, through either the small interfering RNA (siRNA) or microRNA (miRNA) pathways direct de-novo DNA methylation of the original genomic location that produced the RNA. This sort of mechanism is thought to be important in cellular defence against RNA viruses and/or transposons, both of which often form a double-stranded RNA that can be mutagenic to the host genome. By methylating their genomic locations, through an as yet poorly-understood mechanism, they are shut off and are no longer active in the cell, protecting the genome from their mutagenic effect.

In Fungi

It can be seen that many fungi have low levels (0.1 to 0.5%) of cytosine methylation, whereas other fungi have as much as 5% of the genome methylated.

This value seems to vary both among species and among isolates of the same species. There is also evidence that DNA methylation may be involved in state-specific control of gene expression in fungi.

Although brewers' yeast (*Saccharomyces*) and fission yeast (*Schizosaccharomyces*) have very little DNA methylation, the model filamentous fungus *Neurospora crassa* has a well-characterized methylation system. Several genes control methylation in *Neurospora* and mutation of the DNA methyl transferase, *dim-2*, eliminates all DNA methylation but does not affect growth or sexual reproduction. While the *Neurospora* genome has very little repeated DNA, half of the methylation occurs in repeated DNA including transposon relics and centromeric DNA. The ability to evaluate other important phenomena in a DNA methylase-deficient genetic background makes *Neurospora* an important system in which to study DNA methylation.

In Bacteria

Adenine or cytosine methylation is part of the restriction modification system of many bacteria, in which specific DNA sequences are methylated periodically throughout the genome. A methylase is the enzyme that recognizes a specific sequence and methylates one of the bases in or near that sequence.

Foreign DNAs (which are not methylated in this manner) that are introduced into the cell are degraded by sequence-specific restriction enzymes and cleaved. Bacterial genomic DNA is not recognized by these restriction enzymes. The methylation of native DNA acts as a sort of primitive immune system, allowing the bacteria to protect themselves from infection by bacteriophage.

E. coli DNA adenine methyltransferase (Dam) is an enzyme of ~32 kDa that does not belong to a restriction/modification system. The target recognition sequence for *E. coli* Dam is GATC, as the methylation occurs at the N6 position of the adenine in this sequence (G meATC). The three base pairs flanking each side of this site also influence DNA–Dam binding. Dam plays several key roles in bacterial processes, including mismatch repair, the timing of DNA replication, and gene expression.

As a result of DNA replication, the status of GATC sites in the *E. coli* genome changes from fully methylated to hemimethylated. This is because adenine introduced into the new DNA strand is unmethylated. Re-methylation occurs within two to four seconds, during which time replication errors in the new strand are repaired. Methylation, or its absence, is the marker that allows the repair apparatus of the cell to differentiate between the template and nascent strands. It has been shown that altering Dam activity in bacteria results in increased spontaneous mutation rate. Bacterial viability is compromised in dam mutants that also lack certain other DNA repair enzymes, providing further evidence for the role of Dam in DNA repair.

One region of the DNA that keeps its hemimethylated status for longer is the origin of replication, which has an abundance of GATC sites. This is central to the bacterial mechanism for timing DNA replication. SeqA binds to the origin of replication, sequestering it and thus preventing methylation. Because hemimethylated origins of replication are inactive, this mechanism limits DNA replication to once per cell cycle.

Expression of certain genes, for example those coding for pilus expression in *E. coli*, is regulated by the methylation of GATC sites in the promoter region of the gene operon. The cells' environmental conditions just after DNA replication determine whether Dam is blocked from methylating a region proximal to or distal from the promoter region. Once the pattern of methylation has been created, the pilus gene transcription is locked in the on or off position until the DNA is again replicated. In *E. coli*, these pilus operons have important roles in virulence in urinary tract infections. It has been proposed that inhibitors of Dam may function as antibiotics.

On the other hand, DNA cytosine methylase targets CCAGG and CCTGG sites to methylate cytosine at the C5 position (C meC(A/T)GG). The other methylase enzyme, EcoKI, causes mehtylation of adenine in the sequences AAC(N6A)GTGC and GCAC(N6A)GTT.

Most strains used by molecular biologists are derivatives of K-12, and possess both Dam and Dcm, but there are commercially available strains that possess dam-/dcm- activity. In fact, it is possible to unmethylate the DNA extracted from dam+/dcm+ strains by transforming into dam-/dcm- strains. This would help digest sequences that are not being recognized by methylation-sensitive restriction enzymes.

Detection

DNA methylation can be detected by the following assays currently used in scientific research:

- Methylation-Specific PCR (MSP), which is based on a chemical reaction of sodium bisulfite with DNA that converts unmethylated cytosines of CpG dinucleotides to uracil or UpG, followed by traditional PCR. However, methylated cytosines will not be converted in this process, and primers are designed to overlap the CpG site of interest, which allows one to determine methylation status as methylated or unmethylated.
- Whole genome bisulfite sequencing, also known as BS-Seq, which is a high-throughput genome-wide analysis of DNA methylation. It is based on aforementioned sodium bisulfite conversion of genomic DNA, which is then sequencing on a Next-generation sequencing platform. The sequences obtained are then re-aligned to the reference genome to determine methylation states of CpG dinucleotides based on mismatches resulting from the conversion of unmethylated cytosines into uracil.
- The HELP assay, which is based on restriction enzymes' differential ability to recognize and cleave methylated and unmethylated CpG DNA sites.
- ChIP-on-chip assays, which is based on the ability of commercially prepared antibodies to bind to DNA methylation-associated proteins like MeCP2.
- Restriction landmark genomic scanning, a complicated and now rarely-used assay based upon restriction enzymes' differential recognition of methylated and unmethylated CpG sites; the assay is similar in concept to the HELP assay.

- Methylated DNA immunoprecipitation (MeDIP), analogous to chromatin immunoprecipitation, immunoprecipitation is used to isolate methylated DNA fragments for input into DNA detection methods such as DNA microarrays (MeDIP-chip) or DNA sequencing (MeDIP-seq).
- Pyrosequencing of bisulfite treated DNA. This is sequencing of an amplicon made by a normal forward primer but a biatenylated reverse primer to PCR the gene of choice. The Pyrosequencer then analyses the sample by denaturing the DNA and adding one nucleotide at a time to the mix according to a sequence given by the user. If there is a mis-match, it is recorded and the percentage of DNA for which the mis-match is present is noted. This gives the user a percentage methylation per CpG island.
- Molecular break light assay for DNA adenine methyltransferase activity – an assay that relies on the specificity of the restriction enzyme DpnI for fully methylated (adenine methylation) GATC sites in an oligonucleotide labelled with a fluorophore and quencher. The adenine methyltransferase methylates the oligonucleotide making it a substrate for DpnI. Cutting of the oligonucleotide by DpnI gives rise to a fluorescence increase.
- Methyl Sensitive Southern Blotting is similar to the HELP assay, although uses Southern blotting techniques to probe gene-specific differences in methylation using restriction digests. This technique is used to evaluate local methylation near the binding site for the probe.

Demethylating Agent

Demethylating agents are compounds that can inhibit methylation, resulting in the expression of the previously hypermethylated silenced genes. Cytidine analogs such as 5-azacytidine (azacitidine) and 5-azadeoxycytidine (decitabine) are the most commonly used demethylating agents . These compounds work by binding to the enzymes that catalyse the methylation reaction, DNA methyltransferases; and titrate out these enzymes . Both compounds have been approved in the treatment of Myelodysplastic syndrome (MDS) by Food and Drug Administration (FDA) in United States. Azacitidine and decitabine are marketed as Vidaza and Dacogen respectively. Azacitidine is the first drug to be approved by FDA for treating MDS and has been given orphan drug status . Procaine is a DNA-demethylating agent with growth-inhibitory effects in human cancer cells.

Chapter 8

Nitrogenous Base

A nitrogenous (nitrogen-containing) base is a nitrogen-containing molecule having the chemical properties of a base. It is an organic compound that owes its property as a base to the lone pair of electrons of a nitrogen atom.

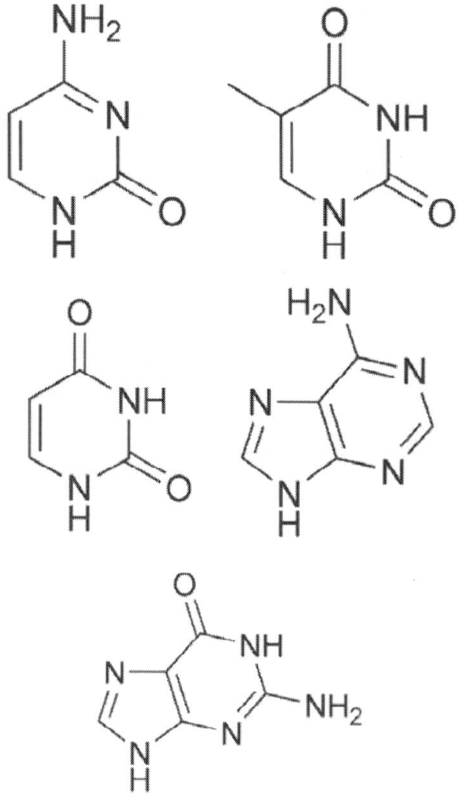

In biological sciences, nitrogenous bases are typically classified as the derivatives of two parent compounds, pyrimidine and purine. They are non-polar and due to their aromaticity, planar. Both pyrimidines and purines resemble pyridine and are thus weak bases and relatively unreactive towards electrophilic aromatic substitution. Their flat shape is particularly important when considering their roles in nucleic acids as nucleobases (building blocks of DNA and RNA): adenine, guanine, thymine, cytosine, and uracil. These nitrogenous bases hydrogen bond between opposing DNA strands to form the rungs of the "twisted ladder" or double helix of DNA or a biological catalyst that is found in the nucleotides. Adenine is always paired with thymine, and guanine is always paired with cytosine.

Microscale Thermophoresis

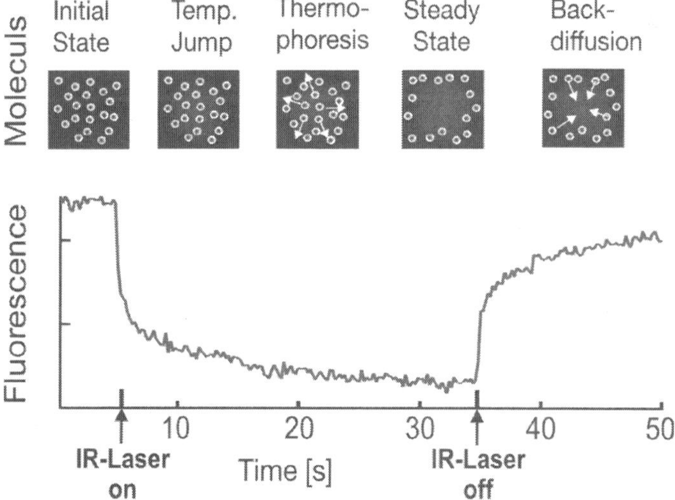

Figure: *Microscale Thermophoresis: The fluorescence inside a capillary is measured and the normalized fluorescence in the IR-Laser heated spot is plotted against time. The IR-laser is switched on at t=5 s and the fluorescence changes as the temperature increases. There are two effects, separated by their time-scales, contributing to the new fluorescence distribution: the fast temperature jump (time scale H" 3 s) and the thermophoretic movement (time scale H" 30 s). Once the IR-laser is switched off (t=35 s) the molecules diffuse back.*

Microscale Thermophoresis (MST) is a technology for the analysis of biomolecules. Microscale Thermophoresis is the directed movement of particles in a microscopic temperature gradient. Any change of the hydration shell of biomolecules due to changes in their structure/ conformation results in a relative change of movement along the temperature gradient and is used to determine binding affinities, binding kinetics and activity kinetics. Events such as the phosphorylation of a protein or the binding of small molecules to a target can be monitored.

MST allows to measure interactions directly in solution without the need of an immobilization to a surface (immobilization-free technology).

Microscale Thermophoresis was developed by the NanoTemper Technologies GmbH, a German high tech company with headquarters in Munich.

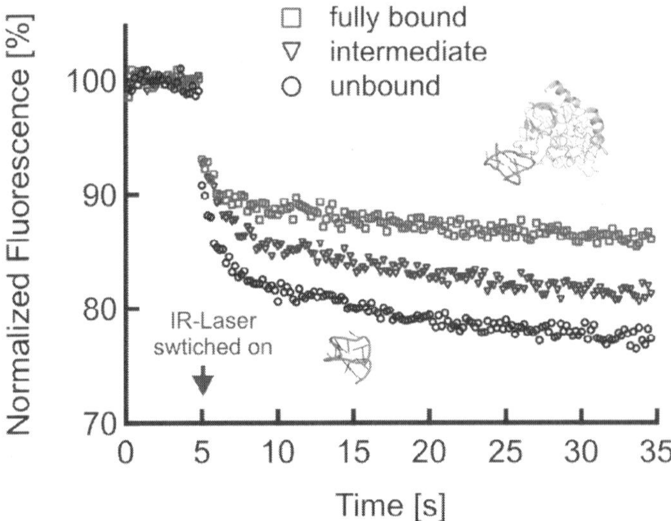

Figure: *Microscale Thermophoresis Aptamer-Protein Binding Assay. The thermophoretic movement of the fluorescently labelled aptamer changes upon binding to the protein*

Applications

Microscale Thermophoresis (MST) can be used to:

- measure affinities between biomolecules including proteins, DNA, RNA, peptides, molecule fragments, small molecules
- measure interactions of biomolecules with nanoparticles, vesicles and nanodiscs
- measure interactions with viruses
- measure the stoichiometry of an interaction
- measure enzyme kinetics
- measure methylation, phosphorylation and the like
- characterize surface modifications of nanoparticles
- measure dissociation constants of multi-component reactions

- recognize different binding sites on a target molecule of interest
- measure the stability of biomolecules in blood, blood serum and blood plasma
- determine binding kinetics (k_{off}/k_{on} rates)
- study the adsorption of small molecules to lipid membranes or plasma proteins
- study protein aggregating compounds/conditions
- perform competition studies including the binding of substrates and inhibitors to an enzyme.

Technology

Microscale thermophoresis (MST) is a new method that enables the quantitative analysis of molecular interactions in solution at the microliter scale. MST is based on the directed movement of molecules along temperature gradients, an effect termed thermophoresis. A spatial temperature difference ΔT leads to a depletion of molecule concentration in the region of elevated temperature, quantified by the Soret coefficient $S_T : c_{hot} / c_{cold} = \exp(-S_T \Delta T)$.

Thermophoresis depends on the interface between molecule and solvent. Under constant buffer conditions, thermophoresis probes the size, charge and solvation entropy of the molecules. The thermophoresis of a fluorescently labelled molecule A typically differs significantly from the thermophoresis of an molecule-target complex AT due to size, charge and solvation entropy differences. This difference in the molecule's thermophoresis is used to quantify the binding in titration experiments under constant buffer conditions.

The thermophoretic movement of the fluorescently labelled molecule is measured by monitoring the fluorescence distribution F inside a capillary . The microscopic temperature gradient is generated by an IR-Laser, which is focused into the capillary and is strongly absorbed by water.

The temperature of the aqueous solution in the laser spot is raised by up to ΔT=5 K. Before the IR-Laser is switched on a homogeneous fluorescence distribution F_{cold} is observed inside the capillary. When the IR-Laser is switched on, two effects, separated by their time-scales, contribute to the new fluorescence distribution F_{hot}. The thermal relaxation time is fast and induces a binding-dependent drop in the fluorescence of the dye due to its local environmental-dependent response to the temperature jump. On the slower diffusive time scale (10 s), the molecules move from the locally heated region to the outer cold regions. The local concentration of molecules decreases in the heated region until it reaches a steady-state distribution. While the mass diffusion D dictates the kinetics of depletion, S_T determines the steady-state concentration ratio $c_{hot} / c_{cold} = \exp(-S_T \Delta T) \approx 1 - S_T \Delta T$ under a temperature increase ΔT. The normalized fluorescence $F_{norm}=F_{hot}/F_{cold}$ measures mainly this concentration ratio, in addition to the temperature jump "F/"T. In the linear approximation we find: $F_{norm} = 1 + (\partial F / \partial T - S_T)\Delta T$. Due to the linearity of the fluorescence intensity and the thermophoretic depletion, the normalized fluorescence from the unbound molecule F_{norm}(A) and the bound complex F_{norm}(AT) superpose linearly. By denoting x the fraction of molecules bound to targets, the changing fluorescence signal during the titration of target T is given by: F_{norm}=(1-x) F_{norm}(A)+x F_{norm}(AT).

Quantitative binding parameters are obtained by using a serial dilution of the binding substrate. By plotting F_{norm} against the logarithm of the different concentrations of the dilution series, a sigmoidal binding curve is obtained. This binding curve can directly be fitted with the nonlinear solution of the law of mass action, with the dissociation constant K_D as result.

Histone Methyltransferase

Histone methyltransferases (HMT) are enzymes, histone-lysine N-methyltransferase and histone-arginine N-methyltransferase, that catalyze the transfer of one to three methyl groups from the cofactor S-Adenosyl methionine to lysine and arginine residues of histone proteins. *These proteins* often contain a SET (Su(var)3-9, Enhancer of Zeste, Trithorax) domain, however the recently discovered HMT Dot1 lacks the characteristic SET domain.

Role in Gene Regulation

Histone methylation serves in epigenetic gene regulation. Methylated histones bind DNA more tightly, which inhibits transcription.

Histone Acetyltransferase

Histone acetyltransferases (HAT) are enzymes that acetylate conserved lysine amino acids on histone proteins by transferring an acetyl group from acetyl CoA to form ε-N-acetyl lysine. In general, histone acetylation is linked to transcriptional activation and associated with euchromatin. When it was first discovered, it was thought that acetylation of lysine neutralizes the positive charge normally present, thus reducing affinity between histone and (negatively charged) DNA, which renders DNA more accessible to transcription factors.

Research has emerged, since, to show that lysine acetylation and other post-translational modifications of histones generate binding sites for specific protein–protein interaction domains, such as the acetyl-lysine-binding bromodomain. Histone acetyltransferases can also acetylate non-histone proteins, such as transcription factors and nuclear receptors to facilitate gene expression.

Examples

Human proteins that possess histone acetyltransferase catalytic activity include:

- CREBBP, CDY1 , CDY2, CDYL1, CLOCK
- ELP3 , EP300
- HAT1
- KAT2A, KAT2B, KAT5
- MYST1, MYST2, MYST3, MYST4
- NCOA1, NCOA3, NCOAT
- TF3C4

Interaction with HDACs

Histone acetyltransferases (HATs) and histone deacetylases (HDACs) are recruited to their target promoters through a physical interaction with a sequence-specific transcription factor (TF). They usually function within a multimolecular complex ('enzymatic complex'), in which the other subunits are necessary for them to modify nucleosomes around the binding site. These enzymes can also modify factors other than histones (protein X).

Acetyltransferase

Acetyltransferase (or transacetylase) is a type of transferase enzyme that transfers an acetyl group. Examples include:

- Histone acetyltransferases including CBP histone acetyltransferase

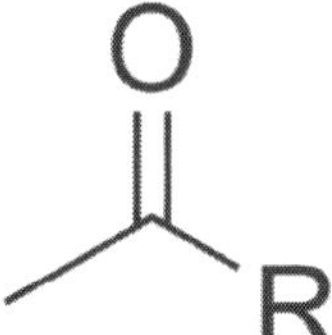

Figure: *Chemical structure of an acetyl group bound to the remainder R of a molecule.*

- Choline acetyltransferase
- Chloramphenicol acetyltransferase
- Serotonin N-acetyl transferase
- NatA Acetyltransferase
- NatB acteyltransferase

Industrial Union Department v. American Petroleum Institute

Industrial Union Department v. American Petroleum Institute (The Benzene Case), 448 U.S. 607 (1980), was a case heard before the United States Supreme Court. This case represented a challenge to the OSHA practice of regulating carcinogens by setting the exposure limit "at the lowest technologically feasible level that will not impair the viability of the industries regulated."

OSHA selected that standard because it believed that (1) it could not determine a safe exposure level and that (2) the authorizing statute did not require it to quantify such a level. A plurality on the Court, led by Justice Stevens, wrote that the authorizing statute did indeed require OSHA to demonstrate a significant risk of harm (albeit not with mathematical certainty) in order to justify setting a particular exposure level.

Perhaps more importantly, the Court noted in dicta that if the government's interpretation of the authorizing statute had been correct, it might violate the Nondelegation doctrine. This line of reasoning may represent the "high-water mark" of recent attempts to revive the doctrine.

Background

The Occupational Safety and Health Act of 1970 delegated broad authority to the Secretary of Labour to promulgate standards to ensure safe and healthful working conditions for the Nation's workers (the Occupational Safety and Health Administration (OSHA) being the agency responsible for carrying out this authority). According to Section 3(8), standards created by the secretary must be "reasonably necessary or appropriate to provide safe or healthful employment and places of employment".

Section 6(b)(5) of the statute sets the principle for creating the safety regulations, directing the Secretary to "set the standard which most adequately assures, to the extent feasible, on the basis of the best available evidence, that no employee will suffer material impairment of health or functional capacity...". At issue in the case, is the Secretary's interpretation of "extent feasible" to mean that if a material is unsafe he must "set an exposure limit at the lowest technologically feasible level that will not impair the viability of the industries regulated."

Opinion of the Court

The Court held the Secretary applied the act inappropriately. To comply with the statute, the secretary must determine 1) that a health risk of a substance exists at a particular threshold and 2) Decide whether to issue the most protective standard, or issue a standard that weighs the costs and benefits. Here, the secretary failed to first determine that a health risk of substance existed for the chemical benzene when workers were exposed at 1 part per million. Data only suggested the chemical was unsafe at 10 parts per million. Thus, the secretary had failed the first step of interpreting the statute, that is, finding that the substance posed a risk at that level.

In its reasoning, the Court noted it would be unreasonable to assume that congress intended to give the Secretary "unprecedented power over American industry". Such a delegation of power would likely be unconstitutional. The court also cited the legislative history of the act, which suggested that congress meant to address major workplace hazards, not hazards with low statistical likelihoods.

Concurring Opinion

In a famous concurrence, Justice Rehnquist argued that the section 6(b)(5) of the statute, which set forth the "extent feasible" principle, should be struck down on the basis of the non-delegation doctrine. The non-delegation doctrine, which has been recognized by the Supreme Court since the era of Chief Justice Marshall, holds that Congress cannot delegate law-making authority to other branches of government. Rehnquist offered three rationales for the application of the non-delegation doctrine. First, ensure Congress makes social policy, not agencies; delegation should only be used when the policy is highly technical or the ground too large to be covered. Second, agencies of the delegated authority require an "intelligible principle" to exercise discretion which was lacking in this case. Third, the intelligible principle must provide judges with a measuring stick for judicial review.

Subsequent Developments

Most scholars have said that the interpretation of statute ignored a foundational principle of statutory interpretation. Generally, specific language governs general language. In this case, the court read the more general provision of Section 3(8) as governing the specific process specified in Section 6(b)(5).

The case also marks the current state of affairs for the non-delegation doctrine. When the court is faced with a provision that appears to be an impermissible delegation of the authority, it will use tools of statutory interpretation to try to narrow the delegation of power.

Friedel–Crafts Reaction

The Friedel–Crafts reactions are a set of reactions developed by Charles Friedel and James Crafts in 1877. There are two main types of Friedel–Crafts reactions: alkylation reactions and acylation reactions. This reaction type is a form of electrophilic aromatic substitution. The general reaction scheme is shown below.

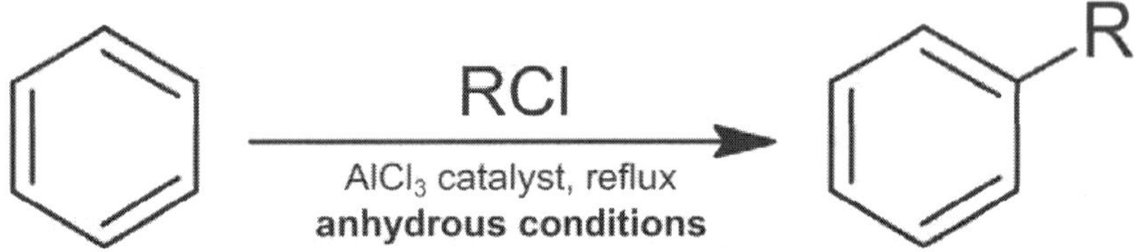

Figure 5: *Several reviews have been written.*

Friedel–Crafts alkylation

Friedel–Crafts alkylation involves the alkylation of an aromatic ring with an alkyl halide using a strong Lewis acid catalyst. With anhydrous ferric chloride as a catalyst, the alkyl group attaches at the former site of the chloride ion. The general mechanism is shown below.

$$R{-}Cl + FeCl_3 \longrightarrow R^+ + FeCl_4^-$$

This reaction has one big disadvantage, namely that the product is more nucleophilic than the reactant due to the electron donating alkyl-chain. Therefore, another hydrogen is substituted with an alkyl-chain, which leads to overalkylation of the molecule. Also, if the chlorine is not on a tertiary carbon, carbocation rearrangement reaction will occur. This is due to the relative stability of the tertiary carbocation over the secondary and primary carbocations. Steric hindrance can be exploited to limit the number of alkylations, as in the *t*-butylation of 1,4-dimethoxybenzene.

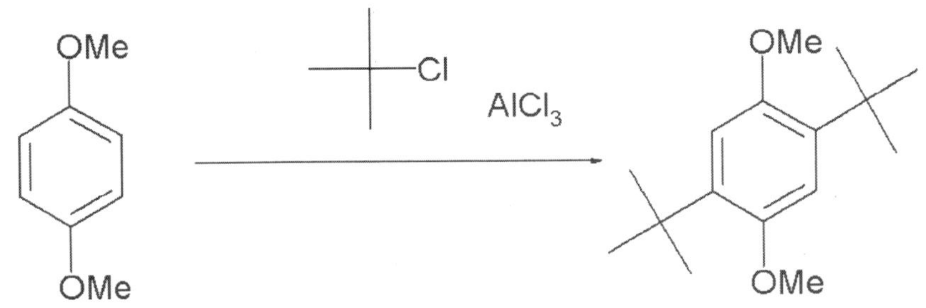

OMe OMe
Cl
$AlCl_3$
OMe OMe

Alkylations are not limited to alkyl halides: Friedel–Crafts reactions are possible with any carbocationic intermediate such as those derived from alkenes and a protic acid, Lewis acid, enones, and epoxides. An example is the synthesis of neophyl chloride from benzene and methallyl chloride:

$$H_2C = C(CH_3)CH_2Cl + C_6H_6 \rightarrow C_6H_5C(CH_3)_2CH_2Cl$$

In one study the electrophile is a bromonium ion derived from an alkene and NBS:

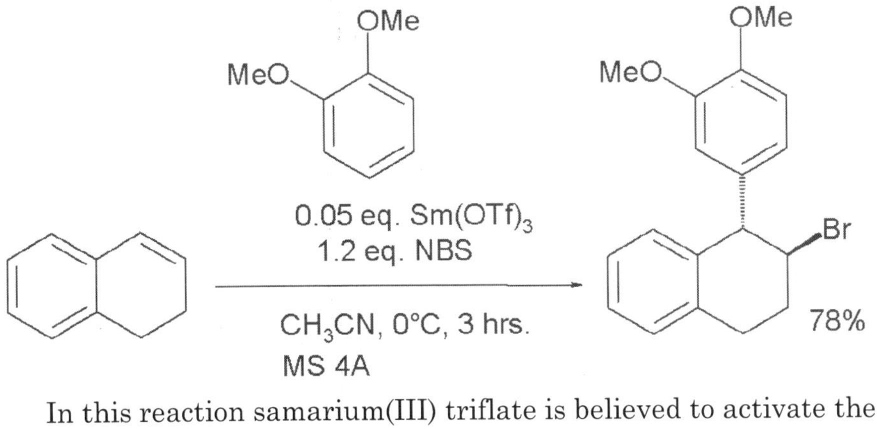

OMe MeO OMe MeO
0.05 eq. $Sm(OTf)_3$
1.2 eq. NBS
Br
CH_3CN, 0°C, 3 hrs.
MS 4A
78%

In this reaction samarium(III) triflate is believed to activate the NBS halogen donor in halonium ion formation.

Friedel–Crafts Dealkylation

Friedel–Crafts alkylation is a reversible reaction. In a reversed Friedel–Crafts reaction or Friedel–Crafts dealkylation, alkyl groups can be removed in the presence of protons and a Lewis acid.

For example, in a multiple addition of ethyl bromide to benzene, *ortho* and *para* substitution is expected after the first monosubstitution step because an alkyl group is an activating group. However, the actual reaction product is 1,3,5-triethylbenzene with all alkyl groups as a *meta* substituent. Thermodynamic reaction control makes sure that thermodynamically favoured *meta* substitution with steric hindrance minimized takes prevalence over less favourable *ortho* and *para* substitution by chemical equilibration. The ultimate reaction product is thus the result of a series of alkylations and dealkylations.

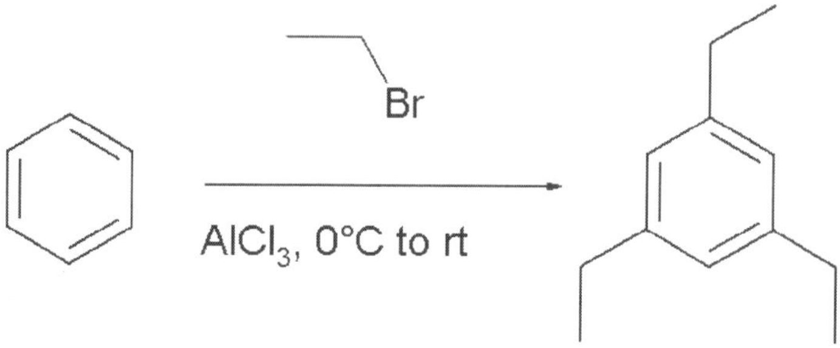

Br

$AlCl_3$, 0°C to rt

Friedel–Crafts Acylation

Friedel–Crafts acylation is the acylation of aromatic rings with an acyl chloride using a strong Lewis acid catalyst. Friedel–Crafts acylation is also possible with acid anhydrides. Reaction conditions are similar to the Friedel–Crafts alkylation mentioned above. This reaction has several advantages over the alkylation reaction. Due to the electron-withdrawing effect of the carbonyl group, the ketone product is always less reactive than the original molecule, so multiple acylations do not occur. Also, there are no carbocation rearrangements, as the carbonium ion is stabilized by a resonance structure in which the positive charge is on the oxygen.

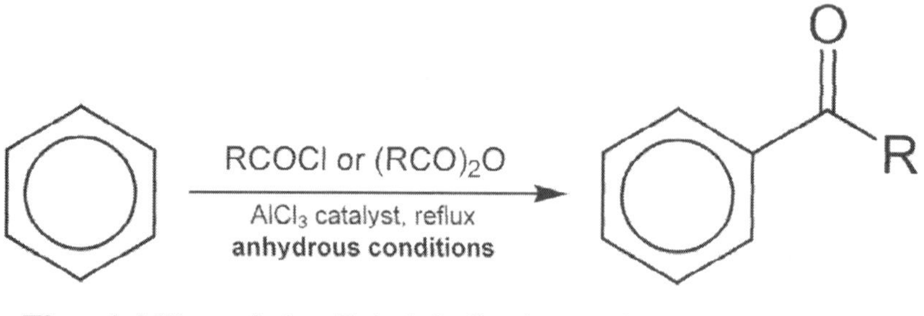

O

RCOCl or $(RCO)_2O$

R

$AlCl_3$ catalyst, reflux
anhydrous conditions

The viability of the Friedel–Crafts acylation depends on the stability of the acyl chloride reagent. Formyl chloride, for example, is too unstable to be isolated. Thus, synthesis of benzaldehyde via the Friedel–Crafts pathway requires that formyl chloride be synthesized *in situ*. This is accomplished via the Gattermann-Koch reaction, accomplished by treating benzene with carbon monoxide and hydrogen chloride under high pressure, catalyzed by a mixture of aluminium chloride and cuprous chloride.

Reaction Mechanism

In a simple mechanistic view, the first step consists of dissociation of a chloride ion to form an acyl cation ("acylium ion"):

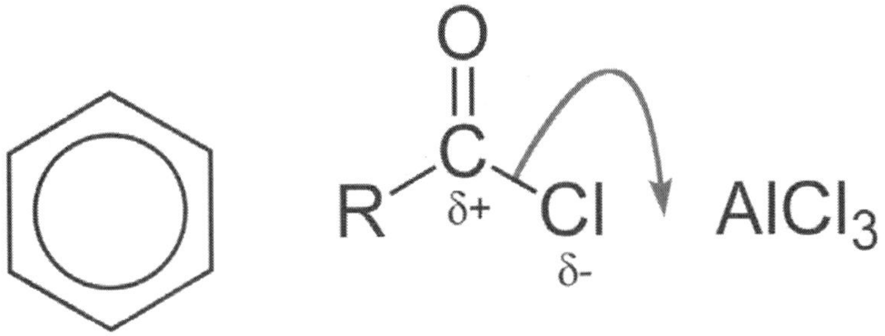

O
C
R δ+ Cl δ- $AlCl_3$

In some cases, the Lewis acid binds to the oxygen of the acyl chloride to form an adduct. Regardless, the resulting acylium ion or a related adduct is subject to nucleophilic attack by the arene:

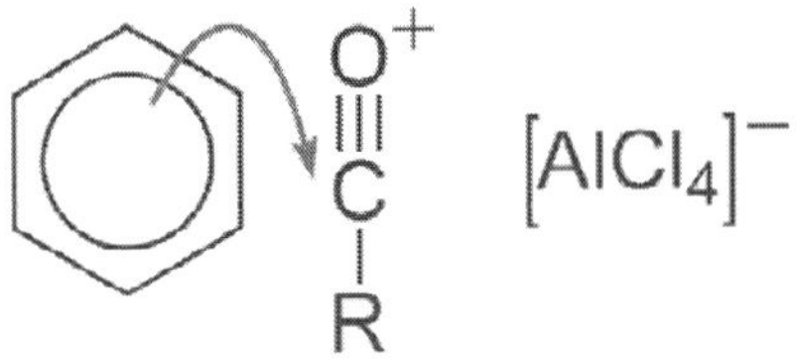

O^+
C
R
$[AlCl_4]^-$

Finally, chloride anion (or $AlCl_4^-$) deprotonates the ring (an "arenium ion") to form HCl, and the $AlCl_3$ catalyst is regenerated:

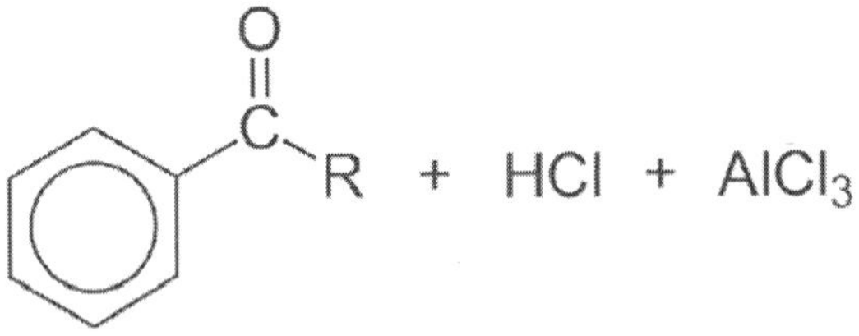

O
C
R + HCl + $AlCl_3$

If desired, the resulting ketone can be subsequently reduced to the corresponding alkane substituent by either Wolff–Kishner reduction or Clemmensen reduction. The net result is the same as the Friedel–Crafts alkylation.

Friedel–Crafts Hydroxyalkylation

Arenes react with certain aldehydes and ketones to form the hydroxyalkylated product for example in the reaction of the mesityl derivative of glyoxal with benzene to form a benzoin with an alcohol rather than a carbonyl group:

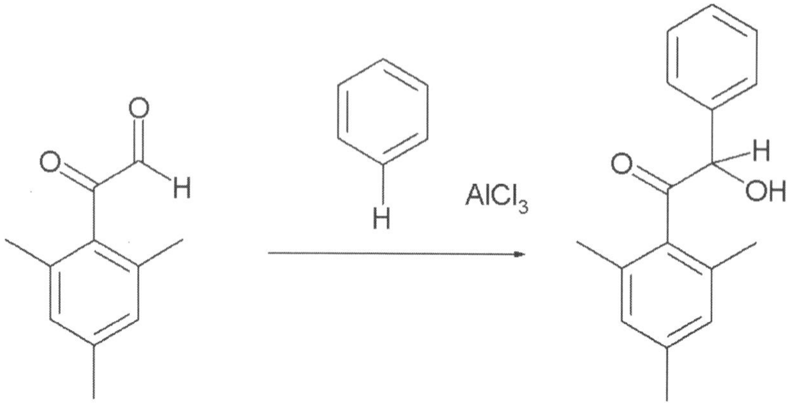

O O H H AlCl$_3$ O H OH

Scope and Variations

This reaction is related to several classic named reactions:

- The acylated reaction product can be converted into the alkylated product via a Clemmensen reduction.
- The Gattermann–Koch reaction can be used to synthesize benzaldehyde from benzene.
- The Gatterman reaction describes arene reactions with hydrocyanic acid
- The Houben–Hoesch reaction describes arene reactions with nitriles
- A reaction modification with an aromatic phenyl ester as a reactant is called the Fries rearrangement.
- In the Scholl reaction two arenes couple directly (sometimes called Friedel–Crafts arylation).
- In the Zincke–Suhl reaction p-cresol is alkylated to a cyclohexadienone with tetrachloromethane
- In the Blanc chloromethylation a chloromethyl group is added to an arene with formaldehyde, hydrochloric acid and zinc chloride.
- The Bogert-Cook Synthesis (1933) involves the dehydration and isomerization of *1-β-phenylethylcyclohexanol* to the octahydro derivative of phenanthrene

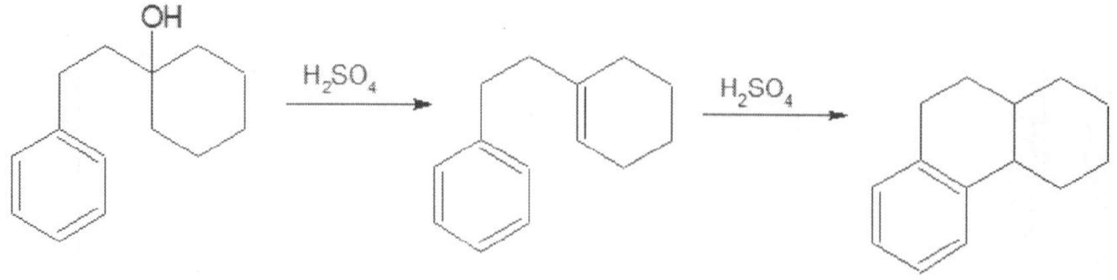

OH H_2SO_4 H_2SO_4

- The Darzens–Nenitzescu Synthesis of Ketones (1910, 1936) involves the acylation of cyclohexene with acetyl chloride to methylcyclohexenylketone.
- In the related Nenitzescu reductive acylation (1936) a saturated hydrocarbon is added making it a reductive acylation to methylcyclohexylketone
- In a green chemistry variation aluminium chloride is replaced by graphite in an alkylation of *p*-xylene with 2-bromobutane. This variation will not work with primary halides from which less carbocation involvement is inferred.

Dyes

Friedel–Crafts reactions have been used in the synthesis of several triarylmethane and xanthene dyes. Examples are the synthesis of thymolphthalein (a pH indicator) from two equivalents of thymol and phthalic anhydride:

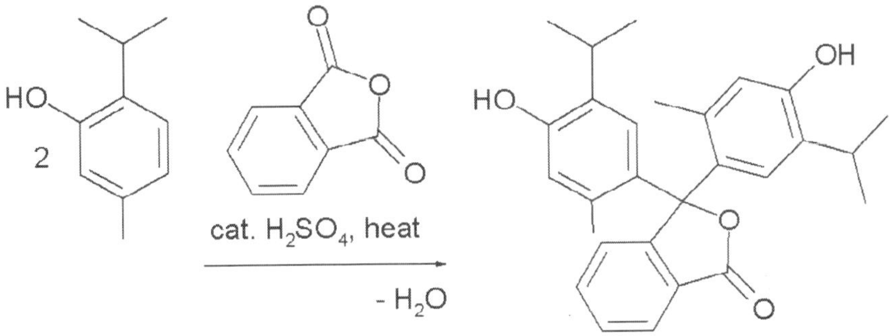

HO 2 — cat. H_2SO_4, heat, $- H_2O$ → HO, OH

A reaction of phthalic anhydride with resorcinol in the presence of zinc chloride gives the fluorophore Fluorescein. Replacing resorcinol by N,N-diethylaminophenol in this reaction gives rhodamine B:

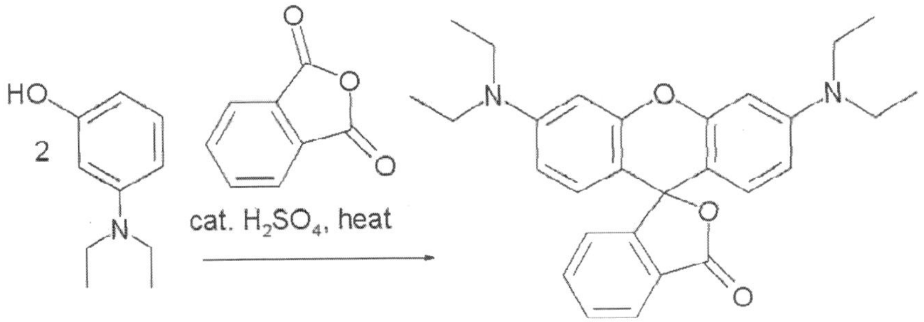

HO 2 N — cat. H_2SO_4, heat →

Haworth Reactions

The Haworth reaction is a classic method for the synthesis of tetralone. In it benzene is reacted with succinic anhydride, the intermediate product is reduced and a second FC acylation takes place with addition of acid.

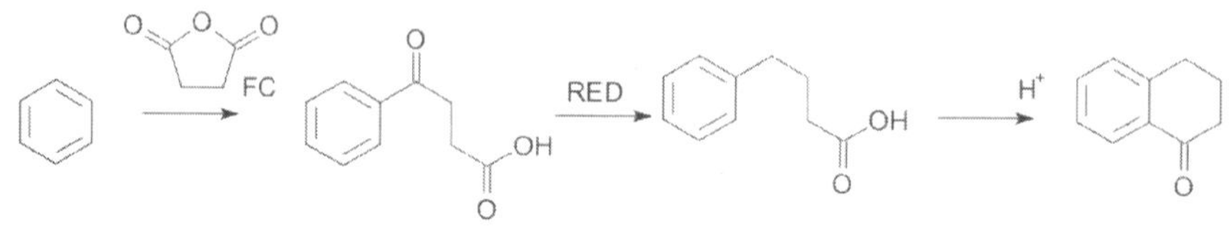

In a related reaction, phenanthrene is synthesized from naphthalene and succinic anhydride in a series of steps.

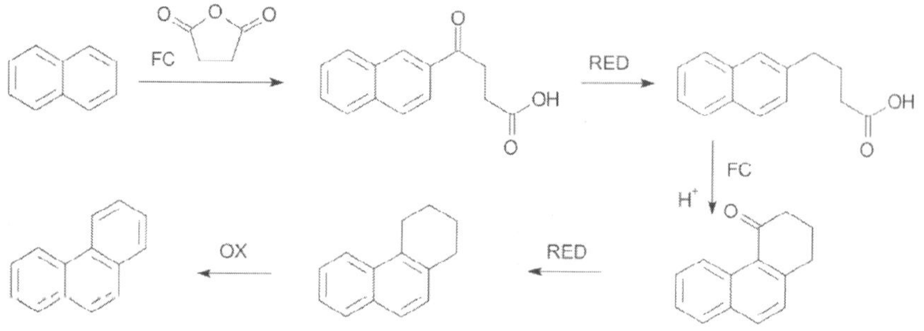

Friedel–Crafts Test for Aromatic Hydrocarbons

Reaction of chloroform with aromatic compounds using an aluminium chloride catalyst gives triarylmethanes, which are often brightly coloured, as is the case in triarylmethane dyes. This is a bench test for aromatic compounds.

Chapter 9

Cyclophosphamide

Cyclophosphamide (INN, trade names Endoxan, Cytoxan, Neosar, Procytox, Revimmune), also known as cytophosphane, is a nitrogen mustard alkylating agent, from the oxazophorines group.

An alkylating agent adds an alkyl group (C_nH_{2n+1}) to DNA. It attaches the alkyl group to the guanine base of DNA, at the number 7 nitrogen atom of the imidazole ring.

It is used to treat various types of cancer and some autoimmune disorders. It is a "prodrug"; it is converted in the liver to active forms that have chemotherapeutic activity.

Uses

The main use of cyclophosphamide is together with other chemotherapy agents in the treatment of lymphomas, some forms of leukemia and some solid tumours. It is a chemotherapy drug that works by slowing or stopping cell growth.

Cyclophosphamide also decreases the immune system's response to various diseases and conditions. Therefore, it has been used in various non-neoplastic autoimmune diseases where disease-modifying antirheumatic drugs (DMARDs) have been ineffective. For example, systemic lupus erythematosus (SLE) with severe lupus nephritis may respond to pulsed cyclophosphamide (in 2005, however, standard treatment for lupus nephritis changed to mycophenolic acid (MMF) from cyclophosphamide). Cyclophosphamide is also used to treat minimal change disease, severe rheumatoid arthritis, Wegener's granulomatosis (with trade name Cytoxan), and multiple sclerosis (with trade name Revimmune).

A 2004 study showed that the biological actions of cyclophosphamide are dose-dependent. At higher doses, it is associated with increased cytotoxicity and immunosuppression, while at low continuous dosage it shows immunostimulatory and antiangiogenic properties. A 2009 study of 17 patients with docetaxel-resistant metastatic hormone refractory prostate cancer showed a Prostate-specific antigen (PSA) decrease in 9 of the 17 patients. Median survival was 24 months for the entire group, and 60 months for those with a PSA response. The study concluded that low-dose cyclophosphamide "might be a viable alternative" treatment for docetaxel-resistant MHRPC and "is an interesting candidate for combination therapies, e.g., immunotherapy, tyrosine kinase inhibitors, and antiangiogenisis."

Pharmacokinetics/Pharmacodynamics

Cyclophosphamide is converted by mixed function oxidase enzymes in the liver to active metabolites. The main active metabolite is 4-hydroxycyclophosphamide, which exists in equilibrium with its tautomer, aldophosphamide. Most of the aldophosphamide is oxidised by the enzyme aldehyde dehydrogenase (ALDH) to make carboxyphosphamide. A small proportion of aldophosphamide is converted into phosphoramide mustard and acrolein. Acrolein is toxic to the bladder epithelium and can lead to hemorrhagic cystitis. This can be prevented through the use of aggressive hydration and/or mesna.

Recent clinical studies have shown that cyclophosphamide induce beneficial immunomodulatory effects in the context of adaptive immunotherapy. Although the mechanisms underlying these effects are not fully understood, several mechanisms have been suggested based on potential modulation of the host environment, including:

1. Elimination of T regulatory cells (CD4+CD25+ T cells) in naive and tumour-bearing hosts
2. Induction of T cell growth factors such as type I IFNs, and/or
3. Enhanced grafting of adoptively transferred tumour-reactive effector T cells by the creation of an immunologic space niche.

Thus, cyclophosphamide pre-conditioning of recipient hosts (for donor T cells) has been used to enhance immunity in naïve hosts, and to enhance adoptive T cell immunotherapy regimens as well as active vaccination strategies, inducing objective anti-tumour immunity.

Mechanism of Action

The main effect of cyclophosphamide is due to its metabolite phosphoramide mustard. This metabolite is only formed in cells that have low levels of ALDH. Phosphoramide mustard forms DNA crosslinks between (interstrand crosslinkages) and within (intrastrand crosslinkages) DNA strands at guanine N-7 positions. This is irreversible and leads to cell death.

Cyclophosphamide has relatively little typical chemotherapy toxicity as ALDHs are present in relatively large concentrations in bone marrow stem cells, liver and intestinal epithelium. ALDHs protect these actively proliferating tissues against toxic effects phosphoramide mustard and acrolein by converting aldophosphamide to carboxyphosphamide that does not give rise to the toxic metabolites (phosphoramide mustard and acrolein).

Side-effects

Many people taking cyclophosphamide do have serious side effects. Side-effects include chemotherapy-induced nausea and vomiting (CINV), bone marrow suppression, stomach ache, diarrhea, darkening of the skin/nails, alopecia (hair loss) or thinning of hair, changes in colour and texture of the hair, and lethargy. Hemorrhagic cystitis is a frequent complication, but this is prevented by adequate fluid intake and Mesna (sodium 2-mercaptoethane sulfonate). Mesna is a sulfhydryl donor and binds acrolein.

Cyclophosphamide is itself carcinogenic, potentially causing transitional cell carcinoma of the bladder as a long-term complication. It can lower the body's ability to fight an infection. It can cause temporary or (rarely) permanent sterility. A serious potential side-effect is Acute Myeloid Leukemia, referred to as secondary AML, due to it occurring secondarily to the primary disease being treated. The risk may be dependent on dose and a number of other factors, including the condition being treated, other agents or treatment modalities used (including radiotherapy), treatment intensity and length of treatment. For some regimens it is a very rare occurrence.

For instance, CMF-therapy for breast cancer (where the cumulative dose is typically less than 20 grams of cyclophosphamide) seems to carry an AML risk of less than 1/2000th, with some studies even finding no increased risk compared to the background population. Other treatment regimens involving higher doses may carry risks of 1-2% or higher, depending on regimen. Cyclophosphamide-induced AML, when it happens, typically presents some years after treatment, with incidence peaking around 3–9 years. After 9 years, the risk has fallen to the level of the regular population. When AML occurs, it is often preceded by a Myelodysplastic syndrome phase, before developing into overt acute leukemia. Cyclophosphamide-induced leukemia will often involve complex cytogenetics, which carries a worse prognosis than de novo AML.

Other (serious) side effects include:

- gross and microscopic hematuria,
- unusual decrease in the amount of urine,
- mouth sores,
- unusual tiredness or weakness,
- joint pain,
- easy bruising/bleeding,
- existing wounds that are slow healing.

History

Cyclophosphamide and the related nitrogen mustard-derived alkylating agent ifosfamide were developed by Norbert Brock and ASTA (now Baxter Oncology). Brock and his team synthesised and screened more than 1,000 candidate oxazaphosphorine compounds. They converted the base nitrogen mustard into a non-toxic "transport form". This transport form was a pro-drug, subsequently actively transported into the cancer cells. Once in the cells, the pro-drug was enzymatically converted into the active, toxic form. The first clinical trials were published at the end of the 1950s.

Ifosfamide

Ifosfamide *(pronounced eye.fos'.fa.mide)* (also marketed as Mitoxana and Ifex) is a nitrogen mustard alkylating agent used in the treatment of cancer.

It is sometimes abbreviated "IFO".

Uses

It is given as a treatment for a variety of cancers, including:

- Testicular cancer
- Breast cancer
- Lymphoma *(Hodgkin's and Non-Hodgkin's)*
- Soft tissue sarcoma
- Osteogenic sarcoma (Bone cancer)
- Lung cancer
- Cervical cancer
- Ovarian cancer.

Administration

It is a white powder which, when prepared for use in chemotherapy becomes a clear, colourless fluid. The delivery is intravenous.

Ifosfamide is often used in conjunction with Mesna to avoid internal bleeding in the patient, in particular hemorrhagic cystitis.

Ifosfamide is given quickly, and in some cases can be given in as little as an hour.

Lipoplatin

Lipoplatin (Liposomal cisplatin) is a nanoparticle of 110 nm average diameter composed of lipids and cisplatin (1). This new drug has successfully finished Phase I, Phase II and Phase III human clinical trials (2,3). It has shown superiority to cisplatin in combination with paclitaxel as a chemotherapy regimen in non-small cell lung cancer (NSCLC) adenocarcinomas. Lipoplatin evades immune surveillance thus escaping clearance from macrophages, circulates for long periods in body fluids after intravenous administration with a half-life of ~120 h, and extravasates through the compromised endothelium of the vasculature in tumours created during the process of neoangiogenesis. Thus, Lipoplatin nanoparticles are concentrated to the primary tumour and metastases. Human studies have shown 40- to 200-fold higher platinum concentration compared to the adjacent normal tissue in specimens from human biopsies 20h post-infusion of the drug (4).

EMEA has given the orphan drug status to Lipoplatin in 2007 in an ongoing registrational Phase II/III study as a first line-treatment in pancreatic cancer (5). Lipoplatin, under the name Nanoplatin, received in 2009 the consent of EMEA to be tested as first line against non-squamous NSCLC mainly composed of adenocarcinomas. In a randomized Phase III it has shown statistically significant reduction of the toxicity of cisplatin, mainly of nephrotoxicity, in an administration regimen that does not require hospitalization of the patients as in the case of cisplatin chemotherapy.

Lomustine

Lomustine (or "CCNU")(Marketed under the name CeeNU in US) is an alkylating nitrosourea compound used in chemotherapy. It is in the same family as streptozotocin. This is a highly lipid soluble drug, and thus crosses the blood brain barrier. This property makes it ideal for treating brain tumours, and is its primary use. Lomustine has a long time to nadir (the time when white blood cells reach their lowest number).

Dose

Adults: 100-150mg/m² PO every 6 weeks (subsequent dose based on WBC) Cats (7,5 kg): 15mg every 6 weeks

Dosage Form

10mg, 15mg, 40mg, 100mg Capsules

Other

It is the “C” in the “PCV” chemotherapy regimen.

Mechlorethamine

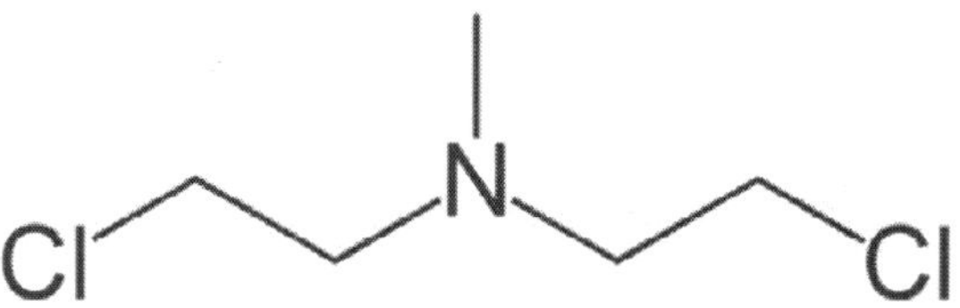

Figure: *Systematic (IUPAC) name*

Mechlorethamine also known as chlormethine, mustine and HN2 is a nitrogen mustard sold under the brand name Mustargen. It is the prototype of alkylating agents, a group of anticancer chemotherapeutic drug. It works by binding to DNA, crosslinking two strands and preventing cell duplication. It binds to the N7 nitrogen on the DNA base guanine. As the chemical is a blister agent, its use is strongly restricted within the Chemical Weapons Convention where it is classified as a Schedule 1 substance. Mechlorethamine belongs to the group of nitrogen mustard alkylating agents.

History

Successful clinical use of mechlorethamine gave birth to the field of anticancer chemotherapy. The drug is an nitrogen-based analogue of mustard gas (which is sulfur-based) and was derived from chemical warfare research. Secret clinical trials of the agent for Hodgkin’s disease and several other lymphomas and leukemias in humans began in December 1942. Because of wartime secrecy restrictions, it was not until 1946 that the results of these trials were published openly.

Uses

It has been derivatized into the estrogen analogue estramustine, used to treat prostate cancer.

It can also be used in chemical warfare where it has the code-name HN2. This chemical is a form of nitrogen mustard gas and a powerful vesicant.

Melphalan

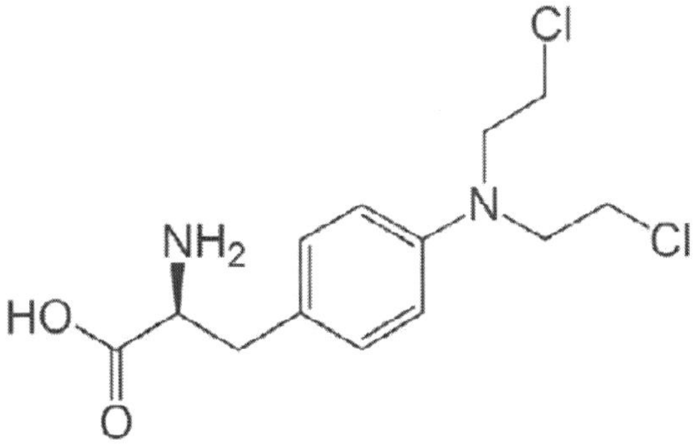

Figure: *Systematic (IUPAC) name*

Melphalan hydrochloride (trade name Alkeran) is a chemotherapy drug belonging to the class of nitrogen mustard alkylating agents. An alkylating agent adds an alkyl group (C_nH_{2n+1}) to DNA. It attaches the alkyl group to the guanine base of DNA, at the number 7 nitrogen atom of the imidazole ring. Otherwise known as L-Phenylalanine Mustard, or L-PAM, melphalan is a phenylalanine derivative of mechlorethamine.

Uses

It is used to treat multiple myeloma and ovarian cancer, and occasionally malignant melanoma. The agent was first investigated as a possible drug for use in melanoma. It was not found to be effective, but has been found to be effective in the treatment of myeloma.

Administration

Oral or intravenous; dosing varies by purpose and route of administration as well as patient weight.

Melphalan Prescribing Information: Alkeran

Melphalan Patient Information: MedlinePlus

Melphalan Material Safety Data Sheet (MSDS): Sequoia Research Products

Side Effects

Common side effects include:

- Nausea and vomiting, and oral ulceration.
- Bone marrow suppression, including
  - o Decreased white blood cell count causing increased risk of infection
  - o Decreased platelet count causing increased risk of bleeding

Less common side effects include:

- Severe allergic reactions
- Pulmonary fibrosis (scarring of lung tissue) including fatal outcomes (usually only with prolonged use)
- Hair loss
- Interstitial pneumonitis
- Rash
- Itching
- Irreversible bone marrow failure due to melphalan not being withdrawn early enough.
- Cardiac arrest.

Nedaplatin

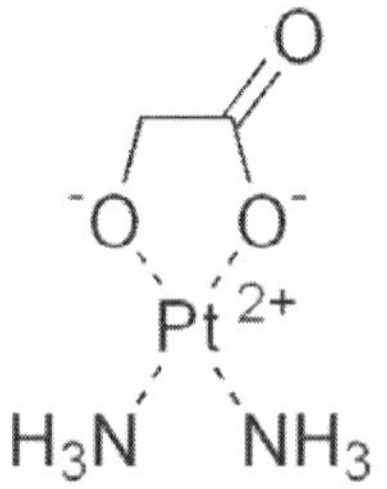

^{-}O O^{-} Pt^{2+} H_3N NH_3

Figure: *Systematic (IUPAC) name*

Nedaplatin (INN, marketed under the tradename Aqupla) is a platinum compound which is used for cancer chemotherapy. It produces less nausea, vomiting and nephrotoxicity than other platinum-containing drugs.

Nimustine

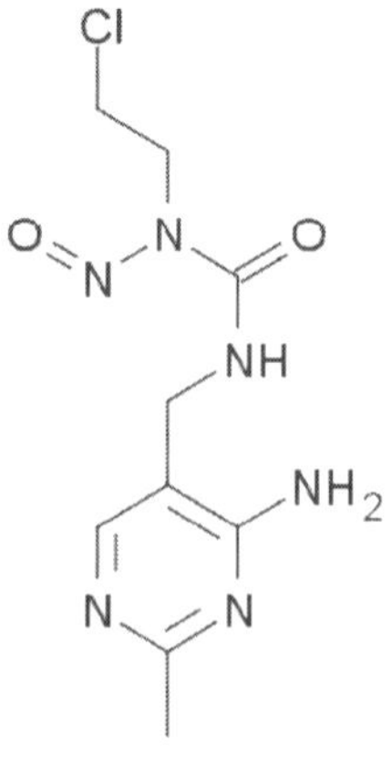

Cl O N N O NH NH_2 N N

Figure: *Nimustine (INN) is a nitrosourea alkylating agent.*

Nitrogen Mustard

Figure: *HN1 (bis(2-chloroethyl)ethylamine)*

Figure: *HN2 (bis(2-chloroethyl)methylamine, mustine)*

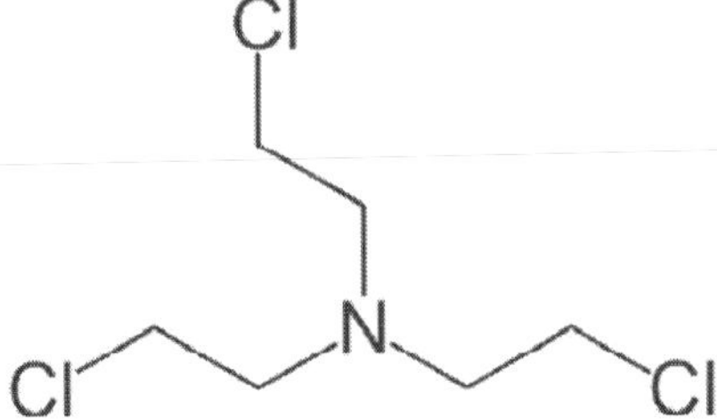

Figure: *HN3 (tris(2-chloroethyl)amine)*

The nitrogen mustards are cytotoxic chemotherapy agents similar to mustard gas. Although their common use is medicinal, in principle these compounds can also be deployed as chemical warfare agents. Nitrogen mustards are nonspecific DNA alkylating agents.

Nitrogen mustard gas was stockpiled by several nations during the Second World War, but it was never used in combat. As with all types of mustard gas, nitrogen mustards are powerful and persistent blister agents and the main examples (HN1, HN2, HN3) are therefore classified as Schedule 1 substances within the Chemical Weapons Convention. Production and use is therefore strongly restricted.

During WWII nitrogen mustards were studied at Yale University and classified human clinical trials of nitrogen mustards for the treatment of lymphoma started in December 1942. Also during WWII, an incident during the air raid on Bari, Italy led to the release of mustard gas that affected several hundred soldiers and civilians.

Medical examination of the survivors showed a decreased number of lymphocytes. After WWII was over, the Bari incident and the Yale group's studies eventually converged prompting a search for other similar compounds. Due to its use in previous studies, the nitrogen mustard known as "HN2" became the first chemotherapy drug mustine.

Examples

The original nitrogen mustard drug, mustine (HN2), is no longer commonly in use. Other nitrogen mustards developed as treatments include cyclophosphamide, chlorambucil, uramustine, ifosfamide, melphalan and bendamustine. Bendamustine has recently reemerged as a viable chemotheraputic treatment.

Nitrogen mustards that can be used for chemical warfare purposes are tightly regulated. Their weapon designations are:

- HN1: Bis(2-chloroethyl)ethylamine
- HN2: Bis(2-chloroethyl)methylamine
- HN3: Tris(2-chloroethyl)amine.

Mechanism of Action

Nitrogen mustards (NMs) form cyclic aminium ions (aziridinium rings) by intramolecular displacement of the chloride by the amine nitrogen. This azidirium group then alkylates DNA by attacking the N-7 nucleophilic centre on the guanine base.

A second attack after the displacement of the second chlorine forms the second alkylation step that results in the formation of interstrand cross-links (ICLs) as it was shown in the early 1960s. At that time it was proposed that the ICLs were formed between N-7 atom of guanine residue in a 5'-d(GC) sequence. These kind of lesion are highly cytotoxic, since they block fundamental metabolic processes such as replication and transcription.

Later was it clearly demonstrated that NMs form a 1,3 ICL in the 5'-d(GNC) sequence.The strong cytotoxic effect caused by the formation of ICLs is what makes NMs an effective chemotherapeutic agent. Other compounds used in cancer chemotherapy that have the ability to form ICLs are cisplatin, mitomycin C, carmustine, psoralen.

Oxaliplatin

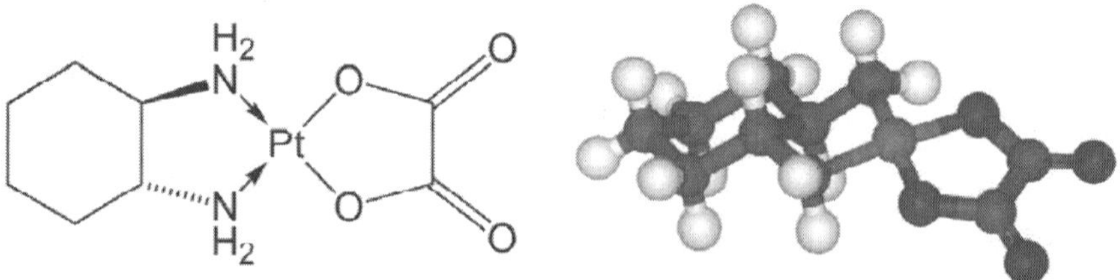

Figure: *Systematic (IUPAC) name*

Oxaliplatin is a coordination complex that is used in cancer chemotherapy. These platinum-based drugs are usually classified as alkylating agents, although they are not actually alkylating groups (they function by a similar mechanism).

Preparation and Structure

Oxaliplatin was discovered in 1976 at Nagoya City University by Professor Yoshinori Kidani, who was granted U.S. Patent 4,169,846 in 1979. Oxaliplatin was subsequently in-licensed by Debiopharm and developed as an advanced colorectal cancer treatment. Debio licensed the drug to Sanofi-Aventis in 1994. Eloxatin gained European approval in 1996 (firstly in France) and approval by the U.S. Food and Drug Administration (FDA) in 2002.

The compound features a square planar platinum(II) centre. In contrast to cisplatin and carboplatin, oxaliplatin features the bidentate ligand 1,2-diaminocyclohexane in place of the two monodentate ammine ligands. It also features a bidentate oxalate group.

Mechanism of Action

The cytotoxicity of platinum compounds is thought to result from inhibition of DNA synthesis in cancer cells. In vivo studies showed that Oxaliplatin has anti-tumour activity against colon carcinoma through its (non-targeted) cytotoxic effects.

Clinical Use

Oxaliplatin is typically administered with fluorouracil and leucovorin in a combination known as FOLFOX for the treatment of colorectal cancer. Oxaliplatin is marketed by Sanofi-Aventis under the trademark Eloxatin or by Medac GmbH under the trademark Oxaliplatin Medac. There are generic equivalents on the market now Oxaliplatin has been compared with other platinum compounds (Cisplatin, Carboplatin) in advanced cancers (gastric, ovarian).

Advanced Colorectal Cancer

In clinical studies, Oxaliplatin by itself has modest activity against advanced colorectal cancer. It has been extensively studied in combination with Fluorouracil and Folinic Acid (a combination known as FOLFOX).

When compared with Fluorouracil and Folinic Acid administered according to the "De Gramont regimen" there was no significant increase in overall survival with the FOLFOX regimen (specifically, FOLFOX4), but progression-free survival, the primary end-point of the phase III randomized trial, was improved with FOLFOX.

Adjuvant Treatment of Colorectal Cancer

After the curative resection of colorectal cancer, chemotherapy based on Fluorouracil and folinic acid reduces the risk of relapse.

The benefit is clinically relevant when cancer has spread to locoregional lymph nodes (stage III, Dukes C).

The addition of Oxaliplatin improves relapse-free survival, but data on overall survival have not yet been published *in extenso*.

When cancer has not spread to the locoregional lymph nodes (stage II, Dukes B) the benefit of chemotherapy is marginal and the decision on whether to give adjuvant chemotherapy should be carefully evaluated by discussing with the patient the realistic benefits and the possible toxic side effects of treatment.

This is even more relevant when the oncologist proposes treatment with Oxaliplatin.

Adverse Effects

Side-effects of oxaliplatin treatment can potentially include:

- Neuropathy, (both an acute, reversible sensitivity to cold and numbness in the hands and feet and a chronic, possibly irreversible foot/leg, hand/arm numbness, often with deficits in proprioception)
- Fatigue
- Nausea, vomiting, and/or diarrhea
- Neutropenia
- Ototoxicity (hearing loss)
- Extravasation if Oxaliplatin leaks from the infusion vein it may cause severe damage to the connective tissues.
- Hypokalemia, which is more common in women than men.

In addition, some patients may experience an allergic reaction to platinum-containing drugs. This is more common in women.

Oxaliplatin has less ototoxicity and nephrotoxicity than cisplatin and carboplatin.

Patent Information

Eloxatin is covered by patent numbers 5338874 (Expiry Apr 07,2013), 5420319 (Expiry Aug 08,2016), 5716988 (Expiry Aug 07,2015) and 5290961 (Expiry Jan 12, 2013). Exclusivity code I-441, which expired on Nov 04, 2007, is for use combination with infusional 5-FU/ LV for adjuvant treatment stage III colon cancer patients who have undergone complete resection primary tumour-based on improvement in disease free survival with no demonstrated benefit overall survival after 4 years.

Exclusivity code NCE, New Chemical Entity, expired on Aug 09, 2007.

Picoplatin

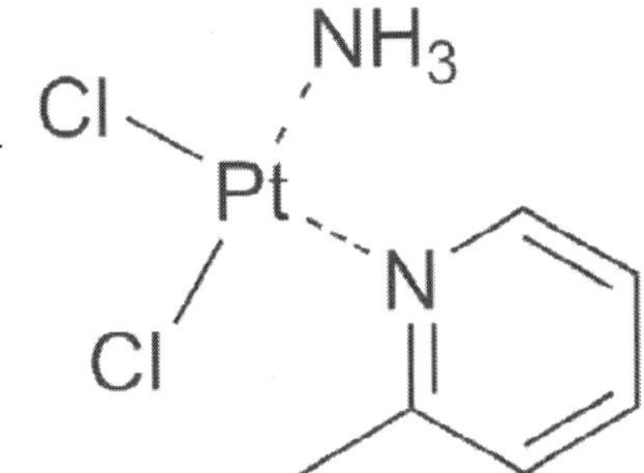

Figure 9: *IUPAC name*

Picoplatin is a cytotoxic platinum compound in clinical development by Poniard Pharmaceuticals (previously NeoRx) for the treatment of patients with solid tumours.

In Phase I and Phase II clinical trials, picoplatin demonstrated activity in a variety of solid tumours, including lung, ovarian, colorectal and hormone-refractory prostate cancer. However, in Phase III trials, picoplatin failed to hit its primary endpoint for advanced small cell lung cancer. Hopes are now pinned on its use for metastatic colorectal cancer.

Prednimustine

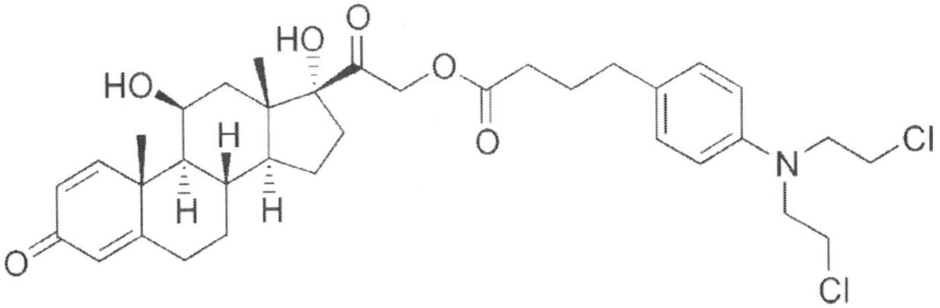

Figure 10: *Systematic (IUPAC) name*

Prednimustine is a drug used in chemotherapy. It is the ester formed from two other drugs, prednisolone and chlorambucil.

It can be associated with myoclonus.Satraplatin

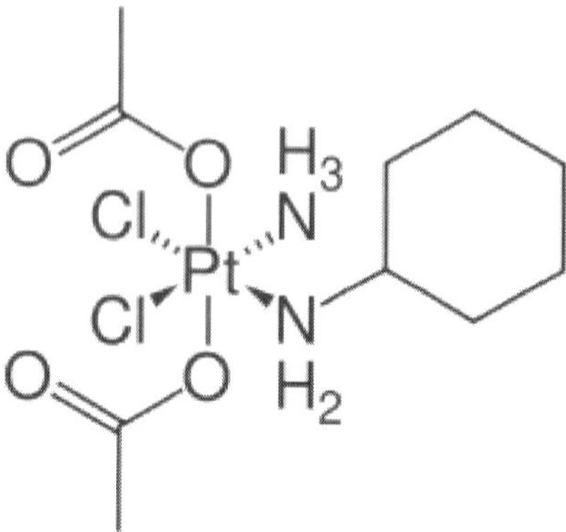

Figure 11: *Systematic (IUPAC) name*

Satraplatin (INN, codenamed JM216) is a platinum compound that is currently under investigation as one treatment of patients with advanced prostate cancer who have failed previous chemotherapy. It has not yet received approval from the U.S. Food and Drug Administration. First mentioned in the medical literature in 1993, satraplatin is the first orally active platinum-based chemotherapeutic drug; other available platinum analogues—cisplatin, carboplatin, and oxaliplatin—must be given intravenously.

It is made available in the United States jointly by Spectrum Pharmaceuticals and GPC Biotech under the name SPERA (Satraplatin Expanded Rapid Access).

The drug has also been used in the treatment of lung and ovarian cancers. The proposed mode of action is that the compound binds to the DNA of cancer cells rendering them incapable of dividing.

Chapter 10

Catalytic Cycle

A catalytic cycle in chemistry is a term for a multistep reaction mechanism that involves a catalyst . The catalytic cycle is the main method for describing the role of catalysts in biochemistry, organometallic chemistry, materials science, etc.

Often such cycles show the conversion of a precatalyst to the catalyst. Since catalysts are regenerated, catalytic cycles are usually written as a sequence of chemical reactions in the form of a loop. In such loops, the initial step entails binding of one or more reactants by the catalyst, and the final step is the release of the product and regeneration of the catalyst. Articles on the Monsanto process, the Wacker process, and the Heck reaction show catalytic cycles.

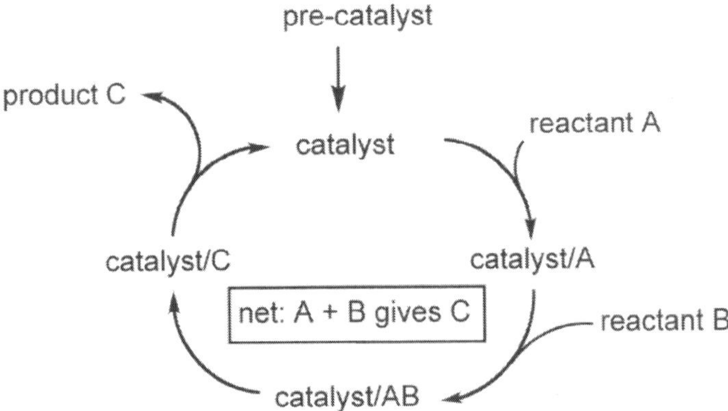

Figure: *Catalytic cycle for conversion of A and B into C.*

A catalytic cycle is not necessarily a full reaction mechanism. For example, it may be that the intermediates have been detected, but it is not known by which mechanisms the actual elementary reactions occur.

Sacrificial Catalysts

Often a so-called sacrificial catalyst is also part of the reaction system with the purpose of regenerating the *true* catalyst in each cycle. As the name implies, the sacrificial catalyst is not regenerated and irreversibly consumed. This sacrificial compound is also known as a stoichiometric catalyst when added in stoichiometric quantities compared to the main reactant. Usually the true catalyst is an expensive and complex molecule and added in quantities as small as possible. The stoichiometric catalyst on the other hand should be cheap and abundant.Reaction mechanism

In chemistry, a reaction mechanism is the step by step sequence of elementary reactions by which overall chemical change occurs. Although only the net chemical change is directly observable for most chemical reactions, experiments can often be designed that suggest the possible sequence of steps in a reaction mechanism. Recently, electrospray ionization mass spectrometry has been used to corroborate the mechanism of several organic reaction proposals.

Reaction Mechanism

Reaction mechanism is the step by step sequence of elementary reactions by which overall chemical change occurs. A chemical mechanism describes in detail exactly what takes place at each stage of an overall chemical reaction (transformation). It also describes each reaction intermediate, activated complex, and transition state, and which bonds are broken (and in what order), and which bonds are formed (and in what order). A complete mechanism must also account for all reactants used, the function of a catalyst, stereochemistry, all products formed and the amount of each, and what the relative rates of the steps are.

Reaction intermediates are chemical species, often unstable and short-lived, which are not reactants or products of the overall chemical reaction, but are temporary products and reactants in the mechanism's reaction steps. Reaction intermediates are often free radicals or ions. Transition states can be unstable intermediate molecular states even in the elementary reactions. Transition states are commonly molecular entities involving an unstable number of bonds and/or unstable geometry which may be at chemical potential maxima.

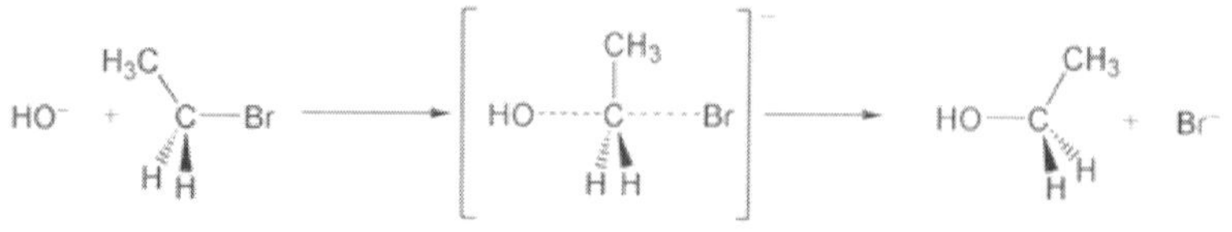

S_N2 reaction mechanism. Note the negatively-charged transition state in brackets in which the central carbon atom in question shows five bonds, an unstable condition.A reaction mechanism must also account for the order in which molecules react. Often what appears to be a single step conversion is in fact a multistep reaction.

Examples

Consider the following reaction:

$$CO + NO_2 \rightarrow CO_2 + NO$$

In this case, it has been experimentally determined that this reaction takes place according to the rate law $r = k[NO_2]^2$. Therefore, a possible mechanism by which this reaction takes place is:

$$2\ NO_2 \rightarrow NO_3 + NO\ \text{(slow)}$$
$$NO_3 + CO \rightarrow NO_2 + CO_2\ \text{(fast)}$$

Each step is called an elementary step, and each has its own rate law and molecularity. The elementary steps should add up to the original reaction. When determining the overall rate law for a reaction, the slowest step is the step that determines the reaction rate. Because the first step (in the above reaction) is the slowest step, it is the rate-determining step. Because it involves the collision of two NO_2 molecules, it is a bimolecular reaction with a rate law of $r = k[NO_2]^2$. If we were to cancel out all the molecules that appear on both sides of the reaction, we would be left with the original reaction.

In organic chemistry, one of the first reaction mechanisms proposed was that for the benzoin condensation, put forward in 1903 by A. J. Lapworth.

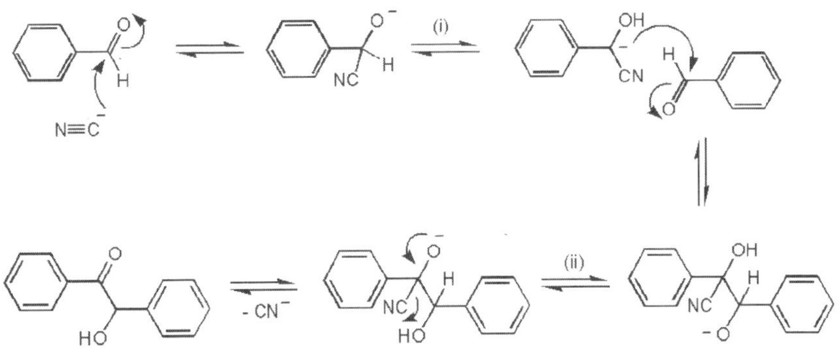

Figure: *Benzoin condensation reaction mechanism. Cyanide ion (CN^-) acts as a catalyst here, entering at the first step and leaving in the last step. Proton (H^+) transfers occur at (i) and (ii). The arrow pushing method is used in some of the steps to show where electron pairs go.*

Modelling

A correct reaction mechanism is an important part of accurate predictive modelling. For many combustion and plasma systems, detailed mechanisms are not available or require development.

Even when information is available, identifying and assembling the relevant data from a variety of sources, reconciling discrepant values and extrapolating to different conditions can be a difficult process without expert help. Rate constants or thermochemical data are often not available in the literature, so computational chemistry techniques or group-additivity methods must be used to obtain the required parameters.

At the different stages of a reaction mechanism's elaboration, appropriate methods must be used.

Molecularity

Molecularity in chemistry is the number of colliding molecular entities that are involved in a single reaction step.

- A reaction involving one molecular entity is called unimolecular.
- A reaction involving two molecular entities is called bimolecular.
- A reaction involving three molecular entities is called termolecular.

Reagent

A reagent is a "substance or compound that is added to a system in order to bring about a chemical reaction or is added to see if a reaction occurs." Although the terms reactant and reagent are often used interchangeably, a reactant is less specifically a "substance that is consumed in the course of a chemical reaction". Solvents and catalysts, although they are involved in the reaction, are usually not referred to as reactants.

In organic chemistry, reagents are compounds or mixtures, usually composed of inorganic or small organic molecules, that are used to affect a transformation on an organic substrate. Examples of organic reagents include the Collins reagent, Fenton's reagent, and Grignard reagent. There are also *analytical reagents* which are used to confirm the presence of another substance. Examples of these are Fehling's reagent, Millon's reagent and Tollens' reagent.

In another use of the term, when purchasing or preparing chemicals, reagent-grade describes chemical substances of sufficient purity for use in chemical analysis, chemical reactions or physical testing. Purity standards for reagents are set by organizations such as

ASTM International. For instance, reagent-quality water must have very low levels of impurities like sodium and chloride ions, silica, and bacteria, as well as a very high electrical resistivity.

Chemical Equilibrium

In a chemical reaction, chemical equilibrium is the state in which the concentrations of the reactants and products have not yet changed with time. It occurs only in reversible reactions, and not in irreversible reactions. Usually, this state results when the forward reaction proceeds at the same rate as the reverse reaction. The reaction rates of the forward and reverse reactions are generally not zero but, being equal, there are no net changes in the concentrations of the reactant and product. This process is called dynamic equilibrium.

The concept of chemical equilibrium was developed after Berthollet (1803) found that some chemical reactions are reversible. For any reaction mixture to exist at equilibrium, the rates of the forward and backward (reverse) reactions are equal. In the following chemical equation with arrows pointing both ways to indicate equilibrium, A and B are reactant chemical species, S and T are product species, and α, β, σ, and τ are the stoichiometric coefficients of the respective reactants and products:

$$\alpha A + \beta B \rightleftharpoons \sigma S + \tau T$$

The equilibrium position of a reaction is said to lie "far to the right" if, at equilibrium, nearly all the reactants are consumed. Conversely the equilibrium position is said to be "far to the left" if hardly any product is formed from the reactants.

Guldberg and Waage (1865), building on Berthollet's ideas, proposed the law of mass action:

$$\text{forward reaction rate} = k_+ A^\alpha B^\beta$$

$$\text{backward reaction rate} = k\_S^\sigma T^\tau$$

where A, B, S and T are active masses and k_+ and k_- are rate constants. Since at equilibrium forward and backward rates are equal:

$$k_+ \{A\}^\alpha \{B\}^\beta = k_- \{S\}^\sigma \{T\}^\tau$$

and the ratio of the rate constants is also a constant, now known as an equilibrium constant.

$$K = \frac{k_+}{k_-} = \frac{\{S\}^\sigma \{T\}^\tau}{\{A\}^\alpha \{B\}^\beta}$$

By convention the products form the numerator. However, the law of mass action is valid only for concerted one-step reactions that proceed through a single transition state and is not valid in general because rate equations do not, in general, follow the stoichiometry of the reaction as Guldberg and Waage had proposed. Equality of forward and backward reaction rates, however, is a necessary condition for chemical equilibrium, though it is not sufficient to explain why equilibrium occurs.

Despite the failure of this derivation, the equilibrium constant for a reaction is indeed a constant, independent of the activities of the various species involved, though it does depend on temperature as observed by the van 't Hoff equation. Adding a catalyst will affect both the forward reaction and the reverse reaction in the same way and will not have an effect on the equilibrium constant. The catalyst will speed up both reactions thereby increasing the speed at which equilibrium is reached.

Although the macroscopic equilibrium concentrations are constant in time reactions do occur at the molecular level. For example, in the case of acetic acid dissolved in water and forming acetate and hydronium ions,

$$CH_3CO_2H + H_2O \rightleftharpoons CH_3CO_2^- + H_3O^+$$

a proton may hop from one molecule of acetic acid on to a water molecule and then on to an acetate anion to form another molecule of acetic acid and leaving the number of acetic acid molecules unchanged. This is an example of dynamic equilibrium. Equilibria, like the rest of thermodynamics, are statistical phenomena, averages of microscopic behavior.

Le Chatelier's principle (1884) gives an idea of the behavior of an equilibrium system when changes to its reaction conditions occur. *If a dynamic equilibrium is disturbed by changing the conditions, the position of equilibrium moves to partially reverse the change.* For example, adding more S from the outside will cause an excess of products, and the system will try to counteract this by increasing the reverse reaction and pushing the equilibrium point backward (though the equilibrium constant will stay the same).

If mineral acid is added to the acetic acid mixture, increasing the concentration of hydronium ion, the amount of dissociation must decrease as the reaction is driven to the left in accordance with this principle. This can also be deduced from the equilibrium constant expression for the reaction:

$$K = \frac{\{CH_3CO_2^-\}\{H_3O^+\}}{\{CH_3CO_2H\}}$$

If $\{H_3O^+\}$ increases $\{CH_3CO_2H\}$ must increase and $\{CH_3CO_2^-\}$ must decrease. The H_2O is left out as it is a pure liquid and its concentration is undefined.

A quantitative version is given by the reaction quotient.

J. W. Gibbs suggested in 1873 that equilibrium is attained when the Gibbs energy of the system is at its minimum value (assuming the reaction is carried out under constant pressure). What this means is that the derivative of the Gibbs energy with respect to reaction coordinate (a measure of the extent of reaction that has occurred, ranging from zero for all reactants to a maximum for all products) vanishes, signalling a stationary point. This derivative is usually called, for certain technical reasons, the Gibbs energy change. This criterion is both necessary and sufficient. If a mixture is not at equilibrium, the liberation of the excess Gibbs energy (or Helmholtz energy at constant volume reactions) is the "driving force" for the composition of the mixture to change until equilibrium is reached. The equilibrium constant can be related to the standard Gibbs energy change for the reaction by the equation

$$\Delta_r G^\ominus = -RT \ln K_{eq}$$

where R is the universal gas constant and T the temperature.

When the reactants are dissolved in a medium of high ionic strength the quotient of activity coefficients may be taken to be constant. In that case the concentration quotient, K_c,

$$K_c = \frac{[S]^\sigma [T]^\tau}{[A]^\alpha [B]^\beta}$$

where [A] is the concentration of A, etc., is independent of the analytical concentration of the reactants. For this reason, equilibrium constants for solutions are usually determined in media of high ionic strength. K_c varies with ionic strength, temperature and pressure (or volume). Likewise K_p for gases depends on partial pressure. These constants are easier to measure and encountered in high-school chemistry courses.

Thermodynamics

The relation between the Gibbs energy and the equilibrium constant can be found by considering chemical potentials. At constant temperature and pressure the function G Gibbs free energy for the reaction, depends only with the extent of reaction: ξ and can only decrease according to the second law of thermodynamics. It means that the derivative of G with ξ must be negative if the reaction happens; at the equilibrium the derivative being equal to zero.

$$\left(\frac{dG}{d\xi}\right)_{T,p} = 0 \ : \ \text{equilibrium}$$

At constant volume, one must consider the Helmholtz free energy for the reaction: A.

In this article only the constant pressure case is considered. The constant volume case is important in geochemistry and atmospheric chemistry where pressure variations are significant. Note that, if reactants and products were in standard state (completely pure), then there would be no reversibility and no equilibrium. The mixing of the products and reactants contributes a large entropy (known as entropy of mixing) to states containing equal mixture of products and reactants. The combination of the standard Gibbs energy change and the Gibbs energy of mixing determines the equilibrium state. In general an equilibrium system is defined by writing an equilibrium equation for the reaction

$$\alpha A + \beta B \rightleftharpoons \sigma S + \tau T$$

In order to meet the thermodynamic condition for equilibrium, the Gibbs energy must be stationary, meaning that the derivative of G with respect to the extent of reaction: ξ, must be zero. It can be shown that in this case, the sum of chemical potentials of the products is equal to the sum of those corresponding to the reactants. Therefore, the sum of the Gibbs energies of the reactants must be the equal to the sum of the Gibbs energies of the products.

$$\alpha\mu_A + \beta\mu_B = \sigma\mu_S + \tau\mu_T$$

where μ is in this case a partial molar Gibbs energy, a chemical potential. The chemical potential of a reagent A is a function of the activity, {A} of that reagent.

$\mu_A = \mu_A^\ominus + RT\ln\{A\}$, ($\mu_A^\ominus$ is the standard chemical potential).

Substituting expressions like this into the Gibbs energy equation:

$dG = Vdp - SdT + \sum_{i=1}^{k} \mu_i dN_i$ in the case of a closed system.

Now

$dN_i = \nu_i d\xi$ (corresponds to the Stoichiometric coefficient and is the differential of the extent of reaction $d\xi$).

At constant pressure and temperature we obtain:

$\left(\frac{dG}{d\xi}\right)_{T,p} = \sum_{i=1}^{k} \mu_i \nu_i = \Delta_r G_{T,p}$ which corresponds to the Gibbs free energy change for the reaction .

This results in:

$$\Delta_r G_{T,p} = \sigma\mu_S + \tau\mu_T - \alpha\mu_A - \beta\mu_B \ .$$

By substituting the chemical potentials:

$$\Delta_r G_{T,p} = (\sigma\mu_S^{\ominus} + \tau\mu_T^{\ominus}) - (\alpha\mu_A^{\ominus} + \beta\mu_B^{\ominus}) + (\sigma RT \ln\{S\} + \tau RT \ln\{T\}) - (\alpha RT \ln\{A\} + \beta RT \ln\{B\}),$$

the relationship becomes:

$$\Delta_r G_{T,p} = \sum_{i=1}^{k} \mu_i^{\ominus} \nu_i + RT \ln \frac{\{S\}^{\sigma}\{T\}^{\tau}}{\{A\}^{\alpha}\{B\}^{\beta}}$$

$\sum_{i=1}^{k} \mu_i^{\ominus} \nu_i = \Delta_r G^{\ominus}$: which is the standard Gibbs energy change for the reaction. It is a constant at a given temperature, which can be calculated, using thermodynamical tables.

$$RT \ln \frac{\{S\}^{\sigma}\{T\}^{\tau}}{\{A\}^{\alpha}\{B\}^{\beta}} = RT \ln Q_r$$

(Q_r is the reaction quotient when the system is not at equilibrium).

Therefore

$$\left(\frac{dG}{d\xi}\right)_{T,p} = \Delta_r G_{T,p} = \Delta_r G^{\ominus} + RT \ln Q_r$$

At equilibrium $\left(\frac{dG}{d\xi}\right)_{T,p} = \Delta_r G_{T,p} = 0$

$Q_r = K_{eq}$; the reaction quotient becomes equal to the equilibrium constant.

leading to:

$$0 = \Delta_r G^{\ominus} + RT \ln K_{eq}$$

and

$$\Delta_r G^{\ominus} = -RT \ln K_{eq}$$

Obtaining the value of the standard Gibbs energy change, allows the calculation of the equilibrium constant

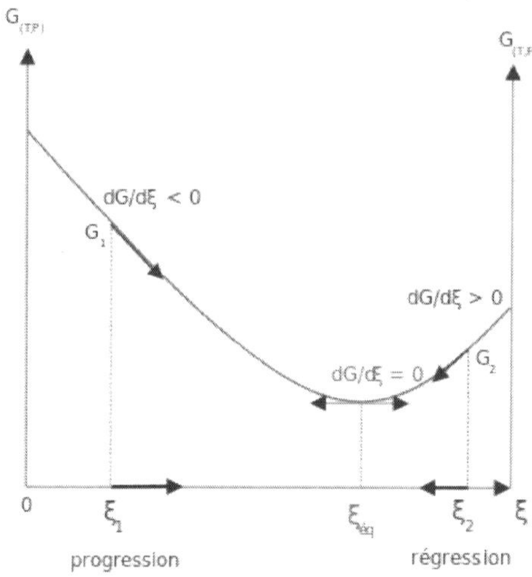

Diagramme G = f (ξ)(T,p)

Addition of Reactants or Products

For a reactional system at equilibrium: $Q_r = K_{(eq)}$;$\xi - \xi_{eq}$.

If are modified activities of constituents, the value of the reaction quotient changes and becomes different from the equilibrium constant: $Q_r \neq K_{(eq)}$

$$\left(\frac{dG}{d\xi}\right)_{T,p} = \Delta_r G^{\ominus} + RT \ln Q_r$$

and

$$\Delta_r G^{\ominus} = -RT \ln K_{(eq)}$$

then

$$\left(\frac{dG}{d\xi}\right)_{T,p} = RT \ln\left(\frac{Q_r}{K_{(eq)}}\right)$$

- If activity of a reagent increases

$Q_r = \dfrac{\prod (a_j)^{\nu_j}}{\prod (a_i)^{\nu_i}}$, the reaction quotient decreases.

then

$Q_r < K_{(eq)}$ and $\left(\dfrac{dG}{d\xi}\right)_{T,p} < 0$: The reaction will shift to the right (i.e. in the forward direction, and thus more products will form).

- If activity of a product increases

then

$Q_r > K_{(eq)}$ and $\left(\frac{dG}{d\xi}\right)_{T,p} > 0$: The reaction will shift to the left (i.e. in the reverse direction, and thus less products will form).

Note that activities and equilibrium constants are dimensionless numbers.

Treatment of Activity

The expression for the equilibrium constant can be rewritten as the product of a concentration quotient, K_c and an activity coefficient quotient, Γ.

$$K = \frac{[S]^\sigma [T]^\tau \ldots}{[A]^\alpha [B]^\beta \ldots} \times \frac{\gamma_S{}^\sigma \gamma_T{}^\tau \ldots}{\gamma_A{}^\alpha \gamma_B{}^\beta \ldots} = K_c \Gamma$$

[A] is the concentration of reagent A, etc. It is possible in principle to obtain values of the activity coefficients, γ. For solutions, equations such as the Debye-Hückel equation or extensions such as Davies equation Specific ion interaction theory or Pitzer equations may be used.. However this is not always possible. It is common practice to assume that Γ is a constant, and to use the concentration quotient in place of the thermodynamic equilibrium constant. It is also general practice to use the term *equilibrium constant* instead of the more accurate *concentration quotient*. This practice will be followed here.

For reactions in the gas phase partial pressure is used in place of concentration and fugacity coefficient in place of activity coefficient. In the real world, for example, when making ammonia in industry, fugacity coefficients must be taken into account. Fugacity, *f*, is the product of partial pressure and fugacity coefficient. The chemical potential of a species in the gas phase is given by

$$\mu = \mu^\ominus + RT \ln\left(\frac{f}{bar}\right) = \mu^\ominus + RT \ln\left(\frac{p}{bar}\right) + RT \ln \gamma$$

so the general expression defining an equilibrium constant is valid for both solution and gas phases.

Concentration Quotients

In aqueous solution, equilibrium constants are usually determined in the presence of an "inert" electrolyte such as sodium nitrate $NaNO_3$ or Potassium perchlorate $KClO_4$. The ionic strength, *I*, of a solution containing a dissolved salt, X^+Y^-, is given by

$$I = \frac{1}{2}\left(c_X z_X^2 + c_Y z_Y^2 + \sum_{i=1}^{n} c_i z_i^2 \right)$$

where c stands for concentration, z stands for ionic charge and the sum is taken over all the species in equilibrium. When the concentration of dissolved salt is much higher than the analytical concentrations of the reagents, the ionic strength is effectively constant. Since activity coefficients depend on ionic strength the activity coefficients of the species are effectively independent of concentration. Thus, the assumption that Γ is constant is justified. The concentration quotient is a simple multiple of the equilibrium constant.

$$K_c = \frac{K}{\Gamma}$$

However, K_c will vary with ionic strength. If it is measured at a series of different ionic strengths the value can be extrapolated to zero ionic strength. The concentration quotient obtained in this manner is known, paradoxically, as a thermodynamic equilibrium constant.

To use a published value of an equilibrium constant in conditions of ionic strength different from the conditions used in its determination, the value should be adjusted.

Metastable Mixtures

A mixture may be appear to have no tendency to change, though it is not at equilibrium. For example, a mixture of SO_2 and O_2 is metastable as there is a kinetic barrier to formation of the product, SO_3.

$$2SO_2 + O_2 \rightleftharpoons 2SO_3$$

The barrier can be overcome when a catalyst is also present in the mixture as in the contact process, but the catalyst does not affect the equilibrium concentrations.

Likewise, the formation of bicarbonate from carbon dioxide and water is very slow under normal conditions

$$CO_2 + 2H_2O \rightleftharpoons HCO_3^- + H_3O^+$$

but almost instantaneous in the presence of the catalytic enzyme carbonic anhydrase.

Pure Compounds

When pure substances (liquids or solids) are involved in equilibria they do not appear in the equilibrium equation

Applying the general formula for an equilibrium constant to the specific case of acetic acid one obtains

$$CH_3CO_2H + H_2O \rightleftharpoons CH_3CO_2^- + H_3O^+$$

$$K_c = \frac{[CH_3CO_2^-][H_3O^+]}{[CH_3CO_2H][H_2O]}$$

It may be assumed that the concentration of water is constant. This assumption will be valid for all but very concentrated solutions. The equilibrium constant expression is therefore usually written as

$$K = \frac{[CH_3CO_2^-][H_3O^+]}{[CH_3CO_2H]}$$

where now

$$K = K_c \cdot [H_2O]$$

a constant factor is incorporated into the equilibrium constant.

A particular case is the self-ionization of water itself

$$H_2O + H_2O \rightleftharpoons H_3O^+ + OH^-$$

The self-ionization constant of water is defined as

$$K_w = [H^+][OH^-]$$

It is perfectly legitimate to write $[H^+]$ for the hydronium ion concentration, since the state of solvation of the proton is constant (in dilute solutions) and so does not affect the equilibrium concentrations. K_w varies with variation in ionic strength and/or temperature.

The concentrations of H^+ and OH^- are not independent quantities. Most commonly $[OH^-]$ is replaced by $K_w[H^+]^{-1}$ in equilibrium constant expressions which would otherwise hydroxide.

Solids also do not appear in the equilibrium equation. An example is the Boudouard reaction:

$$2CO \rightleftharpoons CO_2 + C$$

for which the equation (without solid carbon) is written as:

$$K_c = \frac{[CO_2]}{[CO]^2}$$

Multiple Equilibria

Consider the case of a dibasic acid H_2A. When dissolved in water, the mixture will contain H_2A, HA^- and A^{2-}. This equilibrium can be split into two steps in each of which one proton is liberated.

$$H_2A \rightleftharpoons HA^- + H^+ : K_1 = \frac{[HA^-][H^+]}{[H_2A]}$$

$$HA^- \rightleftharpoons A^{2-} + H^+ : K_2 = \frac{[A^{2-}][H^+]}{[HA^-]}$$

K_1 and K_2 are examples of *stepwise* equilibrium constants. The *overall* equilibrium constant,β_D, is product of the stepwise constants.

$$H_2A \rightleftharpoons A^{2-} + 2H^+ : \beta_D = \frac{[A^{2-}][H^+]^2}{[H_2A]} = K_1K_2$$

Note that these constants are dissociation constants because the products on the right hand side of the equilibrium expression are dissociation products. In many systems, it is preferable to use association constants.

$$A^{2-} + H^+ \rightleftharpoons HA^- : \beta_1 = \frac{[HA^-]}{[A^{2-}][H^+]}$$

$$A^{2-} + 2H^+ \rightleftharpoons H_2A : \beta_2 = \frac{[H_2A]}{[A^{2-}][H^+]^2}$$

β_1 and β_2 are examples of association constants. Clearly $\beta_1 = 1/K_2$ and $\beta_2 = 1/\beta_D$; lg β_1 = pK_2 and lg β_2 = pK_2 + pK_1. For multiple equilibrium systems.

Effect of Temperature

The effect of changing temperature on an equilibrium constant is given by the van 't Hoff equation

$$\frac{d\ln K}{dT} = \frac{\Delta H_m^\ominus}{RT^2}$$

Thus, for exothermic reactions, (ΔH is negative) K decreases with an increase in temperature, but, for endothermic reactions, (ΔH is positive) K increases with an increase temperature. An alternative formulation is

$$\frac{d\ln K}{d(1/T)} = -\frac{\Delta H_m^\ominus}{R}$$

At first sight this appears to offer a means of obtaining the standard molar enthalpy of the reaction by studying the variation of K with temperature. In practice, however, the method is unreliable because error propagation almost always gives very large errors on the values calculated in this way.

Types of Equilibrium

1. In the gas phase. Rocket engines
2. The industrial synthesis such as ammonia in the Haber-Bosch process (depicted right) takes place through a succession of equilibrium steps including adsorption processes.
3. atmospheric chemistry
4. Seawater and other natural waters: Chemical oceanography
5. Distribution between two phases
   * LogD-Distribution coefficient: Important for pharmaceuticals where lipophilicity is a significant property of a drug
   * Liquid-liquid extraction, Ion exchange, Chromatography
   * Solubility product
   * Uptake and release of oxygen by haemoglobin in blood
6. Acid/base equilibria: Acid dissociation constant, hydrolysis, buffer solutions, indicators, acid-base homeostasis
7. Metal-ligand complexation: sequestering agents, chelation therapy, MRI contrast reagents, Schlenk equilibrium
8. Adduct formation: Host-guest chemistry, supramolecular chemistry, molecular recognition, dinitrogen tetroxide
9. In certain oscillating reactions, the approach to equilibrium is not asymptotically but in the form of a damped oscillation .
10. The related Nernst equation in electrochemistry gives the difference in electrode potential as a function of redox concentrations.
11. When molecules on each side of the equilibrium are able to further react irreversibly in secondary reactions, the final product ratio is determined according to the Curtin-Hammett principle.

In these applications, terms such as stability constant, formation constant, binding constant, affinity constant, association/dissociation constant are used. In biochemistry, it is common to give units for binding constants, which serve to define the concentration units used when the constant's value was determined.

Composition of a Mixture

When the only equilibrium is that of the formation of a 1:1 adduct as the composition of a mixture, there are any number of ways that the composition of a mixture can be calculated.

There are three approaches to the general calculation of the composition of a mixture at equilibrium.

1. The most basic approach is to manipulate the various equilibrium constants until the desired concentrations are expressed in terms of measured equilibrium constants (equivalent to measuring chemical potentials) and initial conditions.
2. Minimize the Gibbs energy of the system.
3. Satisfy the equation of mass balance. The equations of mass balance are simply statements that demonstrate that the total concentration of each reactant must be constant by the law of conservation of mass.

Mass-balance Equations

In general, the calculations are rather complicated. For instance, in the case of a dibasic acid, H_2A dissolved in water the two reactants can be specified as the conjugate base, A^{2-}, and the proton, H^+. The following equations of mass-balance could apply equally well to a base such as 1,2-diaminoethane, in which case the base itself is designated as the reactant A:

$$T_A = [A] + [HA] + [H_2A]$$

$$T_H = [H] + [HA] + 2[H_2A] - [OH]$$

With T_A the total concentration of species A. Note that it is customary to omit the ionic charges when writing and using these equations.

When the equilibrium constants are known and the total concentrations are specified there are two equations in two unknown "free concentrations" [A] and [H]. This follows from the fact that $[HA] = \beta_1[A][H]$, $[H_2A] = \beta_2[A][H]^2$ and $[OH] = K_w[H]^{-1}$.

$$T_A = [A] + \beta_1[A][H] + \beta_2[A][H]^2$$

$$T_H = [H] + \beta_1[A][H] + 2\beta_2[A][H]^2 - K_w[H]^{-1}$$

so the concentrations of the "complexes" are calculated from the free concentrations and the equilibrium constants. General expressions applicable to all systems with two reagents, A and B would be

$$T_A = [A] + \sum_i p_i \beta_i [A]^{p_i} [B]^{q_i}$$

$$T_B = [B] + \sum_i q_i \beta_i [A]^{p_i} [B]^{q_i}$$

Polybasic Acids

The composition of solutions containing reactants A and H is easy to calculate as a function of p[H]. When [H] is known, the free concentration [A] is calculated from the mass-balance equation in A. Here is an example of the results that can be obtained.

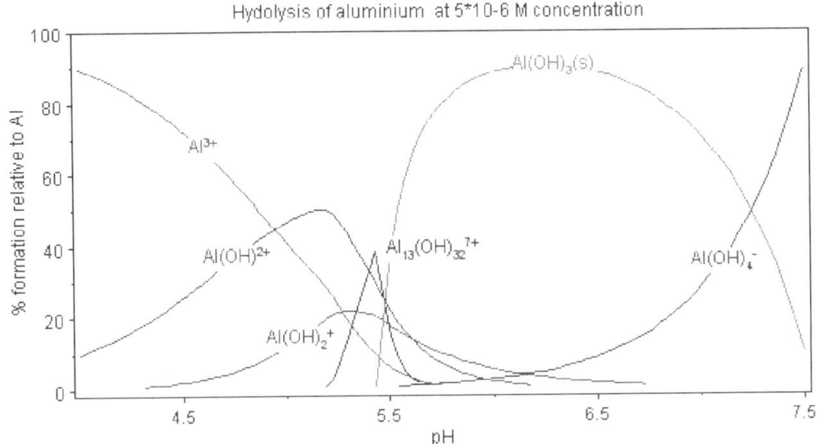

This diagram, for the hydrolysis of the aluminium Lewis acid Al^{3+}_{aq} shows the species concentrations for a 5×10^{-6}M solution of an aluminium salt as a function of pH. Each concentration is shown as a percentage of the total aluminium.

Solution and Precipitation

The diagram above illustrates the point that a precipitate that is not one of the main species in the solution equilibrium may be formed. At pH just below 5.5 the main species present in a 5μM solution of Al^{3+} are aluminium hydroxides $Al(OH)^{2+}$, $Al(OH)_2^+$ and $Al_{13}(OH)_{32}^{7+}$, but on raising the pH $Al(OH)_3$ precipitates from the solution. This occurs because $Al(OH)_3$ has a very large lattice energy. As the pH rises more and more $Al(OH)_3$ comes out of solution.

This is an example of Le Chatelier's principle in action: Increasing the concentration of the hydroxide ion causes more aluminium hydroxide to precipitate, which removes hydroxide from the solution. When the hydroxide concentration becomes sufficiently high the soluble aluminate, $Al(OH)_4^-$, is formed.

Another common instance where precipitation occurs is when a metal cation interacts with an anionic ligand to form an electrically-neutral complex. If the complex is hydrophobic, it will precipitate out of water.

This occurs with the nickel ion Ni^{2+} and dimethylglyoxime, ($dmgH_2$): in this case the lattice energy of the solid is not particularly large, but it greatly exceeds the energy of solvation of the molecule $Ni(dmgH)_2$.

Minimization of Free Energy

At equilibrium, G is at a minimum:

$$dG = \sum_{j=1}^{m} \mu_j \, dN_j = 0$$

For a closed system, no particles may enter or leave, although they may combine in various ways. The total number of atoms of each element will remain constant. This means that the minimization above must be subjected to the constraints:

$$\sum_{j=1}^{m} a_{ij} N_j = b_i^0$$

where a_{ij} is the number of atoms of element i in molecule j and b_i^0 is the total number of atoms of element i, which is a constant, since the system is closed. If there are a total of k types of atoms in the system, then there will be k such equations.

This is a standard problem in optimisation, known as constrained minimisation. The most common method of solving it is using the method of Lagrange multipliers, also known as undetermined multipliers (though other methods may be used).

Define:

$$\mathcal{G} = G + \sum_{i=1}^{k} \lambda_i \left(\sum_{j=1}^{m} a_{ij} N_j - b_i^0 \right) = 0$$

where the λ_i are the Lagrange multipliers, one for each element. This allows each of the N_j to be treated independently, and it can be shown using the tools of multivariate calculus that the equilibrium condition is given by

$$\frac{\partial \mathcal{G}}{\partial N_j} = 0 \quad \text{and} \quad \frac{\partial \mathcal{G}}{\partial \lambda_i} = 0$$

This is a set of *(m+k)* equations in *(m+k)* unknowns (the N_j and the λ_i) and may, therefore, be solved for the equilibrium concentrations N_j as long as the chemical potentials are known as functions of the concentrations at the given temperature and pressure.

This method of calculating equilibrium chemical concentrations is useful for systems with a large number of different molecules. The use of k atomic element conservation equations for the mass constraint is straightforward, and replaces the use of the stoichiometric coefficient equations.

Ethanol

Ethanol, also called ethyl alcohol, pure alcohol, grain alcohol, or drinking alcohol, is a volatile, flammable, colourless liquid. It is a psychoactive drug and one of the oldest recreational drugs. Best known as the type of alcohol found in alcoholic beverages, it is also used in thermometers, as a solvent, and as a fuel. In common usage, it is often referred to simply as alcohol or spirits.

Ethanol is a straight-chain alcohol, and its molecular formula is C_2H_5OH. Its empirical formula is C_2H_6O. An alternative notation is CH_3–CH_2–OH, which indicates that the carbon of a methyl group (CH_3–) is attached to the carbon of a methylene group (–CH_2–), which is attached to the oxygen of a hydroxyl group (–OH). It is a constitutional isomer of dimethyl ether. Ethanol is often abbreviated as EtOH, using the common organic chemistry notation of representing the ethyl group (C_2H_5) with Et.

The fermentation of sugar into ethanol is one of the earliest organic reactions employed by humanity. The intoxicating effects of ethanol consumption have been known since ancient times. In modern times, ethanol intended for industrial use is also produced from ethylene. Ethanol has widespread use as a solvent of substances intended for human contact or consumption, including scents, flavourings, colourings, and medicines. In chemistry, it is both an essential solvent and a feedstock for the synthesis of other products. It has a long history as a fuel for heat and light, and more recently as a fuel for internal combustion engines.

Ethanol is the systematic name defined by the IUPAC nomenclature of organic chemistry for a molecule with two carbon atoms (prefix "eth-"), having a single bond between them (suffix "-ane"), and an attached -OH group (suffix "-ol").

History

Ethanol has been used by humans since prehistory as the intoxicating ingredient of alcoholic beverages. Dried residue on 9,000-year-old pottery found in China imply that Neolithic people consumed alcoholic beverages. Although distillation was well known by the early Greeks and Arabs, the first recorded production of alcohol from distilled wine was by the School of Salerno alchemists in the 12th century. The first to mention absolute alcohol, in contrast with alcohol-water mixtures, was Raymond Lull. In 1796, Johann Tobias Lowitz obtained pure ethanol by filtering distilled ethanol through activated charcoal. Antoine Lavoisier described ethanol as a compound of carbon, hydrogen, and oxygen, and in 1808 Nicolas-Théodore de Saussure determined ethanol's chemical formula. Fifty years later, Archibald Scott Couper published the structural formula of ethanol. It is one of the first structural formulas determined.

Ethanol was first prepared synthetically in 1826 through the independent efforts of Henry Hennel in Great Britain and S.G. Sérullas in France. In 1828, Michael Faraday prepared ethanol by acid-catalyzed hydration of ethylene, a process similar to current industrial ethanol synthesis.

Ethanol was used as lamp fuel in the United States as early as 1840, but a tax levied on industrial alcohol during the Civil War made this use uneconomical. The tax was repealed in 1906. Original Ford Model T automobiles ran on ethanol until 1908. With the advent of Prohibition in 1920, ethanol fuel sellers were accused of being allied with moonshiners, and ethanol fuel fell into disuse until late in the 20th century.

Physical Properties

Ethanol is a volatile, colourless liquid that has a slight odor. It burns with a smokeless blue flame that is not always visible in normal light.

The physical properties of ethanol stem primarily from the presence of its hydroxyl group and the shortness of its carbon chain. Ethanol's hydroxyl group is able to participate in hydrogen bonding, rendering it more viscous and less volatile than less polar organic compounds of similar molecular weight.

Ethanol is a versatile solvent, miscible with water and with many organic solvents, including acetic acid, acetone, benzene, carbon tetrachloride, chloroform, diethyl ether, ethylene glycol, glycerol, nitromethane, pyridine, and toluene. It is also miscible with light aliphatic hydrocarbons, such as pentane and hexane, and with aliphatic chlorides such as trichloroethane and tetrachloroethylene.

Ethanol's miscibility with water contrasts with that of longer-chain alcohols (five or more carbon atoms), whose water miscibility decreases sharply as the number of carbons increases. The miscibility of ethanol with alkanes is limited to alkanes up to undecane, mixtures with dodecane and higher alkanes show a miscibility gap below a certain temperature (about 13 °C for dodecane). The miscibility gap tends to get wider with higher alkanes and the temperature for complete miscibility increases.

Ethanol-water mixtures have less volume than the sum of their individual components at the given fractions. Mixing equal volumes of ethanol and water results in only 1.92 volumes of mixture. Mixing ethanol and water is exothermic. At 298 K, up to 777 J/mol are set free. Mixtures of ethanol and water form an azeotrope at about 89 mole-% ethanol and 11 mole-% water or a mixture of about 96 volume percent ethanol and 4% water at normal pressure and T = 351 K. This azeotropic composition is strongly temperature- and pressure-dependent and vanishes at temperatures below 303 K.

Hydrogen bonding causes pure ethanol to be hygroscopic to the extent that it readily absorbs water from the air. The polar nature of the hydroxyl group causes ethanol to dissolve many ionic compounds, notably sodium and potassium hydroxides, magnesium chloride, calcium chloride, ammonium chloride, ammonium bromide, and sodium bromide. Sodium and potassium chlorides are slightly soluble in ethanol. Because the ethanol molecule also has a nonpolar end, it will also dissolve nonpolar substances, including most essential oils and numerous flavouring, colouring, and medicinal agents.

The addition of even a few percent of ethanol to water sharply reduces the surface tension of water. This property partially explains the "tears of wine" phenomenon. When wine is swirled in a glass, ethanol evaporates quickly from the thin film of wine on the wall of the glass. As the wine's ethanol content decreases, its surface tension increases and the thin film "beads up" and runs down the glass in channels rather than as a smooth sheet.

Mixtures of ethanol and water that contain more than about 50% ethanol are flammable and easily ignited. Alcoholic proof is a widely used measure of how much ethanol (i.e., alcohol) such a mixture contains. In the 18th century, proof was determined by adding a liquor (such as rum) to gunpowder. If the gunpowder still burned, that was considered to be "100 degrees proof" that it was "good" liquor — hence it was called "100 degrees proof".

Ethanol-water solutions that contain less than 50% ethanol may also be flammable if the solution is first heated. Some cooking methods call for wine to be added to a hot pan, causing it to flash boil into a vapour, which is then ignited to burn off excess alcohol. Ethanol is slightly more refractive than water, having a refractive index of 1.36242 (at λ=589.3 nm and 18.35 °C).

Production

Ethanol is produced both as a petrochemical, through the hydration of ethylene, and biologically, by fermenting sugars with yeast. Which process is more economical depends on prevailing prices of petroleum and grain feed stocks.

Ethylene Hydration

Ethanol for use as an industrial feedstock or solvent (sometimes referred to as synthetic ethanol) is often made from petrochemical feed stocks, primarily by the acid-catalyzed hydration of ethylene, represented by the chemical equation

$$C_2H_{4(g)} + H_2O_{(g)} \rightarrow CH_3CH_2OH_{(l)}.$$

The catalyst is most commonly phosphoric acid, adsorbed onto a porous support such as silica gel or diatomaceous earth. This catalyst was first used for large-scale ethanol production by the Shell Oil Company in 1947.

The reaction is carried out with an excess of high pressure steam at 300 °C. In the U.S., this process was used on an industrial scale by Union Carbide Corporation and others; but now only LyondellBasell uses it commercially.

In an older process, first practiced on the industrial scale in 1930 by Union Carbide, but now almost entirely obsolete, ethylene was hydrated indirectly by reacting it with concentrated sulfuric acid to produce ethyl sulfate, which was hydrolysed to yield ethanol and regenerate the sulfuric acid:

$$C_2H_4 + H_2SO_4 \rightarrow CH_3CH_2SO_4H$$

$$CH_3CH_2SO_4H + H_2O \rightarrow CH_3CH_2OH + H_2SO_4$$

Fermentation

Ethanol for use in alcoholic beverages, and the vast majority of ethanol for use as fuel, is produced by fermentation. When certain species of yeast (e.g., *Saccharomyces cerevisiae*) metabolize sugar they produce ethanol and carbon dioxide. The chemical equations below summarize the conversion:

$$C_6H_{12}O_6 \rightarrow 2\ CH_3CH_2OH + 2\ CO_2$$

$$C_{12}H_{22}O_{11} + H_2O \rightarrow 4\ CH_3CH_2OH + 4\ CO_2$$

The process of culturing yeast under conditions to produce alcohol is called fermentation. This process is carried out at around 35–40 °C. Toxicity of ethanol to yeast limits the ethanol concentration obtainable by brewing; higher concentrations, therefore, are usually obtained by fortification or distillation. The most ethanol-tolerant strains of yeast can survive up to approximately 15% ethanol by volume.

To produce ethanol from starchy materials such as cereal grains, the starch must first be converted into sugars. In brewing beer, this has traditionally been accomplished by allowing the grain to germinate, or malt, which produces the enzyme amylase. When the malted grain is mashed, the amylase converts the remaining starches into sugars. For fuel ethanol, the hydrolysis of starch into glucose can be accomplished more rapidly by treatment with dilute sulfuric acid, fungally produced amylase, or some combination of the two.

Cellulosic Ethanol

Sugars for ethanol fermentation can be obtained from cellulose. Until recently, however, the cost of the cellulase enzymes capable of hydrolyzing cellulose has been prohibitive. The Canadian firm Iogen brought the first cellulose-based ethanol plant on-stream in 2004. Its primary consumer so far has been the Canadian government, which, along with the United States Department of Energy, has invested heavily in the commercialization of cellulosic ethanol. Deployment of this technology could turn a number of cellulose-containing agricultural by-products, such as corncobs, straw, and sawdust, into renewable energy resources. Other enzyme companies are developing genetically engineered fungi that produce large volumes of cellulase, xylanase, and hemicellulase enzymes. These would convert agricultural residues such as corn stover, wheat straw, and sugar cane bagasse and energy crops such as switchgrass into fermentable sugars.

Cellulose-bearing materials typically also contain other polysaccharides, including hemicellulose. When undergoing hydrolysis, hemicellulose decomposes into mostly five-carbon sugars such as xylose. *S. cerevisiae*, the yeast most commonly used for ethanol production, cannot metabolize xylose. Other yeasts and bacteria are under investigation to ferment xylose and other pentoses into ethanol.

On January 14, 2008, General Motors announced a partnership with Coskata, Inc. The goal is to produce cellulosic ethanol cheaply, with an eventual goal of US$1 per US gallon ($0.30/L) for the fuel. The partnership plans to begin producing the fuel in large quantity by the end of 2008.

In June 2009, this goal is still ahead of the firm. By 2011 a full-scale plant will come on line, capable of producing 50 million US gallons (190,000 m^3) to 100 million US gallons (380,000 m^3) of ethanol a year (200–400 ML/a).

Prospective Technologies

The anaerobic bacterium *Clostridium ljungdahlii*, discovered in commercial chicken wastes, can produce ethanol from single-carbon sources including synthesis gas, a mixture of carbon monoxide and hydrogen that can be generated from the partial combustion of either fossil fuels or biomass. Use of these bacteria to produce ethanol from synthesis gas has progressed to the pilot plant stage at the BRI Energy facility in Fayetteville, Arkansas. The BRI technology has been purchased by INEOS.

Another prospective technology is the closed-loop ethanol plant. Ethanol produced from corn has a number of critics who suggest that it is primarily just recycled fossil fuels because of the energy required to grow the grain and convert it into ethanol. There is also the issue of competition with use of corn for food production.

However, the closed-loop ethanol plant attempts to address this criticism. In a closed-loop plant, renewable energy for distillation comes from fermented manure, produced from cattle that have been fed the DDSG by-products from grain ethanol production.

The concentrated compost nutrients from manure are then used to fertilize the soil and grow the next crop of grain to start the cycle again. Such a process is expected to lower the fossil fuel consumption used during conversion to ethanol by 75%.

An alternative technology allows for the production of biodiesel from distillers grain as an additional value product. Though in an early stage of research, there is some development of alternative production methods that use feed stocks such as municipal waste or recycled products, rice hulls, sugarcane bagasse, small diameter trees, wood chips, and switchgrass.

Testing

Breweries and biofuel plants employ two methods for measuring ethanol concentration. Infrared ethanol sensors measure the vibrational frequency of dissolved ethanol using the CH band at 2900 cm^{-1}. This method uses a relatively inexpensive solid state sensor that compares the CH band with a reference band to calculate the ethanol content.

The calculation makes use of the Beer-Lambert law. Alternatively, by measuring the density of the starting material and the density of the product, using a hydrometer, the change in specific gravity during fermentation indicates the alcohol content. This inexpensive and indirect method has a long history in the beer brewing industry.

Purification

Ethylene hydration or brewing produces an ethanol–water mixture. For most industrial and fuel uses, the ethanol must be purified. Fractional distillation can concentrate ethanol to 95.6% by volume (89.5 mole%). This mixture is an azeotrope with a boiling point of 78.1 °C, and cannot be further purified by distillation.

Common methods for obtaining absolute ethanol include desiccation using adsorbents such as starch, corn grits, or zeolites, which adsorb water preferentially, as well as azeotropic distillation and extractive distillation. Most ethanol fuel refineries use an adsorbent or zeolite to desiccate the ethanol stream.

In another method to obtain absolute alcohol, a small quantity of benzene is added to rectified spirit and the mixture is then distilled. Absolute alcohol is obtained in the third fraction, which distills over at 78.3 °C (351.4 K). Because a small amount of the benzene used remains in the solution, absolute alcohol produced by this method is not suitable for consumption, as benzene is carcinogenic.

There is also an absolute alcohol production process by desiccation using glycerol. Alcohol produced by this method is known as spectroscopic alcohol—so called because the absence of benzene makes it suitable as a solvent in spectroscopy.

Grades of Ethanol

Denatured Alcohol: Pure ethanol and alcoholic beverages are heavily taxed, but ethanol has many uses that do not involve consumption by humans. To relieve the tax burden on these uses, most jurisdictions waive the tax when an agent has been added to the ethanol to render it unfit to drink. These include bittering agents such as denatonium benzoate and toxins such as methanol, naphtha, and pyridine. Products of this kind are called *denatured alcohol.*

Absolute Ethanol

Absolute or anhydrous alcohol refers to ethanol with a low water content. There are various grades with maximum water contents ranging from 1% to ppm levels. Absolute alcohol is not intended for human consumption. If azeotropic distillation is used to remove water, it will contain trace amounts of the material separation agent (e.g. benzene).

Absolute ethanol is used as a solvent for laboratory and industrial applications, where water will react with other chemicals, and as fuel alcohol. Spectroscopic ethanol is an absolute ethanol with a low absorbance in ultraviolet and visible light, fit for use as a solvent in ultraviolet-visible spectroscopy.

Pure ethanol is classed as 200 proof in the USA, equivalent to 175 degrees proof in the UK system.

Rectified Spirits

Rectified spirit, an azeotropic composition containing 4% water, is used instead of anhydrous ethanol for various purposes. Wine spirits are about 188 proof. The impurities are different from those in 190 proof laboratory ethanol.

Reactions

Ethanol is classified as a primary alcohol, meaning that the carbon its hydroxyl group attaches to has at least two hydrogen atoms attached to it as well. Many ethanol reactions occur at its hydroxyl group.

Ester Formation

In the presence of acid catalysts, ethanol reacts with carboxylic acids to produce ethyl esters and water:

$$RCOOH + HOCH_2CH_3 \rightarrow RCOOCH_2CH_3 + H_2O$$

This reaction, which is conducted on large scale industrially, requires the removal of the water from the reaction mixture as it is formed. Esters react in the presence of an acid or base to give back the alcohol and a salt. This reaction is known as saponification because it is used in the preparation of soap. Ethanol can also form esters with inorganic acids. Diethyl sulfate and triethyl phosphate are prepared by treating ethanol with sulfur trioxide and phosphorus pentoxide respectively. Diethyl sulfate is a useful ethylating agent in organic synthesis. Ethyl nitrite, prepared from the reaction of ethanol with sodium nitrite and sulfuric acid, was formerly a widely used diuretic.

Dehydration

Strong acid desiccants cause the dehydration of ethanol to form diethyl ether and other byproducts. If the dehydration temperature exceeds around 160 °C, ethylene will be the main product. Millions of kilograms of diethyl ether are produced annually using sulfuric acid catalyst:

$$2\ CH_3CH_2OH \rightarrow CH_3CH_2OCH_2CH_3 + H_2O \text{ (on 120 °C)}$$

Combustion

Complete combustion of ethanol forms carbon dioxide and water vapour:

C_2H_5OH (l) + 3 O_2 (g) → 2 CO_2 (g) + 3 H_2O (g); (ΔH_c = -1371 kJ/mol)
specific heat = 2.44 kJ/(kg·K)

Acid-base Chemistry

Ethanol is a neutral molecule and the pH of a solution of ethanol in water is nearly 7.00. Ethanol can be quantitatively converted to its conjugate base, the ethoxide ion ($CH_3CH_2O^-$), by reaction with an alkali metal such as sodium:

$$2\ CH_3CH_2OH + 2\ Na \rightarrow 2\ CH_3CH_2ONa + H_2$$

or a very strong base such as sodium hydride:

$$CH_3CH_2OH + NaH \rightarrow CH_3CH_2ONa + H_2$$

The acidity of water and ethanol are nearly the same, as indicated by their pKa of 15.7 and 16 respectively. Thus, sodium ethoxide and sodium hydroxide exist in an equilbrium that is closely balanced:

$$CH_3CH_2OH + NaOH \rightleftharpoons CH_3CH_2ONa + H_2O$$

Halogenation

Ethanol is not used industrially as a precursor to ethyl halides, but the reactions are illustrative. Ethanol reacts with hydrogen halides to produce ethyl halides such as ethyl chloride and ethyl bromide via an S_N2 reaction:

$$CH_3CH_2OH + HCl \rightarrow CH_3CH_2Cl + H_2O$$

These reactions require a catalyst such as zinc chloride. HBr requires refluxing with a sulfuric acid catalyst. Ethyl halides can, in principle, also be produced by treating ethanol with more specialized halogenating agents, such as thionyl chloride or phosphorus tribromide.

$$CH_3CH_2OH + SOCl_2 \rightarrow CH_3CH_2Cl + SO_2 + HCl$$

Upon treatment with halogens in the presence of base, ethanol gives the corresponding haloform (CHX_3, where X = Cl, Br, I). This conversion is called the haloform reaction. " An intermediate in the reaction with chlorine is the aldehyde called chloral:

$$4\ Cl_2 + CH_3CH_2OH \rightarrow CCl_3CHO + 5\ HCl$$

Oxidation

Ethanol can be oxidized to acetaldehyde and further oxidized to acetic acid, depending on the reagents and conditions. This oxidation is of no importance industrially, but in the human body, these oxidation reactions are catalyzed by the enzyme liver alcohol dehydrogenase. The oxidation product of ethanol, acetic acid, is a nutrient for humans, being a precursor to acetyl CoA, where the acetyl group can be spent as energy or used for biosynthesis.

Uses

As a Fuel

Energy content of some fuels compared with ethanol:

Fuel type	MJ/L	MJ/kg	Research octane number
Dry wood (20% moisture)		~19.5	
Methanol	17.9	19.9	108.7
Ethanol	21.2	26.8	108.6
E85(85% ethanol, 15% gasoline)	25.2	33.2	105
Liquefied natural gas	25.3	~55	
Autogas (LPG)(60% propane + 40% butane)	26.8	50.	
Aviation gasoline(high-octane gasoline, not jet fuel)	33.5	46.8	100/130 (lean/rich)
Gasohol(90% gasoline + 10% ethanol)	33.7	47.1	93/94
Regular gasoline	34.8	44.4	min. 91
Premium gasoline			max. 104
Diesel	38.6	45.4	25
Charcoal, extruded	50	23	

The largest single use of ethanol is as a motor fuel and fuel additive. Brazil has the largest national fuel ethanol industry. Gasoline sold in Brazil contains at least 25% anhydrous ethanol. Hydrous ethanol (about 95% ethanol and 5% water) can be used as fuel in more than 90% of new cars sold in the country. Brazilian ethanol is produced from sugar cane and noted for high carbon sequestration. The US uses Gasohol (max 10% ethanol) and E85 (85% ethanol) ethanol/gasoline mixtures. Ethanol may also be utilized as a rocket fuel, and is currently in lightweight rocket-powered racing aircraft.

Australian law limits of the use of pure Ethanol sourced from Sugarcane waste to up to 10% in automobiles. It has been recommended that older cars (and vintage cars designed to use a slower burning fuel) have their valves upgraded or replaced.

Ethanol as a fuel reduces harmful tailpipe emissions of carbon monoxide, particulate matter, oxides of nitrogen, and other ozone-forming pollutants. Argonne National Laboratory analyzed the greenhouse gas emissions of many different engine and fuel combinations. Comparing ethanol blends with gasoline alone, they showed reductions of 8% with the biodiesel/petrodiesel blend known as B20, 17% with the conventional E85 ethanol blend, and that using cellulosic ethanol lowers emissions 64%.

Ethanol combustion in an internal combustion engine yields many of the products of incomplete combustion produced by gasoline and significantly larger amounts of formaldehyde and related species such as acetaldehyde. This leads to a significantly larger photochemical reactivity that generates much more ground level ozone.

These data have been assembled into The Clean Fuels Report comparison of fuel emissions and show that ethanol exhaust generates 2.14 times as much ozone as does gasoline exhaust. When this is added into the custom *Localised Pollution Index (LPI)* of The Clean Fuels Report the local pollution (pollution that contributes to smog) is 1.7 on a scale where gasoline is 1.0 and higher numbers signify greater pollution. The California Air Resources Board formalized this issue in 2008 by recognizing control standards for formaldehydes as an emissions control group, much like the conventional NOx and Reactive Organic Gases (ROGs).

World production of ethanol in 2006 was 51 gigalitres (1.3×10^{10} US gal), with 69% of the world supply coming from Brazil and the United States. More than 20% of Brazilian cars are able to use 100% ethanol as fuel, which includes ethanol-only engines and flex-fuel engines. Flex-fuel engines in Brazil are able to work with all ethanol, all gasoline or any mixture of both. In the US flex-fuel vehicles can run on 0% to 85% ethanol (15% gasoline) since higher ethanol blends are not yet allowed or efficient. Brazil supports this population of ethanol-burning automobiles with large national infrastructure that produces ethanol from domestically grown sugar cane. Sugar cane not only has a greater concentration of sucrose than corn (by about 30%), but is also much easier to extract. The bagasse generated by the process is not wasted, but is used in power plants as a surprisingly efficient fuel to produce electricity.

The United States fuel ethanol industry is based largely on corn. According to the Renewable Fuels Association, as of October 30, 2007, 131 grain ethanol bio-refineries in the United States have the capacity to produce 7.0 billion US gallons (26,000,000 m^3) of ethanol per year. An additional 72 construction projects underway (in the U.S.) can add 6.4 billion US gallons (24,000,000 m^3) of new capacity in the next 18 months. Over time, it is believed that a material portion of the H"150-billion-US-gallon (570,000,000 m^3) per year market for gasoline will begin to be replaced with fuel ethanol.

One problem with ethanol is its high miscibility with water, which means that it cannot be efficiently shipped through modern pipelines, like liquid hydrocarbons, over long distances. Mechanics also have seen increased cases of damage to small engines, in particular, the carburetor, attributable to the increased water retention by ethanol in fuel.

Alcoholic Beverages

Ethanol is the principal psychoactive constituent in alcoholic beverages, with depressant effects on the central nervous system. It has a complex mode of action and affects multiple systems in the brain, the most notable one being its agonistic action on the GABA receptors. Similar psychoactives include those that also interact with GABA receptors, such as gamma-hydroxybutyric acid (GHB). Ethanol is metabolized by the body as an energy-providing nutrient, as it metabolizes into acetyl CoA, an intermediate common with glucose and fatty acid metabolism that can be used for energy in the citric acid cycle or for biosynthesis.

Alcoholic beverages vary considerably in ethanol content and in foodstuffs they are produced from. Most alcoholic beverages can be broadly classified as fermented beverages, beverages made by the action of yeast on sugary foodstuffs, or distilled beverages, beverages whose preparation involves concentrating the ethanol in fermented beverages by distillation. The ethanol content of a beverage is usually measured in terms of the volume fraction of ethanol in the beverage, expressed either as a percentage or in alcoholic proof units.

Fermented beverages can be broadly classified by the foodstuff they are fermented from. Beers are made from cereal grains or other starchy materials, wines and ciders from fruit juices, and meads from honey. Cultures around the world have made fermented beverages from numerous other foodstuffs, and local and national names for various fermented beverages abound.

Distilled beverages are made by distilling fermented beverages. Broad categories of distilled beverages include whiskeys, distilled from fermented cereal grains; brandies, distilled from fermented fruit juices; and rum, distilled from fermented molasses or sugarcane juice. Vodka and similar neutral grain spirits can be distilled from any fermented material (grain, tomatoes or potatoes are most common); these spirits are so thoroughly distilled that no tastes from the particular starting material remain. Numerous other spirits and liqueurs are prepared by infusing flavors from fruits, herbs, and spices into distilled spirits.

A traditional example is gin, which is created by infusing juniper berries into a neutral grain alcohol. In a few beverages, ethanol is concentrated by means other than distillation. Applejack is traditionally made by freeze distillation, by which water is frozen out of fermented apple cider, leaving a more ethanol-rich liquid behind. Ice beer (also known by the German term *Eisbier* or *Eisbock*) is also freeze-distilled, with beer as the base beverage. Fortified wines are prepared by adding brandy or some other distilled spirit to partially fermented wine. This kills the yeast and conserves some of the sugar in grape juice; such beverages not only are more ethanol-rich but are often sweeter than other wines.

Alcoholic beverages are sometimes used in cooking, not only for their inherent flavors but also because the alcohol dissolves hydrophobic flavour compounds, which water cannot. Just as industrial ethanol is used as feedstock for the production of industrial acetic acid, alcoholic beverages are made into culinary/household vinegar: Wine and cider vinegar are both named for their respective source alcohols, whereas malt vinegar is derived from beer.

Feedstock

Ethanol is an important industrial ingredient and has widespread use as a base chemical for other organic compounds. These include ethyl halides, ethyl esters, diethyl ether, acetic acid, ethyl amines, and to a lesser extent butadiene.

Antiseptic

Ethanol is used in medical wipes and in most common antibacterial hand sanitizer gels at a concentration of about 62% v/v as an antiseptic. Ethanol kills organisms by denaturing their proteins and dissolving their lipids and is effective against most bacteria and fungi, and many viruses, but is ineffective against bacterial spores.

Treatment for Poisoning by Other Alcohols

Ethanol is sometimes used to treat poisoning by other, more toxic alcohols, in particular methanol and ethylene glycol. Ethanol competes with other alcohols for the alcohol dehydrogenase enzyme, lessening metabolism into toxic aldehyde and carboxylic acid derivatives, and reducing one of the more serious toxic effect of the glycols to crystallize in the kidneys.

Solvent

Ethanol is miscible with water and is a good general purpose solvent. It is found in paints, tinctures, markers, and personal care products such as perfumes and deodorants. It may also be used as a solvent in cooking, such as in vodka sauce.

Historical Uses

Before the development of modern medicines, ethanol was used for a variety of medical purposes. It has been known to be used as a truth drug (as hinted at by the maxim *"in vino veritas"*), as medicine for depression and as an anesthetic.

Ethanol was commonly used as fuel in early bipropellant rocket (liquid propelled) vehicles, in conjunction with an oxidizer such as liquid oxygen. The German V-2 rocket of World War II, credited with beginning the space age, used ethanol, mixed with 25% of water to reduce the combustion chamber temperature.

The V-2's design team helped develop U.S. rockets following World War II, including the ethanol-fueled Redstone rocket, which launched the first U.S. satellite. Alcohols fell into general disuse as more efficient rocket fuels were developed.

Pharmacology

Ethanol binds to acetylcholine, GABA, serotonin, and NMDA receptors.

The removal of ethanol through oxidation by alcohol dehydrogenase in the liver from the human body is limited. Hence, the removal of a large concentration of alcohol from blood may follow zero-order kinetics. This means that alcohol leaves the body at a constant rate, rather than having an elimination half-life.

Also, the rate-limiting steps for one substance may be in common with other substances. For instance, the blood alcohol concentration can be used to modify the biochemistry of methanol and ethylene glycol. In this way the oxidation of methanol to the toxic formaldehyde and formic acid in the (human body) can be prevented by giving an appropriate amount of ethanol to a person having ingested methanol. Note that methanol is very toxic and causes blindness and death. A person having ingested ethylene glycol can be treated in the same way.

Drug Effects

Pure ethanol will irritate the skin and eyes. Nausea, vomiting and intoxication are symptoms of ingestion. Long-term use by ingestion can result in serious liver damage. Atmospheric concentrations above one in a thousand are above the European Union Occupational exposure limits.

Short-term

BAC (g/L)	*BAC(% v/v)*	*Symptoms*
0.5	0.05%	Euphoria, talkativeness, relaxation
1	0.1 %	Central nervous system depression, nausea, possible vomiting, impaired motor and sensory function, impaired cognition
>1.4	>0.14%	Decreased blood flow to brain
3	0.3%	Stupefaction, possible unconsciousness
4	0.4%	Possible death
>5.5	>0.55%	Death

Effects on the Central Nervous System

Ethanol is a central nervous system depressant and has significant psychoactive effects in sublethal doses; for specifics, see "Effects of alcohol on the body by dose". Based on its abilities to change the human consciousness, ethanol is considered a psychoactive drug. Death from ethyl alcohol consumption is possible when blood alcohol level reaches 0.4%. A blood level of 0.5% or more is commonly fatal. Levels of even less than 0.1% can cause intoxication, with unconsciousness often occurring at 0.3–0.4%.

The amount of ethanol in the body is typically quantified by blood alcohol content (BAC), which is here taken as weight of ethanol per unit volume of blood. Small doses of ethanol, in general, produce euphoria and relaxation; people experiencing these symptoms tend to become talkative and less inhibited, and may exhibit poor judgment. At higher dosages (BAC > 1 g/L), ethanol acts as a central nervous system depressant, producing at progressively higher dosages, impaired sensory and motor function, slowed cognition, stupefaction, unconsciousness, and possible death.

Ethanol acts in the central nervous system by binding to the GABA-A receptor, increasing the effects of the inhibitory neurotransmitter GABA (i.e., it is a positive allosteric modulator).

In USA, about half of the deaths in car accidents occur in alcohol-related crashes. The risk of a fatal car accident increases exponentially with the level of alcohol in the driver's blood. Most drunk driving laws governing the acceptable levels in the blood while driving or operating heavy machinery set typical upper limits of blood alcohol content (BAC) between 0.05% and 0.08%.

Discontinuing consumption of alcohol after several years of heavy drinking can also be fatal. Alcohol withdrawal can cause anxiety, autonomic dysfunction, seizures, and hallucinations. Delirium tremens is a condition that requires people with a long history of heavy drinking to undertake an alcohol detoxification regimen.

Effects on Metabolism

Ethanol within the human body is converted into acetaldehyde by alcohol dehydrogenase and then into the acetyl in acetyl CoA by acetaldehyde dehydrogenase. Acetyl CoA is the final product of both carbohydrate and fat metabolism, where the acetyl can be further used to produce energy or for biosynthesis. As such, ethanol is a nutrient. However, the product of the first step of this breakdown, acetaldehyde, is more toxic than ethanol. Acetaldehyde is linked to most of the clinical effects of alcohol. It has been shown to increase the risk of developing cirrhosis of the liver, multiple forms of cancer, and alcoholism.

Drug Interactions

Ethanol can intensify the sedation caused by other central nervous system depressant drugs such as barbiturates, benzodiazepines, opioids, phenothiazines, and anti-depressants.

Magnitude of Effects

Some individuals have less effective forms of one or both of the metabolizing enzymes, and can experience more severe symptoms from ethanol consumption than others. However, those having acquired alcohol tolerance have a greater quantity of these enzymes, and metabolize ethanol more rapidly.

Long-term

Birth Defects: Ethanol is classified as a teratogen.

Other Effects

Frequent drinking of alcoholic beverages has been shown to be a major contributing factor in cases of elevated blood levels of triglycerides.

Ethanol is not a carcinogen. However, the first metabolic product of ethanol, acetaldehyde, is toxic, mutagenic, and carcinogenic.

Natural Occurrence

Ethanol is a byproduct of the metabolic process of yeast. As such ethanol will be present in any yeast habitat. Ethanol can commonly be found in overripe fruit. Ethanol produced by symbiotic yeast can be found in Bertam Palm blossoms. Although some species such as the Pentailed Treeshrew exhibit ethanol seeking behaviours, most show no interest or avoidance of food sources containing ethanol. Ethanol is also produced during the germination of many plants as a result of natural anerobiosis. Ethanol has been detected in outer space, forming an icy coating around dust grains in interstellar clouds.

2,2,2-Trichloroethanol

2,2,2-Trichloroethanol is an organic compound related to ethanol, except the hydrogen atoms at position 2 are replaced with chlorine atoms. In humans, its pharmacological effects are similar to those of chloral hydrate: it is used as a sedative or a hypnotic. The hypnotic drug triclofos (2,2,2-trichloroethyl phosphate) is metabolized *in vivo* to 2,2,2-trichloroethanol. Chronic exposure may result in kidney and liver damage.1,1,1-Trichloroethane

The organic compound 1,1,1-trichloroethane, also known as methyl chloroform, is a chloroalkane. This colourless, sweet-smelling liquid was once produced industrially in large quantities for use as a solvent. It is regulated by the Montreal Protocol as an ozone-depleting substance and its use is being rapidly phased out.

Production

1,1,1-Trichloroethane was first reported by Henri Victor Regnault in 1840. Industrially, it is usually produced in a two-step process from vinyl chloride. In the first step, vinyl chloride reacts with hydrogen chloride at 20-50 °C to produce 1,1-dichloroethane:

$$CH_2{=}CHCl + HCl \rightarrow CH_3CHCl_2$$

This reaction is catalyzed by a variety of Lewis acids, mainly aluminium chloride, iron(III) chloride, or zinc chloride. The 1,1-dichloroethane is then converted to 1,1,1-trichloroethane by reaction with chlorine under ultraviolet irradiation:

$$CH_3CHCl_2 + Cl_2 \rightarrow CH_3CCl_3 + HCl$$

This reaction proceeds at 80-90% yield, and the hydrogen chloride byproduct can be recycled to the first step in the process. The major side-product is the related compound 1,1,2-trichloroethane, from which the 1,1,1-trichloroethane can be separated by distillation.

A somewhat smaller amount of 1,1,1-trichloroethane is produced from the reaction of vinylidene chloride and hydrogen chloride in the presence of an iron(III) chloride catalyst:

$$CH_2{=}CCl_2 + HCl \rightarrow CH_3CCl_3$$

1,1,1-Trichloroethane is marketed with stabilizers since it is unstable with respect to dehydrochlorination and attacks some metals.

Stabilizers comprise up to 8% of the formulation, including acid scavengers (epoxides, amines) and complexants. The Montreal Protocol targeted 1,1,1-trichloroethane as one of those compounds responsible for ozone depletion and banned its use beginning in 1996. Since then, its manufacture and use has been phased out throughout most of the world.

Uses

1,1,1-Trichloroethane is generally considered as a nonpolar solvent. Owing to its unsymmetrical structure, it is a superior solvent for organic compounds that do not dissolve well in hydrocarbons such as hexane. It is an excellent solvent for many organic materials and also one of the least toxic of the chlorinated hydrocarbons. Prior to the Montreal Protocol, it was widely used for cleaning metal parts and circuit boards, as a photoresist solvent in the electronics industry, as an aerosol propellant, as a cutting fluid additive, and as a solvent for inks, paints, adhesives and other coatings.

It was also the standard cleaner for photographic film (movie/slide/negatives, etc.). Other commonly available solvents damage emulsion, and thus are not suitable for this application. The standard replacement, Forane 141 is much less effective, and tends to leave a residue. 1,1,1-Trichloroethane was used as a thinner in Correction fluid products such as Liquid paper. Many applications for 1,1,1-trichloroethane (including film cleaning) were previously done with carbon tetrachloride, which was banned in 1970. In the laboratory, 1,1,1-trichloroethane is being been replaced by other solvents.

1,1,1-Trichloroethane is used as an insecticidal fumigant.

Bibliography

Antwerpen, F.J. : *The Origins of Chemical Engineering, History of Chemical Engineering*, Washington D.C.: ACS, 1980.

Beizer, Boris: *Software Testing Techniques,* International Thomson Computer Press, Boston, MA., 2004.

Bernard S. Matisoff: *Handbook of Electronics Manufacturing Engineering,* Chapman & Hall, New York, 1997.

Chopey, Nicholas P.: *Handbook of Chemical Engineering Calculations,* McGraw-Hill, New York, 2004.

Conaway, Charles F.: *The Petroleum Industry: A Nontechnical Guide,* PennWell Pub. Co., Tulsa, OK, 1999.

Daniel Yergin, *The Prize: The Epic Quest for Oil, Money, and Power*, Simon and Schuster 1991.

Datta-Barua, Lohit: *Natural Gas Measurement and Control: A Guide for Operators and Engineers,* McGraw-Hill Professional Book Group, London, 1991.

Eisner, A.: *Hydrogenation and Petrography of Some Low-rank Coals from the Western United States,* Bureau of Mines, Washington, D.C., 1940.

Engel, M.H., and S. Macko: *Organic Geochemistry; Principles and Applications,* Plenum Press, New York, 1993.

Frank, Alison Fleig: *Oil Empire: Visions of Prosperity in Austrian Galicia,* Harvard University Press, 2005.

George B. Hogaboom: *Principles of Electroplating and Electroforming (Electrotyping)*, McGraw-Hill, New York, 1949.

Gomes, K.J., Braden, J.F.: *Mechanical Fastening of Plastics - An Engineering Handbook,* Marcel Dekker, New York, 1984.

Henley, E. J. and J. D. Seader: *Equilibrium Stage Operations in Chemical Engineering,* John Wiley and Sons, New York, 1981.

Hill, J. W.; Petrucci, R. H.; McCreary, T. W.; Perry, S. S.: *General Chemistry*, Pearson Prentice Hall, Upper Saddle River, New Jersey, 2005.

Kondolf, G. Mathias; Hervé Piégay: *Tools in Fluvial Geomorphology*, Wiley, New York, 2003.

Laurence M. Harwood, Christopher J. Moody: *Experimental Organic Chemistry: Principles and Practice*, WileyBlackwell, Oxford, 1989.

Manthorp, P.A.: *Estuarine Circulation and Sediment Transport in the Saco River estuary*, Boston University, Boston, MA, 1995.

March, Jerry: *Advanced Organic Chemistry: Reactions, Mechanisms, and Structure*, New York: Wiley, 1985.

N.Y. Krylov, A.A. Bokserman, E.R.Stavrovsky: *The Oil Industry of the Former Soviet Union*, CRC Press, 1998.

O'Keeffe, M.; Hyde, B.G.: *Crystal Structures; I. Patterns and Symmetry*, Mineralogical Society of America, Washington, DC, 1996.

Rabek, J.: *Experimental Methods in Physical Chemistry and Photophysics*, New York: Oxford University Press, 1989.

Ruthven, D. M.: *Principles of Adsorption and Adsorption Processes,* John Wiley and Sons, New York, 1984)

Sachanen, A. N.: *Conversion of Petroleum: Production of Motor Fuels by Thermal and Catalytic Processes*, Reinhold Pub. Corp., New York , 1940.

Saferstein, Richard: *Criminalistics: An Introduction to Forensic Science.* Upper Saddle River, NJ: Prentice Hall, 1998.

Thomas, L.: *Handbook of Practical Coal Geology*: John Wiley & Sons, Chichester, 1992.

Tissot, B., and D.H. Welte: *Petroleum Formation and Occurrence*, Springer-Verlag, New York, 1984.

Tolf, Robert W.: *The Russian Rockefellers: The Saga of the Nobel Family and the Russian Oil Industry,* Hoover Press, 1976.

Torben Smith Sørensen: *Surface Chemistry and Electrochemistry of Membranes*, CRC Press, NY, 1999.

Tyson, R.V.: *Sedimentary Organic Matter; Organic Facies and Palynofacies*, Chapman & Hall, New York, 1995.

Wachtman, John B.: *Mechanical Properties of Ceramics*, Wiley-Interscience, John Wiley & Son's, New York, 2002.

Walas, S. M.: *Phase Equilibria in Chemical Engineering,* Butterworths, Reading, MA, 1985.

Wang, Y.: *Physical Chemistry and Photobiology of Nucleic Acids: Chemistry*, Athens: U of Georgia Press, 2005.

Wothers, P.: *Molecular Quantum Mechanics*, Lexington, KY: University Press of Kentucky, 1996.

Yang, Y., R. Zou, Z. Shi, and R. Jiang: *Atlas for Coal Petrography of China*, China University of Mining and Technology Press, Beijing, 1996.

Zim, Herbert S. and Paul R. Shaffer: *Rocks and Minerals: A Guide to Familiar Mnierals, Gems, Ores and Rocks,* Golden, New York, 1957.

Index

G

H

I

L

M

N

O

P

R

S

T

V

W

❑❑❑